MicroSim PSpice for Windows

2nd Edition

Volume II

Operational Amplifiers and Digital Circuits

A Circuit Simulation Primer

Roy W. Goody

Mission College, Santa Clara, CA

Prentice Hall

Upper Saddle River, New Jersey Columbus, Ohio

Library of Congress Cataloging-in-Publication Data
Goody, Roy W.
MicroSim PSpice for Windows / Roy W. Goody. — 2nd ed.
p. cm.
Previous ed. published under title: PSpice for Windows.
Includes index.
Contents: v. 1. A circuit simulation primer
ISBN 0-13-655804-6
1. PSpice for Windows. 2. Electric circuits—Computer simulation.
3. Electronic circuits—computer simulation. I. Goody, Roy W.
PSpice for Windows. II. Title
TK454.G66 1998 97-7764
621.3815'01'135369—dc21 CIP

Cover photo: M. Angelo, Westlight
Editor: Linda Ludewig
Production Editor: Christine M. Harrington
Design Coordinator: Karrie M. Converse
Cover Designer: Rod Harris
Production Manager: Laura Messerly
Marketing Manager: Debbie Yarnell

This book was printed and bound by Courier/Kendallville, Inc. The cover was printed by Phoenix Color Corp.

Simon & Schuster/A Viacom Company
Upper Saddle River, New Jersey 07458

Printed in the United States of America

10 9 8 7 6 5 4 3 2

ISBN: 0-13-655804-6

Prentice-Hall International (UK) Limited, *London*
Prentice-Hall of Australia Pty. Limited, *Sydney*
Prentice-Hall of Canada, Inc., *Toronto*
Prentice-Hall Hispanoamericana, S. A., *Mexico*
Prentice-Hall of India Private Limited, *New Delhi*
Prentice-Hall of Japan, Inc., *Tokyo*
Simon & Schuster Asia Pte. Ltd., *Singapore*
Editora Prentice-Hall do Brasil, Ltda., *Rio de Janeiro*

Contents

Part III — Filter Synthesis

Part IV — Applications

Part V — PC Boards

Appendices

Preface

We know what we are,
but know not what we may be.
WILLIAM SHAKESPEARE, *HAMLET,* 1600

This text is a follow-up to *PSpice for Windows, 2nd Edition, Volume I.* The experimental activities assume that you have acquired a working knowledge of the PSpice techniques presented in Volume I.

Take all that you have learned in Volume I, covering DC/AC and devices and circuits, and fundamental PSpice techniques, and add to that the feedback, digital, filter design, and advanced techniques of this text. Mix all this together, blend in experience, skill, and artistry—and step into an amazing world of design and innovation where the only limitation is your imagination.

OVERVIEW

PSpice for Windows is offered in two volumes. Volume I covers DC/AC and devices and circuits. This text, Volume II, continues our studies into operational amplifiers, digital, advanced filter design, and printed circuit board development.

The major features of *PSpice for Windows* are:

- It is based on the popular professional-level PSpice software from MicroSim Corporation—not a scaled-down version intended primarily for education.
- It will operate on any computer that runs Windows 95, 3.1, or NT and has 8M of extended memory.
- It combines both circuit simulation and electronic theory.
- Most of the experimental activities can be done hands-on using conventional equipment as well as PSpice.

- It is designed to supplement or replace a laboratory or theory text in a conventional op amp or digital course.
- It is comprehensive, covering nearly every available feature of PSpice.
- It requires no previous knowledge of circuit simulation or op amp and digital theory.
- It is aimed at the technology student, but it is entirely appropriate for technicians and engineers.

THE SECOND EDITION

In format and content, this second edition is very similar to the first. Although new chapters have been added and major improvements have been made in graphics and layout, its major difference lies in the use of PSpice version 7.1. Version 7.1 is more powerful, is easier to use, and offers several new features. If you are presently using the first edition, you will find it very easy to adjust to the changes.

In fact, the timely release of new editions based on updated versions of PSpice is a major feature of this text series. Because all text and graphics is done on the computer, manuscripts are submitted to the publisher in camera-ready format—greatly reducing publication costs and allowing a faster than normal edition update. Therefore, the promise and commitment of the author and publisher is to ensure the continuous availability of a state-of-the-art PSpice experience.

THE EVALUATION VERSION

Fortunately for those of us in education, MicroSim Corporation has made *evaluation* software available at no cost—with copying of the software "welcome and encouraged." Evaluation version 7.1 on CD-ROM comes with the Instructor's Guide to this text. The evaluation software is also available directly from MicroSim or from their web site—from which floppy disks can be created.

The two filter synthesis chapters of Part III use special filter design software (*fseval63*) provided with the Instructor's Guide. It can also be downloaded from MicroSim's web site.

Newer versions of the PSpice software are constantly being released and chances are good that they will work with this manual. In general, you should use the latest version; if any adjustments are necessary, they should be minor. *However, to be perfectly safe, you may wish to stay with version 7.1 until the next edition of PSpice for Windows is released.*

All the activities in this book are based on the evaluation version. In features and performance, it is identical to the professional-level version used within industry. Its only major limitation is the number of symbols and components that can be placed on the schematic, and that all circuits must fit on a single schematic page. We can live easily within these limitations, and for the most part they will be completely invisible.

For those who may wish to go beyond the evaluation version, MicroSim Corporation also gives very generous academic discounts on their full-fledged software packages.

SYSTEM REQUIREMENTS

- PSpice version 7.1 (used with this second edition) requires 8Mb of RAM (16Mb recommended); a 486 or Pentium computer; and Windows 3.1, 95, or NT. It is most easily installed via CD-ROM (but can be installed from MicroSim's website, or by way of floppy disks created from the website). Version 7.1 is the last release that will support Windows 3.1.
- Version 6.0 (used in the first edition) requires only 4MEG of RAM, can easily run on a 386 computer, can be used with Windows 3.1 (but <u>not</u> Windows 95), and is contained on only three floppy disks.

If the equipment required for version 7.1 presents a problem, you may wish to temporarily stay with the 6.0-based first edition.

ORGANIZATION

The first ten chapters of Part I (Feedback) emphasize the operational amplifier in the closed loop configuration. The ten chapters of Part II (Digital Circuits) cover all aspects of basic digital, from the simple gates to random-access memory. The two chapters of Part III (Filter Synthesis) cover advanced filter design using special filter synthesis software. The four chapters of Part IV (Applications) make use of all the previous material to design and test more complex analog/digital projects, and the final chapter of Part V introduces us to printed circuit board design.

A SUGGESTION

As with Volume I, we recommend that students also receive hands-on experience by prototyping actual circuits and troubleshooting with conventional instruments.

A useful approach when computers are limited is to divide students into two or more groups and alternate between PSpice and hands-on. It is especially instructive to perform the same activity using both and to compare the results. *In this regard, most of the experimental activities outlined in this text can be performed using both techniques.*

PREREQUISITES

Besides the ability to perform simple mathematical operations, the only prerequisite needed is a basic understanding of the theory and techniques presented in Volume I.

CREDITS

A very special thanks to copy editor Marianne L'Abbate for her profound knowledge of the English language, her ability to root out the most entrenched errors, and her willingness to make many useful suggestions concerning organization and substance.

I also wish to express my sincere gratitude to production editor Christine Harrington and administrative editor Linda Ludewig of Prentice Hall Publishing. Under their careful guidance, the project steadily moved forward and was released right on time.

Of course, MicroSim Corporation deserves special credit for making the evaluation disk available at no cost. Their foresight makes it possible for colleges and universities to teach circuit simulation at the professional level without breaking their ever-shrinking budgets.

Thank you for adopting *PSpice for Windows*; good luck and good success.

Roy W. Goody
Mission College

Introduction

Dreams are true while they last,
and do we not live in dreams?
LORD TENNYSON, 1869

In many ways, our modern civilization is a product of our dreams, for every advance we now enjoy—from agriculture to space flight—first came alive in our imagination. PSpice is the perfect dream machine, for it is truly a link between the fantasy world of our mind where ideas are born, and the material world of our everyday lives where ideas are realized.

With PSpice circuit simulation, our wildest dreams can come true.

MOUSE CONVENTIONS

The same mouse/keyboard conventions introduced by Volume I also apply here.

- **CLICKL** or **BOLD PRINT** (*click left once*) to select an item.
- **DCLICKL** (*double click left*) to end a mode or edit a selection.
- **CLICKR** (*click right once*) to abort a mode.
- **DCLICKR** (*double click right*) to repeat an action.
- **CLICKLH** (*click left, hold down, and move mouse*) to drag a selected item. Release left button when placed.
- **DRAG** (*no clicks, move mouse*) to move an item.

GETTING STARTED

For those who may not have ready access to Volume I, this section is reproduced in its entirety.

The evaluation version 7.1 on CD-ROM comes with the Instructor's Guide to this text. You may also contact MicroSim Corporation and ask for a copy, order a copy from MicroSim's website, or you can download the software from the website. The website will also allow you to make a floppy disk set.

> Be aware that the busy folks at MicroSim are constantly releasing new versions of PSpice, and chances are good that the new versions will work with this manual. However, to be perfectly safe, we recommend the use of version 7.1.

MicroSim Corporation 20 Fairbanks Irvine, CA 92618	
GENERAL:	**phone: (714) 770-3022 fax: (714) 455-0554 autofax: (714) 454-3296 WWW: http://www.microsim.com FTP: ftp://ftp.microsim.com**
SALES:	**phone: (800) 245-3022 E-mail: sales@microsim.com**
TECH SUPPORT:	**phone: (714) 837-0790 E-mail: tech.support@microsim.com**

INSTALLING THE SOFTWARE

PSpice

There are three methods available for installing PSpice software: From the evaluation version CD-ROM, by downloading from their website, or by using floppy disks (originally made from the website).

CD-ROM

Windows 95

1. Place the CD-ROM in drive.
2. Wait for the MicroSim AutoPlay screen to appear.
3. Follow the directions, accepting all default options.

If the AutoPlay function is not enabled:

1. Select Run from the Start Menu.
2. Enter D:\SETUP.EXE.
3. Follow the directions, accepting all default options.

Windows NT

1. Place the CD-ROM in drive.
2. Select Run from the File menu of the Program Manager.
3. Enter D:\SETUP.EXE.
4. Follow the directions, accepting all default options.

Windows 3.1

1. Place the CD-ROM in drive.
2. Enter the File Manager and **CLICKL** on drive D.
3. **DCLICKL** on SETUP.EXE.
4. Follow the directions, accepting all default options.

DOWNLOAD FROM WEBSITE / CREATE FLOPPIES

1. Go to MicroSim's website: **http://www.microsim.com**
2. CLICKL on **Download**.
3. CLICKL on *Evaluation requirements, installation* to bring up the *About the MicroSim DesignLab Evaluation Version* page. Either print or make a copy of the *How to Install the Windows Evaluation Versions* section.

4. Follow steps 1-5 of the install instructions. Be aware that all files are self-extracting ZIP files. (If your intent is to create the 13-disk floppy set, be sure to download and type the 13 pieces.)

Filter design

FSEVAL63 is available on the CD-ROM that comes with the Instructor's Guide or from MicroSim's website. *FSEVAL63* requires at least 425K of free base-memory to install and execute. Type *mem* from C: to determine available memory.

1. First, create directory *fseval63* on drive C.

2. If you are using the website, download compressed file *filter.exe* and place in directory *fseval63*. If you are using the CD-ROM, copy compressed file *filter.exe* to directory *fseval63*.

3. Restart Windows in the MS-DOS mode and enter directory *fseval63*. Type *filter* to extract files. Type *install.* (***C*** is boot drive, ***Yes*** *C:\FSEVAL63* is installation directory, ***Yes*** to overwrite files, ***Yes*** to add to path statement, ***Go ahead and modify***.)

4. Type *filter* to enter the program, and you are ready to go!

ENTERING SCHEMATICS

Windows 95

1. **CLICKL** on *Start*, **Drag** to *Programs*, **Drag** to *MicroSim71* (or other), **Schematics**. The Schematics window appears.

2. Turn to Chapter 1; you are ready to go!

Windows 3.1

1. Bring up the Program Manager window and the *MicroSim Eval* group (**CLICKL** on appropriate icons, as necessary). **DCLICKL** on the *Schematics* icon to bring up the *Schematics* window.

2. Turn to Chapter 1; you are ready to go!

EVALUATION VERSION LIMITATIONS

If you were able to download the *Evaluation requirements, installation* page from MicroSim's Website, refer to the "Limitations" section. Otherwise, refer to the abbreviated limitations listed below:

- Maximum of 50 symbols on a single A-size schematic page.
- Maximum of 30 components can be placed on a PCB layout.
- Maximum of 64 nodes, 10 transistors, 2 op amps, or 65 digital primitive devices, or a combination thereof.
- Optimizer limited to one goal, parameter, and constraint.

Special Note

Many times throughout this text you will be asked to bring up the *output file*. Inability to do so is typically caused by an incompatibility between the MicroSoft IntelliPoint mouse drive (pointer.exe) and Windows 95/NT.

To correct the problem, bring up file WIN.INI in directory WINDOWS and look for either of the following:

[WINDOWS]
LOAD=C:\mouse\pointer.exe
OR
[WINDOWS]
LOAD=C:\MSINPUT\pointer.exe

Modify either line as follows and reset your computer:

[WINDOWS]
LOAD=

If this does not fix the problem, try any of the following:

- If you are using *Logitech's* Smart mouse driver, obtain a later version.
- Do not use *AfterDark* version 3.0.
- Reload the latest versions of Windows 95 Service Pack and OLE upgrade. These can be obtained from:

http://www.microsoft.com

PART I
Feedback

In Part I, we concentrate on the science of *cybernetics*, where the power of *feedback* gives conventional circuits amazing properties. We will find that the *differential amplifier* is the ideal circuit to implement feedback.

A precision differential amplifier is known as an *operational amplifier*, and is one of the most powerful analog devices available to the designer. When the operational amplifier is combined with negative and positive feedback, we create such devices as amplifiers, buffers, integrators, differentiators, oscillators, and filters.

CHAPTER 1

The Differential Amplifier
Common-Mode Rejection

OBJECTIVES

- To analyze the differential amplifier.
- To compare the differential mode with the common mode.

DISCUSSION

A differential amplifier has two inputs. It strongly amplifies the *difference* between the two inputs, but it rejects signals that are *common* to both inputs. Such an amplifier is shown in Figure 1.1.

- Signals common to both inputs are forced to pass through the large 5kΩ emitter bias resistor and are rejected.
- Signals differential to both inputs pass easily through two small re' values and are amplified.

As an added bonus, the differential amplifier of Figure 1.1 requires no coupling or bypass capacitors and is therefore a *DC amplifier* (one with no low-frequency roll-off). The split power supply eliminates the input coupling capacitor by placing the input leads at ground, and the symmetric arrangement of the transistor pair eliminates the bypass capacitor because each transistor bypasses the other.

As we will see in the following chapters, the differential amplifier provides a far more important function than simply amplifying differential-mode DC signals: it provides a convenient method of implementing *negative or positive feedback*.

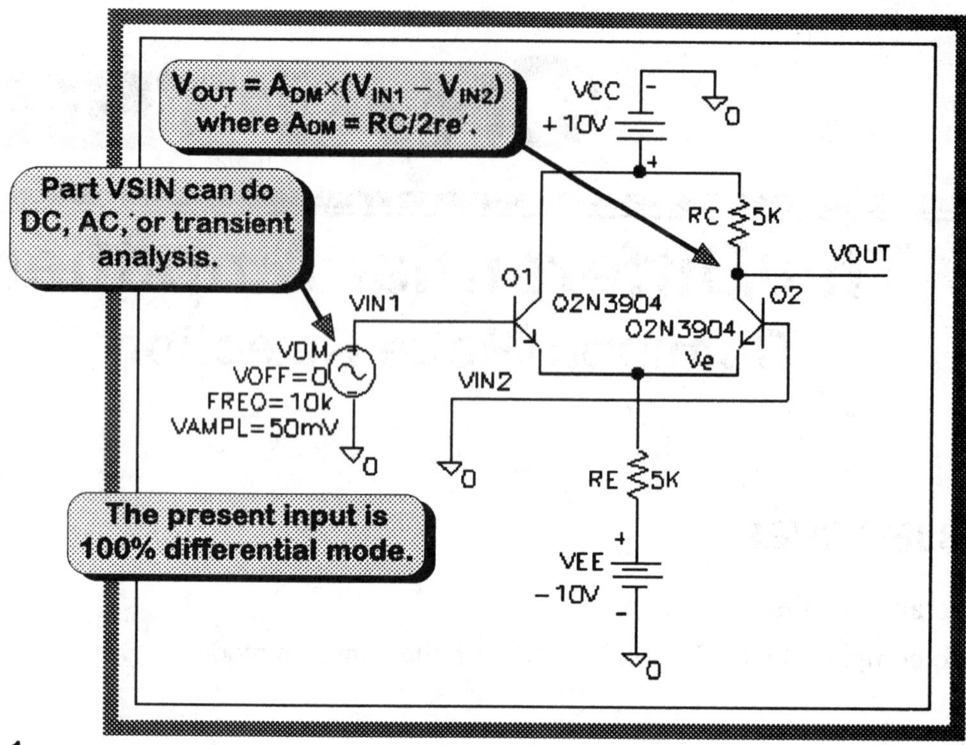

FIGURE 1.1
The differential amplifier

SIMULATION PRACTICE

1. Draw the circuit of Figure 1.1 and set the attributes as shown. (V_{DM} is a differential mode input.)

2. Using both hand calculations and PSpice (bias point analysis), determine the Q-point of transistor Q2 and enter your answers into the table below.

   Do theory and experiment approximately match?

   Yes **No**

	Calculations	**PSpice**
I_{CQ}	____________	____________
V_{CEQ}	____________	____________

Differential mode

In most of the analysis steps to follow, you are free to use either the AC Sweep or the Transient mode—or both. It's your choice!

Reminders

- When using AC analysis, trace variables may be plotted directly in *combinational* form (such as V_{OUT}/V_{IN}). (All AC trace variables automatically display polar-form peak magnitudes by default, and all bias point (DC) components are automatically rejected.)
- When using transient analysis, trace variables must be plotted *separately* in order to determine each peak value. All transient trace variables display instantaneous values, and bias point values are not rejected. Use peak-to-peak/2 to factor out the DC value and average the + and – peaks.

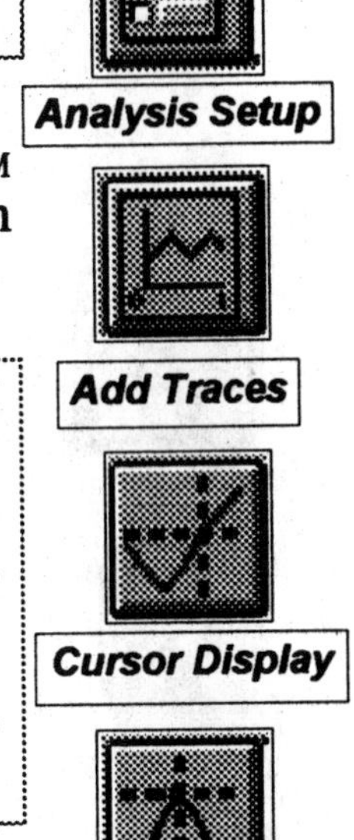

3. Using hand calculations and PSpice, determine midband A_{DM} (differential mode gain), Z_{IN}, and Z_{OUT}, and place your answers in the table below. (Assume *Beta* equals 175 and re′ ≈ $25mV/I_{EQ}$.)

Reminder: To determine Z_{OUT}, add a capacitor to the output, and use either of the following methods:

(a) **Algebra**: Measure V_{OUT} with and without a load. (Without a load, $V_{OUT} = V_{THEVENIN}$). Use algebra to determine Z_{OUT}.

(b) **Injection**: Ground the inputs, move VSIN to the output, and measure V_{OUT} and I_{OUT} directly.

	Calculations	PSpice
A_{DM}	________	________
Z_{IN}	________	________
Z_{OUT}	________	________

Use peak-to-peak/2 where appropriate to factor out DC offset.

Do theory and experiment approximately match (within 30%)?

Yes **No**

4. Based on the phase relationship between the input and output voltages, circle the appropriate label for V_{IN1}. (In the AC Sweep mode, use *P(V(Vout)/V(Vin1))*.)

   Inverting **Noninverting**

5. Move V_{DM} from V_{IN1} to V_{IN2}, again measure the phase relationship between input and output, and circle the appropriate label for V_{IN2}.

   Inverting **Noninverting**

6. Using the symbols for *inverting* (–) and *noninverting* (+), label the two inputs of Figure 1.1.

7. Determine the bandwidth of the amplifier of Figure 1.1.

   BW = ________________

   Is it a DC amplifier?

   Yes **No**

Common mode

8. Connect the circuit in the *common mode*, as shown in Figure 1.2.

9. Using the test mode of your choice, measure the midband common-mode voltage gain (V_{OUT}/V_{IN}) and enter the result below.

   A_{CM} = __________

   Comparing A_{DM} to A_{CM}, does the differential amplifier tend to reject common-mode input signals?

   Yes **No**

10. As a measure of how much better differential-mode signals are amplified over common-mode signals, determine the *common-mode rejection ratio* (CMRR).

$$\text{CMRR} = \frac{A_{DM}}{A_{CM}} = \_\_\_\_\_\_\_\_\_\_\_\_$$

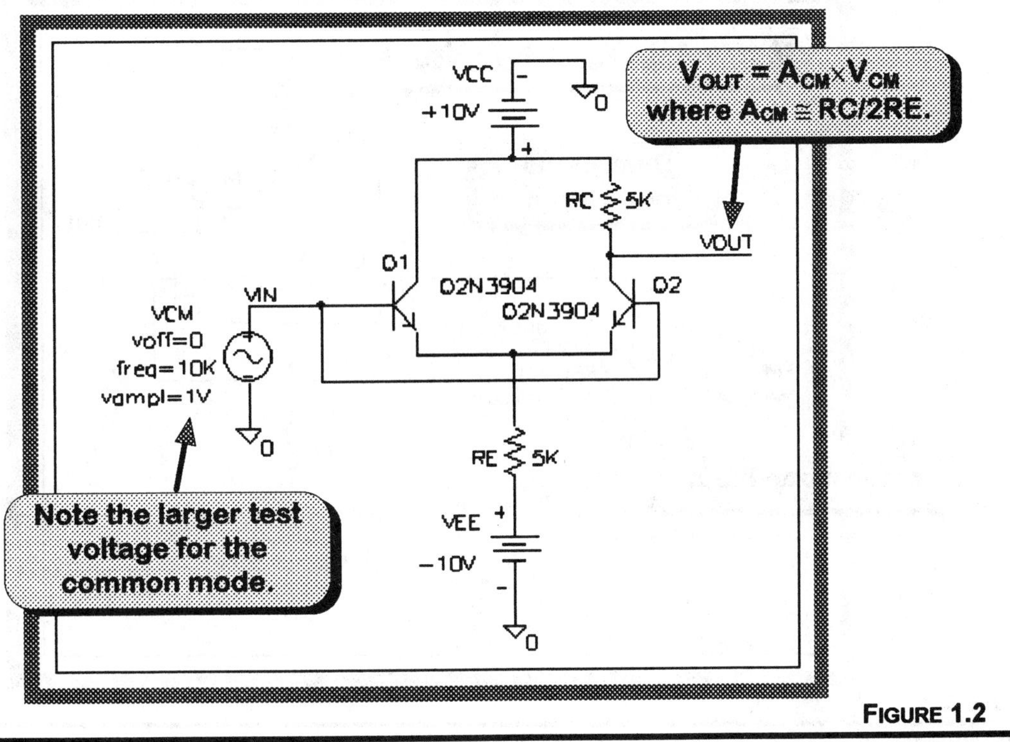

FIGURE 1.2
Common mode input

Advanced activities

11. Based on the values of A_{DM} and A_{CM} determined earlier, sketch a time-domain graph of V_{OUT} for the circuit of Figure 1.3. Compare your answer with the measured (PSpice) results.

12. The circuit of Figure 1.4 uses *current-source* biasing to improve CMRR. Measure the circuit's CMRR and compare your results to those of Step 10. Did the CMRR go up?

    Yes **No**

13. Measure the harmonic distortion of any of the amplifiers of this chapter. Would you say it is high or low? (Assume that any value above 10% is high.)

    High **Low**

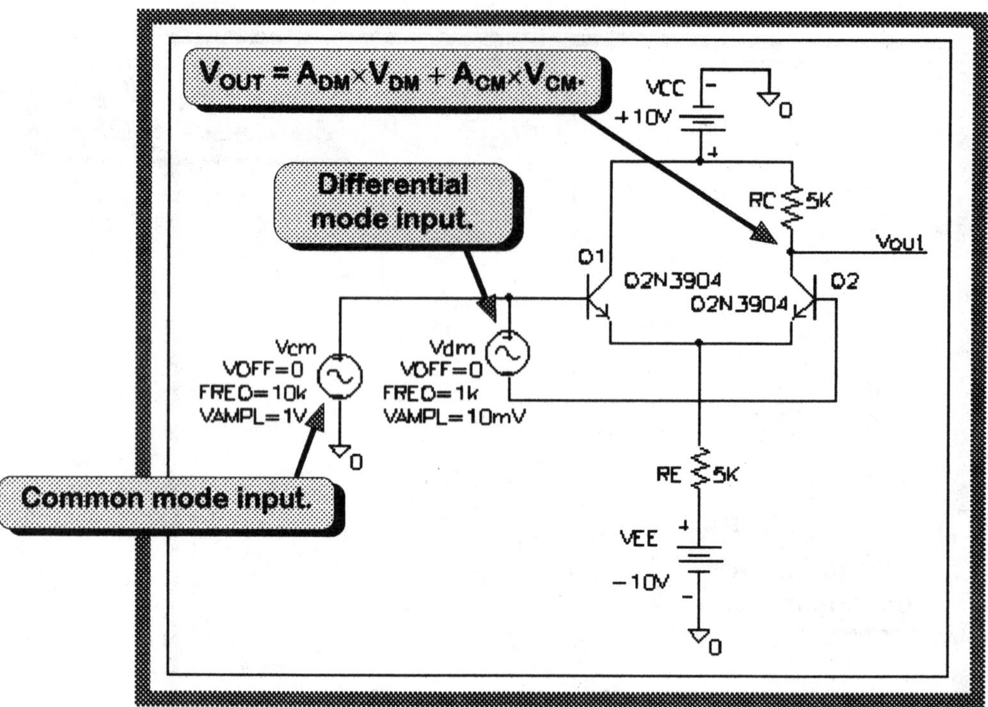

FIGURE 1.3
Combined circuit

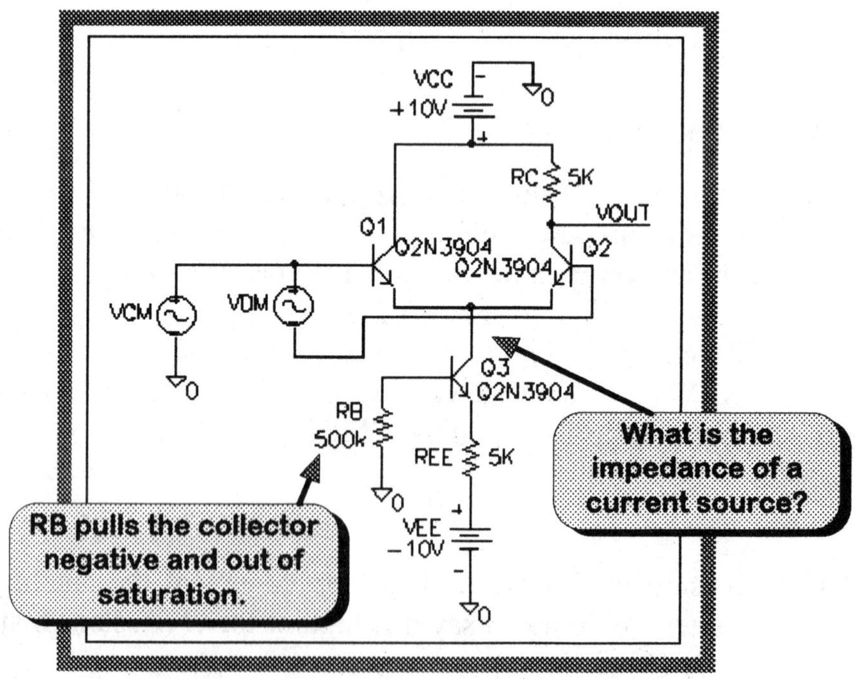

FIGURE 1.4
Using current-source biasing

EXERCISE

- Design a differential amplifier that gives a differential (balanced) *output*. (Hint: Add a 5k resistor to the collector of Q1 in Figure 1.4 and tap the voltage off both Q1 and Q2.) How does the gain compare to that of Figure 1.4?

QUESTIONS AND PROBLEMS

1. Why is a *bypass* capacitor not required in the differential amplifier?

2. Why is there a "2" in each of the gain equations below?

Differential-mode gain	Common-mode gain
$A_{DM} = RC/2re'$	$A_{CM} \cong RC/2RE$

3. How could you swamp the circuit of Figure 1.1 in order to reduce the harmonic distortion? What would happen to the voltage gain (A_{DM})?

4. When the circuit was modified as in Figure 1.4, why did the CMRR go up?

5. Given the following, determine V_{OUT}(peak) as accurately as you can. (Note: Assume that V_{IN1} and V_{IN2} are peak values, and are the same frequency and phase.)

V_{IN1} = 200mV, V_{IN2} = 215mV, A_{DM} = 100, A_{CM} = 1

V_{OUT} = ____________

CHAPTER 2

Negative Feedback

Open Loop Versus Closed Loop

OBJECTIVES

- To design a three-stage amplifier.
- To compare the properties of an open-loop and closed-loop amplifier.

DISCUSSION

In this chapter, we show how a magical drop of negative feedback can greatly enhance the properties of an amplifier. We start with the conventional three-stage amplifier of Figure 2.1—an amplifier that presently does not employ feedback (the *open loop* configuration).

Then, for just 5 cents in added components (two resistors), we add negative feedback (*close the loop*) and observe the amazing improvement in design ease, stability, distortion, bandwidth, Z_{IN} and Z_{OUT}. We choose the differential amplifier of Chapter 1 for the first stage because it is ideally suited to carry out negative feedback.

Referring to Figure 2.1, we have a differential input stage, a common-emitter middle stage, and a Class B buffer final stage. We note the complete absence of coupling and bypass capacitors, making the amplifier a DC amplifier.

- Coupling capacitors are not required because of the split power supply, and because the middle stage is also a *level-shifter* that allows for *direct coupling* between stages.
- Bypass capacitors are not required because the first stage is a differential amplifier, the second stage is a level-shifter, and the third stage is a buffer.

SIMULATION PRACTICE

Open loop (without feedback)

1. Draw the three-stage open-loop (OL) amplifier of Figure 2.1 and set the attributes as shown. (We tap the first stage signal off Q1 to compensate for the phase inversion introduced by Q3.)

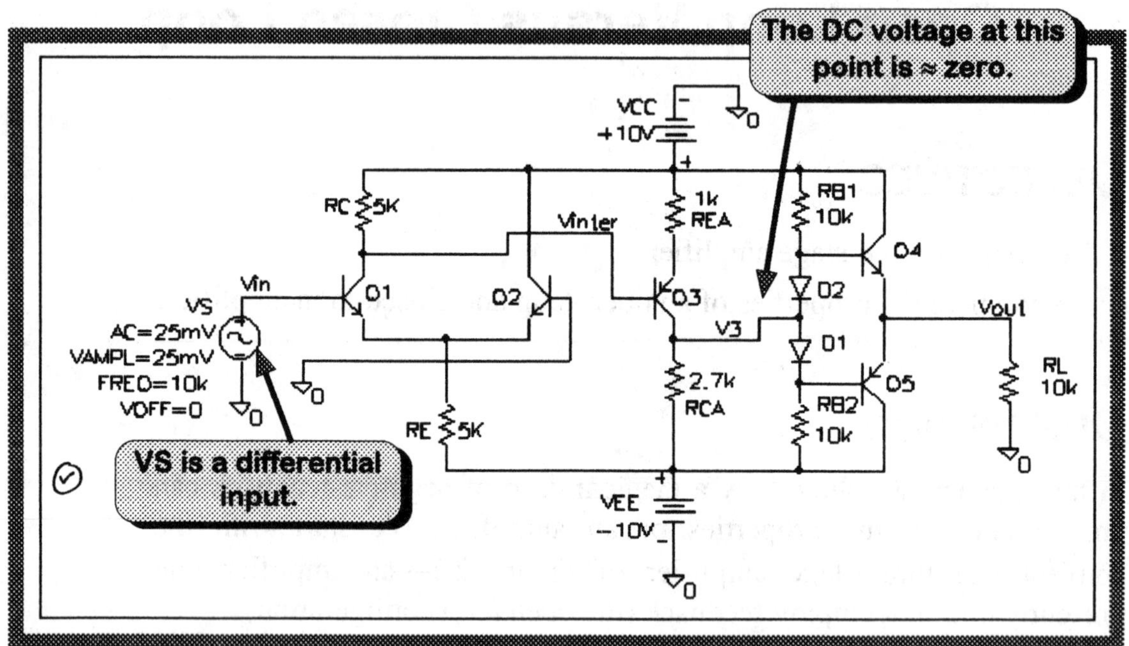

FIGURE 2.1
Three-stage amplifier (open loop)

2. Using PSpice (transient, Fourier, and AC modes as needed), determine the following open-loop (OL) values. (<u>Note</u>: When using AC Sweep mode, all values are midband.)

A (gain) = ________ A_{dB} = ________

Z_{IN} = ________

Z_{OUT} = ________

BW (bandwidth) = ________

HD (distortion) = ________

SNR (signal/noise) = ________

Hint: Review Volume I as needed.

Closed loop (with feedback)

3. Add a negative feedback path to the circuit of Figure 2.1, and create the closed-loop (CL) amplifier of Figure 2.2.

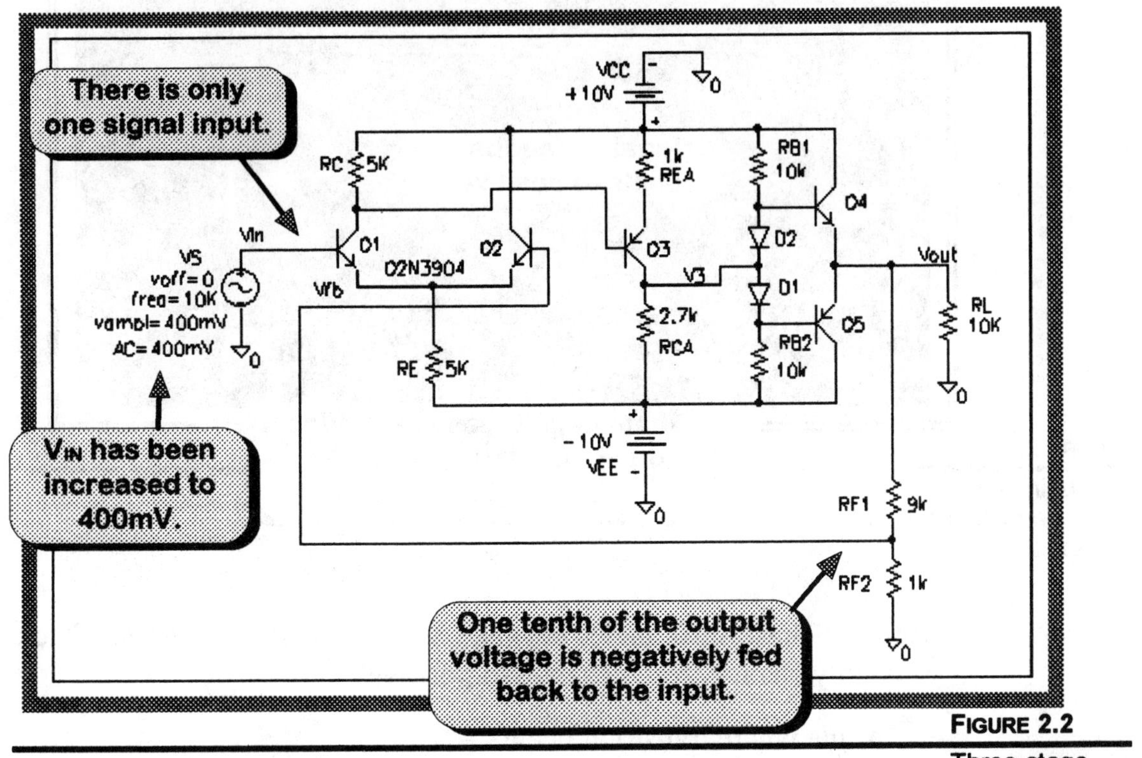

FIGURE 2.2
Three-stage amplifier (closed loop)

4. First, use the transient mode to look at the input and output waveforms. Is the output "well-behaved" (no clipping or distortion), and is the voltage gain approximately 10?

Yes **No**

5. Switch to the AC Sweep mode and generate the gain and phase plots of Figure 2.3.

> The dramatic spike at approximately 20MHz results from: (1) negative feedback turning into positive feedback due to high-frequency phase shifts, and (2) overall gain > 1.

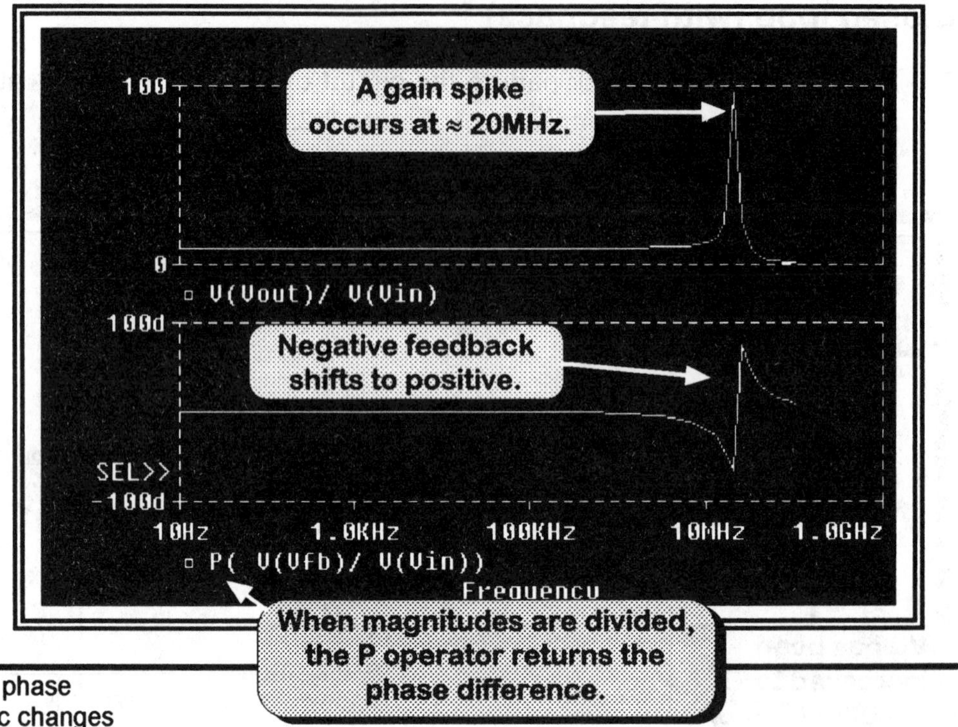

FIGURE 2.3

The gain and phase show dramatic changes around 20MHz.

6. At what phase shift angle (between V_{FB} and V_{IN}) does the feedback suddenly change from negative to positive?

   Angle where transition occurs = ____________

7. To remove the high-frequency gain spike, add the .5nF *compensation capacitor* (CC) shown in Figure 2.4.

   > The compensation capacitor creates a low-pass filter that reduces the overall gain below 1 when the feedback turns positive at high frequencies. This prevents the output signal from growing and producing a spike.

8. We again generate plots of gain and phase shift and find that the high-frequency gain spike and sudden phase shift has been removed (Figure 2.5).

   Is the low-frequency gain approximately the same as before compensation?

   Yes **No**

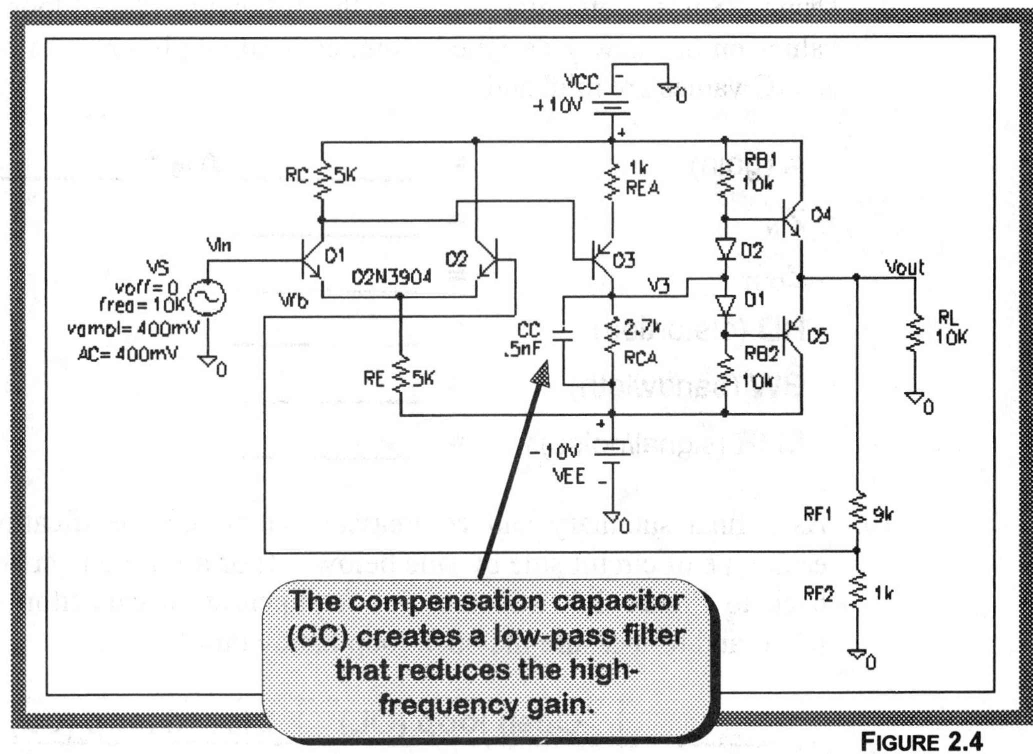

FIGURE 2.4
Compensation capacitor added

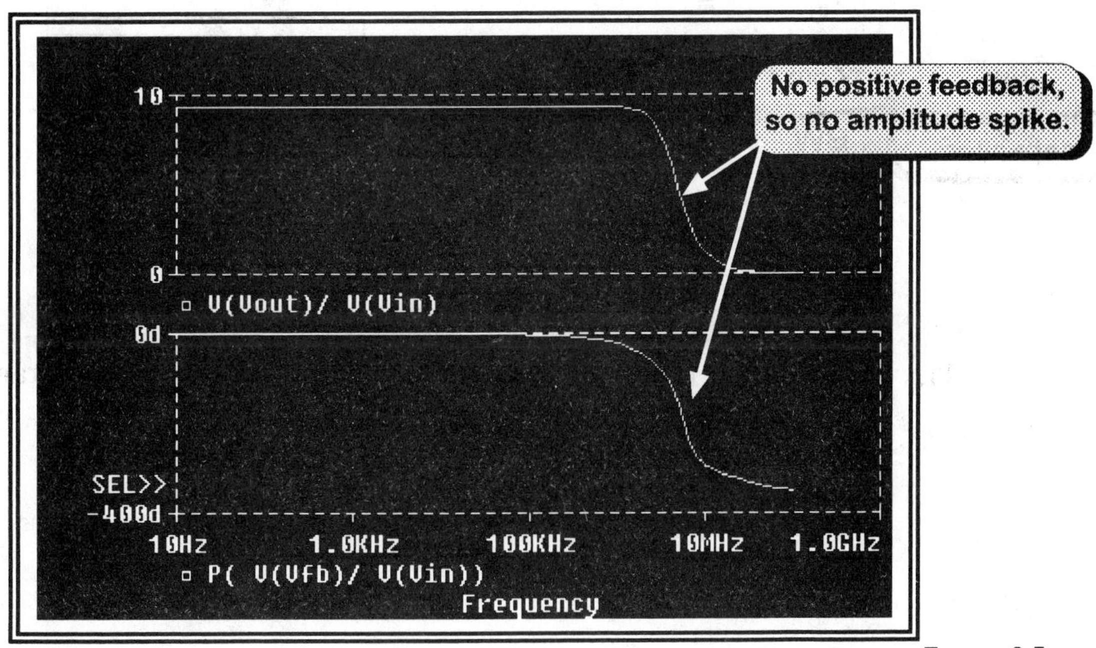

FIGURE 2.5
Waveforms with compensation

9. Using PSpice, determine each of the following closed-loop (CL) values on our newly designed, compensated amplifier. (As before, all AC values are midband.)

A (gain)	=	________	A_{dB} =	________
Z_{IN}	=	________		
Z_{OUT}	=	________		
HD (distortion)	=	________		
BW (bandwidth)	=	________		
SNR (signal/noise)	=	________		

10. As a final summary and comparison, enter the specifications of each type of circuit side by side below. (For a fair comparison, go back to Figure 2.1, add the .5nF compensation capacitor across RCA, and re-measure the open loop bandwidth.)

		Open loop (OL)	Closed loop (CL)
A(reg)	=	________	________
Z_{IN}	=	________	________
Z_{OUT}	=	________	________
BW	=	________	________
HD	=	________	________
SNR	=	________	________

With compensation capacitor. (→ BW)

11. Based on the results of Step 10, what conclusions can you draw about the effects of negative feedback? (What do we profit by sacrificing voltage gain?)

Advanced activities

12. Reduce the load (RL) of both amplifiers (with and without feedback) from 10k to 100Ω and explain the results.

13. Set 10% device tolerances to all the resistors of the feedback amplifier of Figure 2.4, except the two feedback resistors (RF1 and RF2). Run a worst case analysis and note the results. Then reverse the configuration (only RF1 and RF2 have 10% tolerances) and note the results. By comparing the two cases, what conclusions can you draw? (To improve sensitivity, would it be a waste of money to upgrade all resistors to 1% tolerance?)

14. Change the value of the compensation capacitor (CC) and explain the results.

EXERCISE

- Design a negative feedback amplifier that uses 100% feedback. What are its major characteristics? (Would it make a good buffer?)

QUESTIONS AND PROBLEMS

1. To achieve negative feedback, the feedback signal must be:

   (a) in phase with the input signal.
   (b) 180° out of phase with the input signal.

2. What is a *direct-coupled* amplifier?

3. Why is the final stage of the amplifier of Figure 2.1 a Class B buffer?

4. How is the gain of the closed-loop amplifier of Figure 2.4 related to the values of the feedback resistors (RF1 and RF2)? How is it related to the percentage of feedback?

5. Why is the closed-loop amplifier of Figure 2.4 a better *buffer* than the open loop amplifier of Figure 2.1?

6. Did the addition of a compensation capacitor to the closed-loop configuration (Figure 2.4) cost us any significant bandwidth? Can the same be said of the open-loop configuration of Figure 2.1?

7. What is the difference between *negative* and *positive* feedback?

CHAPTER 3

Subcircuits

Creating Symbols

OBJECTIVES

- To create a subcircuit.
- To design and test a schematic that uses a subcircuit.

DISCUSSION

The three-stage amplifier of Figure 3.1 (reproduced from Chapter 2) might be called a standard circuit. It is a combination of fundamental components that performs a basic function (differential amplification and buffering).

In a large-scale schematic, such a circuit combination may appear many times. Rather than clutter the main schematic with the circuit details, it would be most convenient if we could represent the circuit as a single functional block (a *subcircuit*) and store it within a symbol library. Then, whenever we needed to incorporate such a device within our design, we could fetch it from the library as easily as we do a diode or transistor. Creating a subcircuit is the purpose of this chapter.

To create a ready-to-use subcircuit, we follow a three-step sequence:

Step 1: Create the subcircuit MODEL and make it available *locally* (valid only within the schematic presently being edited).

Step 2: Create the subcircuit PART by designing the PART symbol, assigning the symbol a PART name, and storing it in a symbol library.

Step 3: Associate the MODEL with the PART.

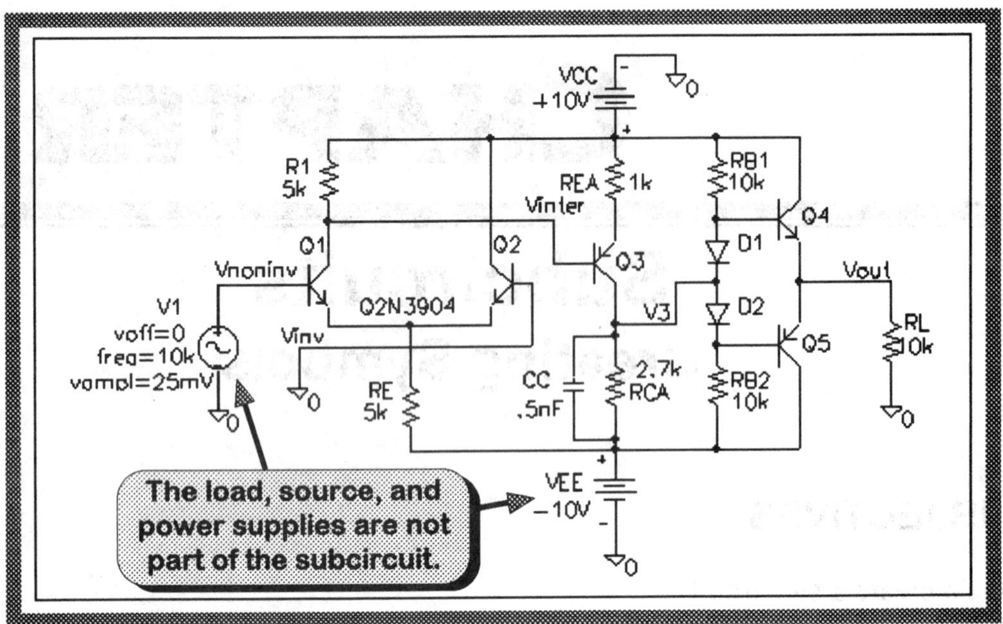

FIGURE 3.1
Three-stage amplifier (open loop)

Note: Some of the steps to follow may not be necessary if previously performed by another student on the same computer.

SIMULATION PRACTICE

Step 1: Create the MODEL

Open File

1. Bring back the three-stage differential amplifier/buffer of Figure 3.1 (from Chapter 2).

Create File

2. As shown in Figure 3.2, replace the input/output lines with the *interface port symbols* IF_IN and IF_OUT (from *port.slb*). To label each port symbol as shown, **DCLICKL** on each symbol and fill in.

3. Save the schematic to file *opamp.sch*.

Save File

4. Perform **Tools, Create Subcircuit** to generate the subcircuit MODEL definition of Table 3.1. This model definition is automatically stored as *opamp.sub* in the same directory as *opamp.sch*.

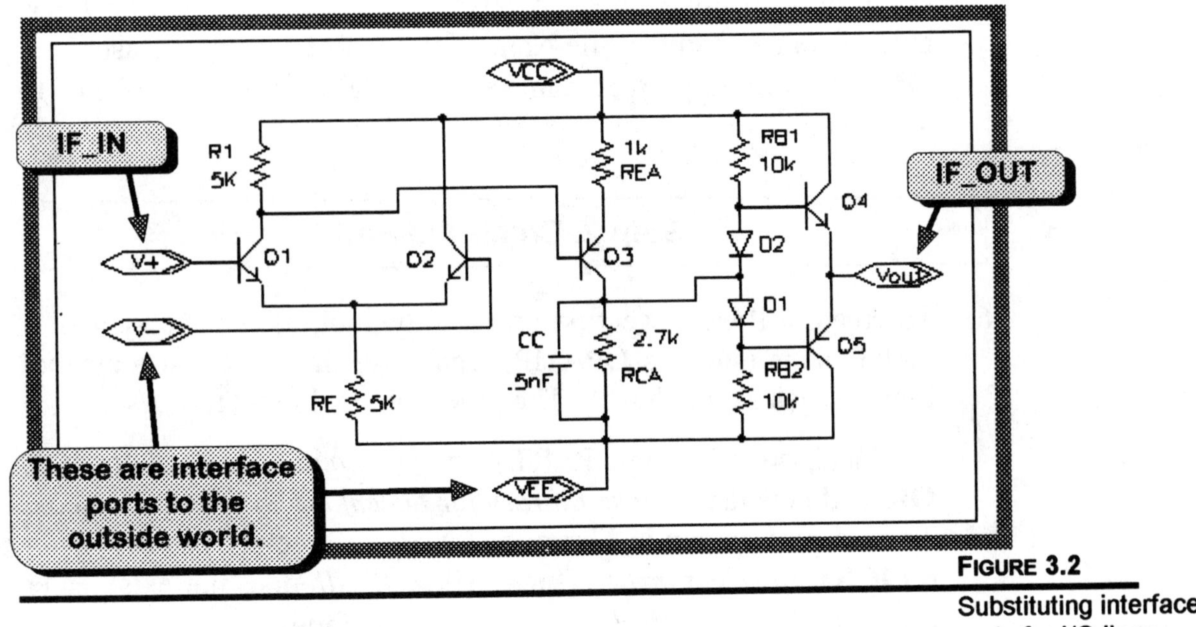

FIGURE 3.2
Substituting interface ports for I/O lines

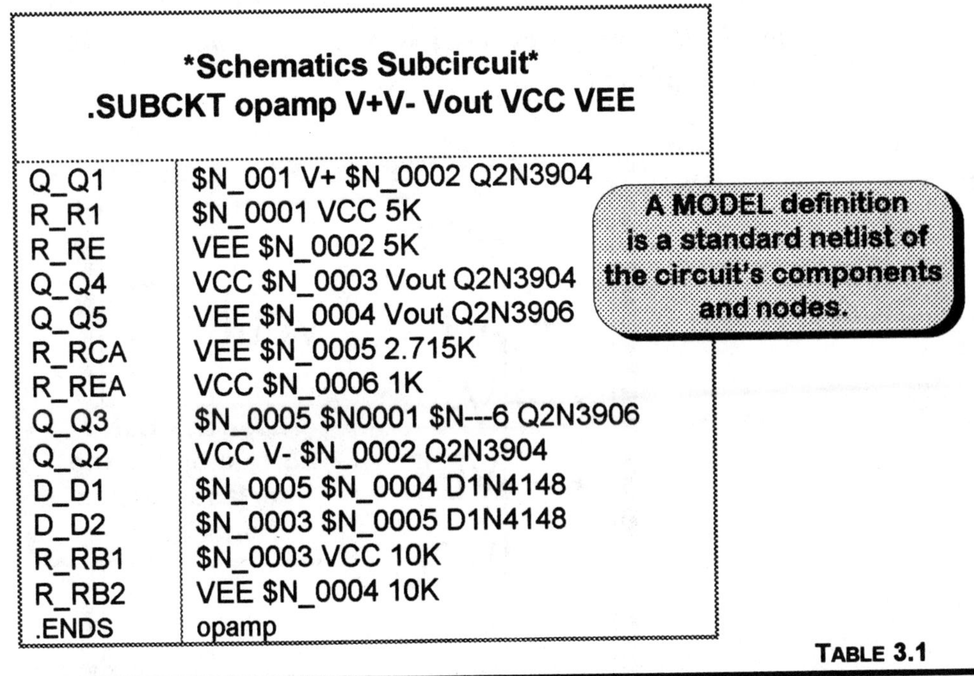

Schematics Subcircuit .SUBCKT opamp V+V- Vout VCC VEE	
Q_Q1	$N_001 V+ $N_0002 Q2N3904
R_R1	$N_0001 VCC 5K
R_RE	VEE $N_0002 5K
Q_Q4	VCC $N_0003 Vout Q2N3904
Q_Q5	VEE $N_0004 Vout Q2N3906
R_RCA	VEE $N_0005 2.715K
R_REA	VCC $N_0006 1K
Q_Q3	$N_0005 $N0001 $N---6 Q2N3906
Q_Q2	VCC V- $N_0002 Q2N3904
D_D1	$N_0005 $N_0004 D1N4148
D_D2	$N_0003 $N_0005 D1N4148
R_RB1	$N_0003 VCC 10K
R_RB2	VEE $N_0004 10K
.ENDS	opamp

TABLE 3.1
Subcircuit MODEL definition

5. To make the MODEL definition (*opamp.sub*) available *locally* (valid only to *opamp.sch*): **Analysis, Library and Include Files,** enter *opamp.sub* in the File Name field, **Add Library** (no asterisk), **OK**. (Add Library* would make the model global and available to all schematics.)

Step 2: Create the PART

6. To create a PART, we first create a symbol, assign the symbol a PART name (such as OPAMP), and store the PART in a symbol library (such as *mylib.slb*). The procedure to do this follows:

   File, Symbolize, type PART name *opamp* in *Save As* dialog box, **OK**, and note the *Choose Library for Schematic Symbol* dialog box. To save part *opamp* in library *mylib.slb* in directory *msimev71\lib*, **CLICKL** to select *mylib*, **Open**. (If *mylib.slb* does not exist, enter *c:\msimev71\lib\mylib.slb* in file name box, **Open**.)

7. The next step is to edit the new PART symbol by bringing up the Symbol Editor: **File, Edit Library, File, Open, DCLICKL** on *mylib.slb*, **Part, Get, DCLICKL** on *opamp* to bring up the initial symbol of Figure 3.3.

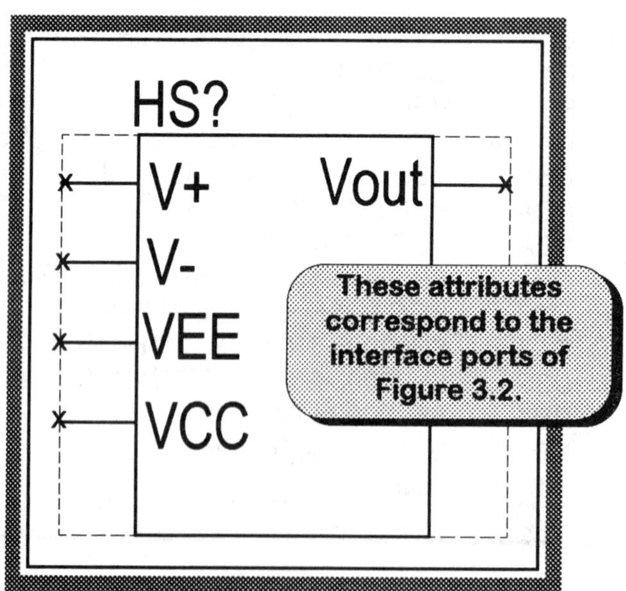

FIGURE 3.3
The OPAMP PART symbol

8. If you wish, modify the initial symbol by moving the pins and attributes by **select**, **DRAG**. Also, **DCLICKL** on any attribute or pin to place pin numbers and reorient pin names. Because this edit symbol process can be quite involved, you may wish to remain with the initial symbol of Figure 3.3.

9. To save the edited PART symbol again in the symbol library (*mylib.slb*): **File**, **Save** (and if necessary, **Yes** add to list of configured libraries).

Step 3: Associate the MODEL with the PART

10. To associate the MODEL definition (*opamp.sub*) with the PART symbol (in *mylib.slb*): **Part**, **Attributes**, **CLICKL** on PART=, enter OPAMP in Value field, **Save Attr, CLICKL** on MODEL=, enter OPAMP in Value field, **Save Attr**, **OK**. (**DRAG** part name OPAMP to desired position.)

11. To return to Schematics, **File**, **Close**, (and if necessary, **Yes** save changes to Part, **Yes** save changes to library).

12. **File**, **Close** to exit the *opamp.sch* window (and if necessary, **Yes** save all changes).

Amplifier design using the OPAMP subcircuit

13. Create a new schematic screen

14. Fetch subcircuit PART *OPAMP* from *mylib.slb*.

15. Add feedback components to create the circuit of Figure 3.4.

16. Store the circuit in a file of your choosing, run PSpice (transient mode), and test the circuit. Is the voltage gain close to the expected value of 10?

    Yes **No**

17. If you wish, determine other amplifier characteristics (Z_{IN}, Z_{OUT}, bandwidth, etc.) and compare them to Chapter 2.

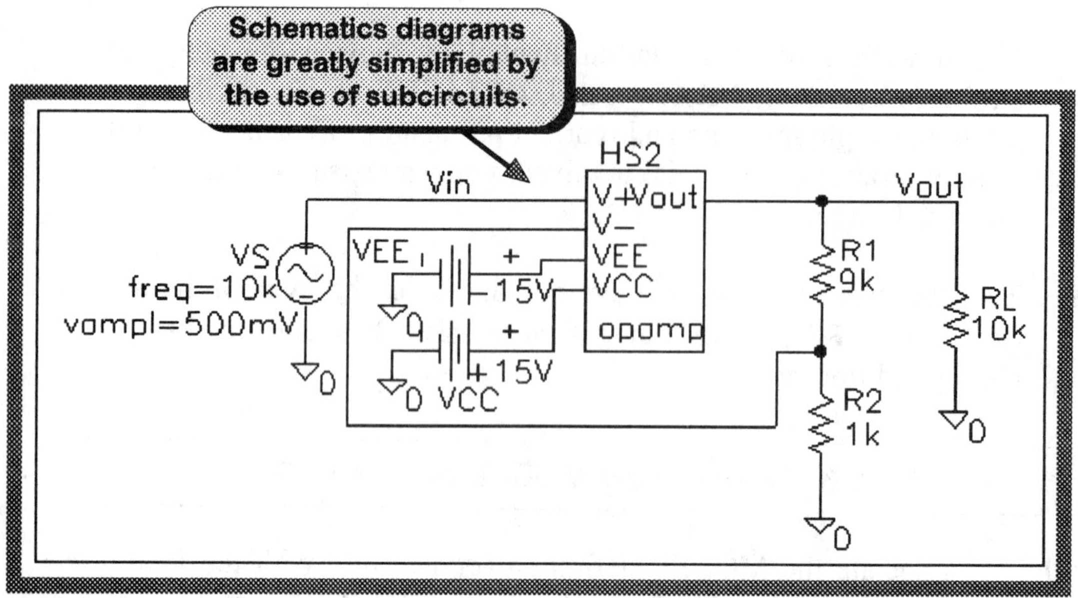

FIGURE 3.4

VCVS using subcircuit opamp

Advanced activities

18. To show that a subcircuit block is really a hierarchical module, select block OPAMP, **DCLICKL** (or **Navigate**, **Push**), and note that the block contents are displayed. (**Navigate**, **Pop** to return.)

19. To view an arbitrary waveform (such as the collector of Q3), **Navigate**, **Push** and drop a voltage marker at this location (or find and select alias variable *V(HS1.Q3:c)*).

20. Cascade two op amp circuits (of the type shown in Figure 3.4) to generate an overall gain of approximately 100.

EXERCISES

- Create and test a power supply subcircuit based on the design of Chapter 9 in Volume I.

- Based on the design of Figure 3.5, create and test a subcircuit called INVERTER. Determine its input/output characteristics and its propagation delay.

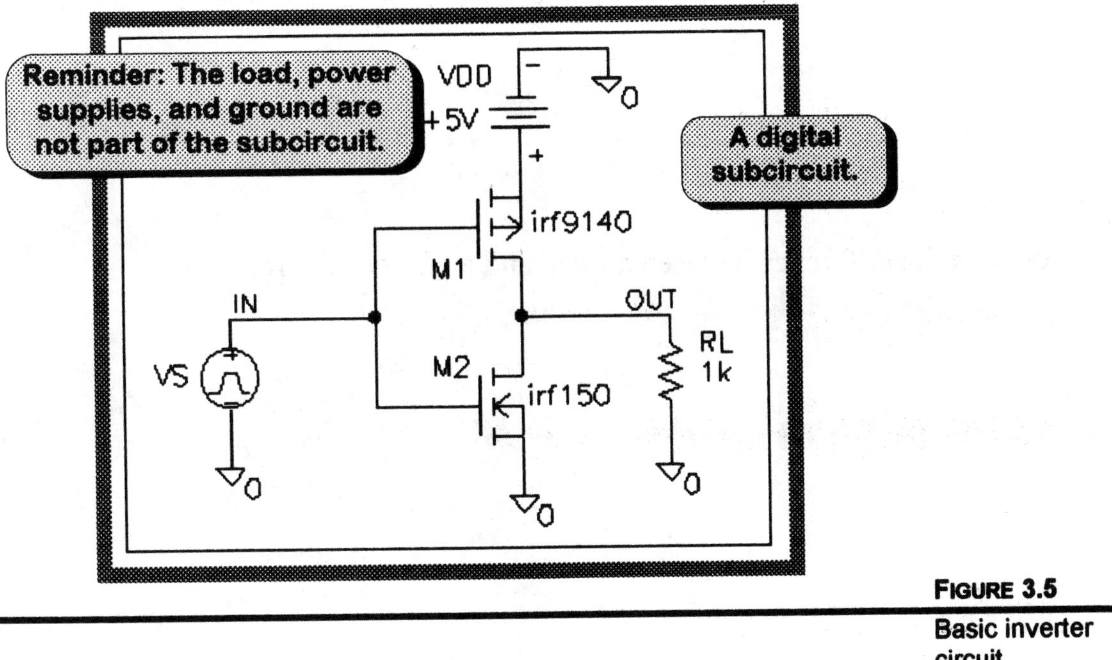

FIGURE 3.5
Basic inverter circuit

QUESTIONS AND PROBLEMS

1. List several advantages of using subcircuits.

2. Can a circuit be constructed entirely of subcircuits?

3. Can we push into a subcircuit?

4. Referring to the trace variable *V(HS1.Q3:c)*, what does the dot mean?

5. What is the difference between a *local* and a *global* subcircuit?

CHAPTER 4

The Operational Amplifier
Specifications

OBJECTIVE

- To compare test results of the 741 op amp with its major specifications.

DISCUSSION

The homemade op amp (subcircuit) of the last chapter has a number of drawbacks. First, the gain and input impedance are too low, and the output impedance is too high. There are many other secondary characteristics we wish to improve.

In this chapter, we will examine a commercially available operational amplifier and see if it offers the improved characteristics we desire. Our choice is the 741 workhorse op amp, supplied within PSpice library *eval.slb*.

Specifications

The characteristics of any circuit, such as the 741 op amp, are known as *specifications.* The important specifications of the 741 op amp are listed in Table 4.1. Using PSpice we will verify each of these one at a time. (If time is short, verify those of primary interest first.)

Of all the specifications listed, perhaps the most surprising is the very low value of the break frequency (10Hz). However, as we learned in Chapter 2, this is done deliberately in order to force the overall closed-loop op amp gain below 1 before phase shifts cause oscillations by way of positive feedback.

741 specifications	
• Input offset voltage	1mV
• Maximum/minimum output voltages	14V (15V supplies)
• Open loop voltage gain	200,000 (106dB)
• Input bias current	80nA
• Input offset current	20nA
• Short circuit output current	25mA
• Input impedance	2MEG
• Output impedance	75ohms
• CMRR	90DB (30,000)
• Slew rate	.5V/μs
• Break frequency	10Hz
• Frequency rolloff	20dB/Dec
• Gain-BW product	1MEGHz

TABLE 4.1
The 741 op amp specifications

SIMULATION PRACTICE

1. Draw the test circuit of Figure 4.1 and set the attributes as shown.

Input offset voltage - *The voltage at either input that drives V_{OUT} to zero.*

2. Perform a DC Sweep of Vni between –200μV and +200μV, in steps of .1μV, and display the V_{OUT} curve of Figure 4.2. Determine V_{OFF}, the value of Vni that drives V_{OUT} to zero. Compare your result with the spec sheet value from Table 4.1.

If you wish, set V_{OFF} to this value in the future.

V_{OFF} **(PSpice) = ________** V_{OFF} **(spec) = 1 mV**

Maximum/minimum V_{OUT} values - *Also known as the rail voltages, the maximum and minimum output voltages allowed for a particular power supply level (15V, in our case).*

3. Using the graphical results of Figure 4.2, enter the rail voltages.

$V_{RAIL}+$ **(PSpice) = ________** $V_{RAIL}+$**(spec) = +14V**

$V_{RAIL}-$ **(PSpice) = ________** $V_{RAIL}-$**(spec) = –14V**

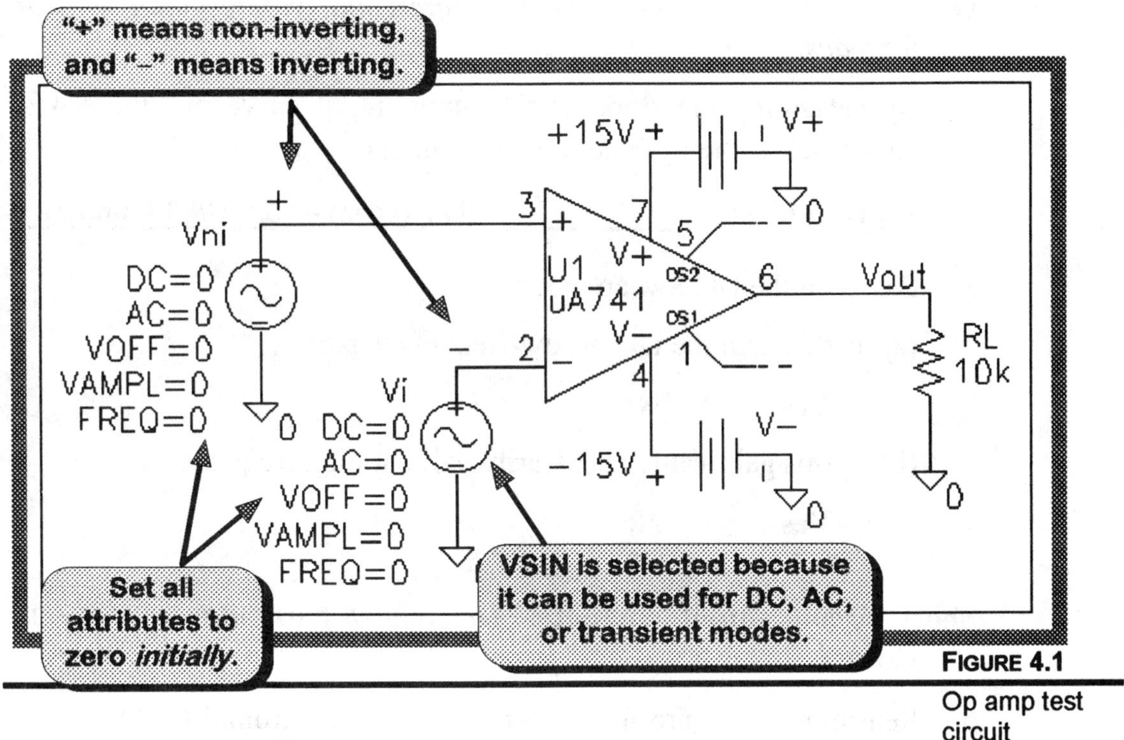

FIGURE 4.1

Op amp test circuit

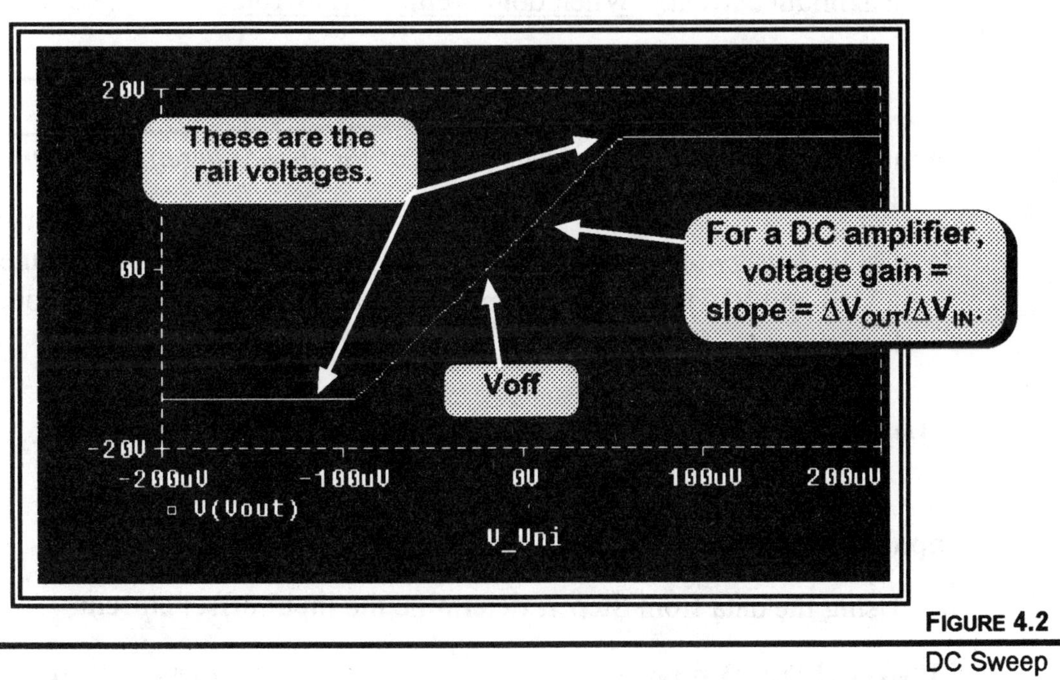

FIGURE 4.2

DC Sweep curves

Open loop voltage gain - *The DC differential voltage gain without feedback.*

4. By measuring the slope of the input/output curve of Figure 4.2, record below the open-loop voltage gain ($\Delta V_{OUT}/\Delta V_{IN}$).

A_{OL} (PSpice) = ____________ **A_{OL} (spec) = 200,000 (106dB)**

5. Based on the DC Sweep mode:

(a) Is this gain the DC or low-frequency gain?

Yes **No**

(b) Is this gain value necessarily valid at high frequencies?

Yes **No**

Short circuit output current - *The output current when V_{OUT} is shorted.*

6. Returning to Figure 4.1, short the output to ground (set RL to a very small value) and plot the output current (I(RL)) using the DC Sweep methods of Step 2. The short circuit current is the absolute maximum current. When done, return RL to 10k.

I_{SHORT} (PSpice) = _______ **I_{SHORT} (spec) = 25mA**

Input bias current - *The average value of the input currents (Vni and Vi grounded).*

7. Using *bias point* techniques (IPROBE or output file), determine I(Vni) and I(Vi), and use the equation below to calculate the input bias current. (Make sure that DC = 0 for both inputs.)

I_{BIAS} (PSpice) = $\dfrac{I_{NI} + I_I}{2}$ = _______ **I_{BIAS} (spec) = 80nA**

Input offset current - *The difference between the input bias currents.*

8. Using the data from Step 7, determine the input offset current.

I_{OFFSET} (PSpice) = $I_{NI} - I_I$ = _______ **I_{OFFSET} (spec) = 20nA**

Input impedance - *The total resistance between the inputs.*

9. Set Vni's VAMPL to 1mV and FREQ to 10k (leave Vi grounded). Using transient analysis, measure I(Vni) and determine Z_{IN} using Ohm's law. (Hint: Use peak-to-peak to cancel out the DC bias current offset.)

Z_{IN} (PSpice) = __________ **Z_{IN} (spec) = 2MEG**

Output impedance - *The resistance viewed from the output terminal.*

10. Using the same transient analysis as Step 9, plot V_{OUT} with and without a load (i.e., RL = 100Ω and 100MEGΩ.) Using algebra, determine Z_{OUT}. (If necessary, see Chapter 1, Step 3.)

Z_{OUT} (PSpice) = __________ **Z_{OUT} (spec) = 75Ω**

Common-mode rejection ratio (CMRR) - *Open-loop differential gain (A_{DM}) divided by common mode gain (A_{CM}).*

11. Short both inputs to Vni, as shown in Figure 4.3. Perform a DC Sweep between +5V and −5V, display V_{OUT}, and determine A_{CM}. (Hint: A_{CM} equals the slope of the output curve.)

A_{CM} = __________

12. Using the data gathered in Steps 4 and 11, determine CMRR.

CMRR (PSpice) = __________ **CMRR (spec) = 90dB (30,000)**

Slew rate - *The slope of V_{OUT} to a step input signal.*

13. Set up the circuit as shown in Figure 4.4. Program part VPULSE for a fast rise-time step input of 0 to .1V, and run a transient analysis from 0 to 50μs. (1pF would be a fast rise time.)

14. Display V_{OUT} (to 50μsec) and determine its slew rate (slope) in volts/μsec.

Slew rate (PSpice) = __________ **Slew rate (spec) = .5V/μs**

15. Do you think the slew rate factor might produce output signal distortions at high frequencies?

Yes **No**

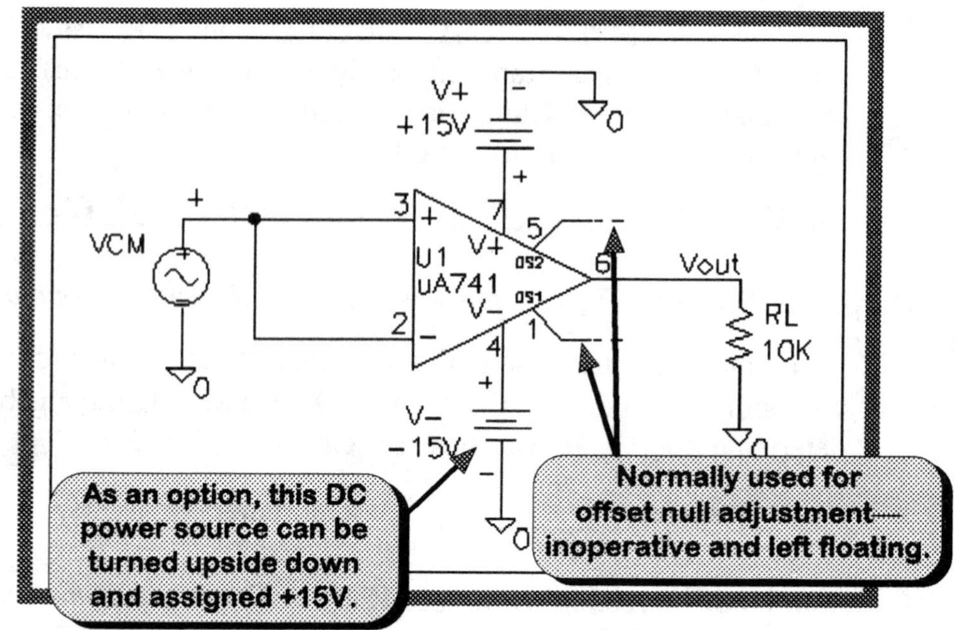

FIGURE 4.3
Common-mode configuration

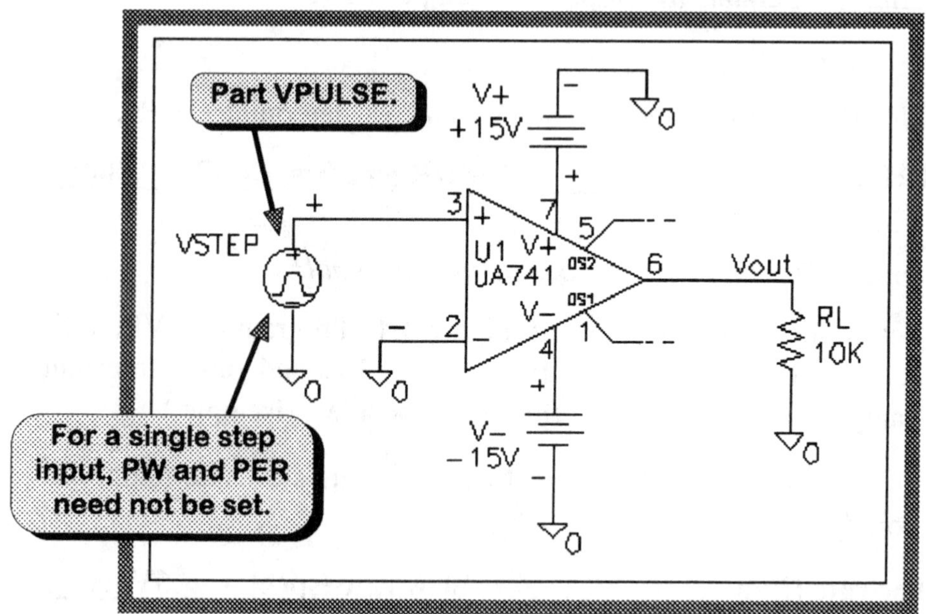

FIGURE 4.4
Slew rate test setup

Frequency response - *The gain versus the frequency (Bode plot).*

16. Using Figure 4.4, set Vni to AC = 50μV (the use of part VPULSE is okay) and leave Vi grounded. Perform a logarithmic AC sweep from 1Hz to 10MEGHz, and generate the graph of Figure 4.5. Answer the following:

    (a) What is the break frequency?

    F_B (PSpice) = _________ **F_B (spec) = 10Hz**

    (b) What is the rolloff? (Hint: Measure ΔVdB between 1kHz and 10kHz.)

    Rolloff (PSpice) = _________ **Rolloff (spec) = 20dB/Dec**

    (c) What is the *gain-bandwidth product*? (Hint: Measure the *unity gain frequency*, the frequency at which the gain is 1, or 0dB).

    Gain-BW (PSpice) = _________ **Gain-BW (spec) = 1MEGHz**

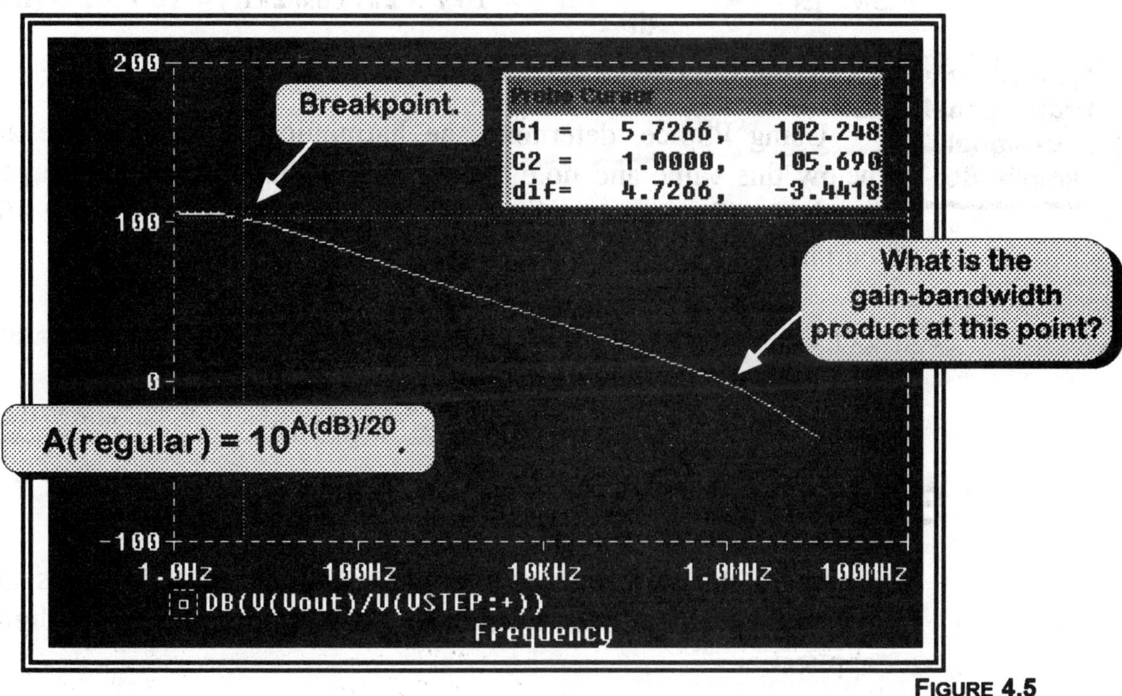

FIGURE 4.5
Open-loop Bode plot

17. Pick several points along the curve (well past the breakpoint) and determine A × f (gain times frequency). What generalization can you make about this value (the same *gain/bandwidth product* defined in Step 16(c))?

Advanced activities

18. Repeat any or all of the specification steps using the LM324 op amp. Compare and summarize your results.

19. Following up on Step 15, assume we have a 10V output signal from a 741 op amp. Theoretically, what is the highest frequency possible without slew rate distortion?

    Hint: As shown below, we obtain the slope of a sine wave by differentiation. The frequency at which the maximum slope equals the slew rate is the highest frequency possible without slew rate distortion. (Hint: The maximum value of cos( ) = 1.)

$$.5V/\mu s = \frac{d(10V\sin(2\pi ft))}{dt} = 10V \times 2\pi f\cos(2\pi ft) = 10V \times 2\pi f(\max)$$

    Solving for f gives = 7958Hz

    f is inversely proportional to signal amplitude.

    Using PSpice, determine the harmonic distortion above and below this value and note the waveform shape. Summarize your results. (Be sure to adjust the input to always generate a 10V output.)

20. For the amplifier of Figure 4.1, determine the signal-to-noise ratio at a midband (very low) frequency.

EXERCISES

- Compare the slew rate of the 741, 324, and 711 op amps and determine which one would be most appropriate for digital applications.

- Design a *comparator* circuit in which V_{OUT} = +15V for all V_{IN} greater than +5V and V_{OUT} = −15V for all V_{IN} +5V or less.

QUESTIONS AND PROBLEMS

1. Is the circuit of Figure 4.1 in the *open-loop* or *closed-loop* configuration?

2. For many applications, why is it possible to ignore many of the specifications of this experiment (such as *input offset current*)?

3. How do the specifications for A, Z_{IN}, and Z_{OUT} for the 741 differ from the characteristics of the amplifier of the previous experiment?

4. What is a *compensated* op amp? (Is the 741 compensated?)

5. To verify the voltage gain of Step 4 using transient methods, what range of frequencies would we have to use? Why?

6. Could we use the DC sweep method of Step 4 to determine voltage gain if the amplifier were not a DC amplifier? Why or why not?

CHAPTER 5

The Non-Inverting Configuration
VCVS and VCIS

OBJECTIVE

- To determine the characteristics of the VCVS and VCIS op amp configurations.

DISCUSSION

A common form of negative feedback is the VCVS (voltage-controlled-voltage-source) mode of Figure 5.1.

> ***A voltage-controlled-voltage-source is a device in which the output voltage is directly proportional to the input voltage, regardless of the load.***

It is useful to think of the VCVS configuration as a *voltage-to-voltage transducer*; that is, there is a linear one-to-one correspondence between the input voltage and the output voltage—regardless of load value. In other words, give me an input voltage, and I will give you the corresponding output voltage—the load (or output current) doesn't matter.

The transfer function

The *transfer function* is the ratio of output over input. For the VCVS configuration of Figure 5.1, both the input and output are voltage, and therefore the transfer function is its *voltage gain (A)*. Of special importance to the designer, the VCVS transfer function depends primarily on the feedback resistors (RF1 and RF2).

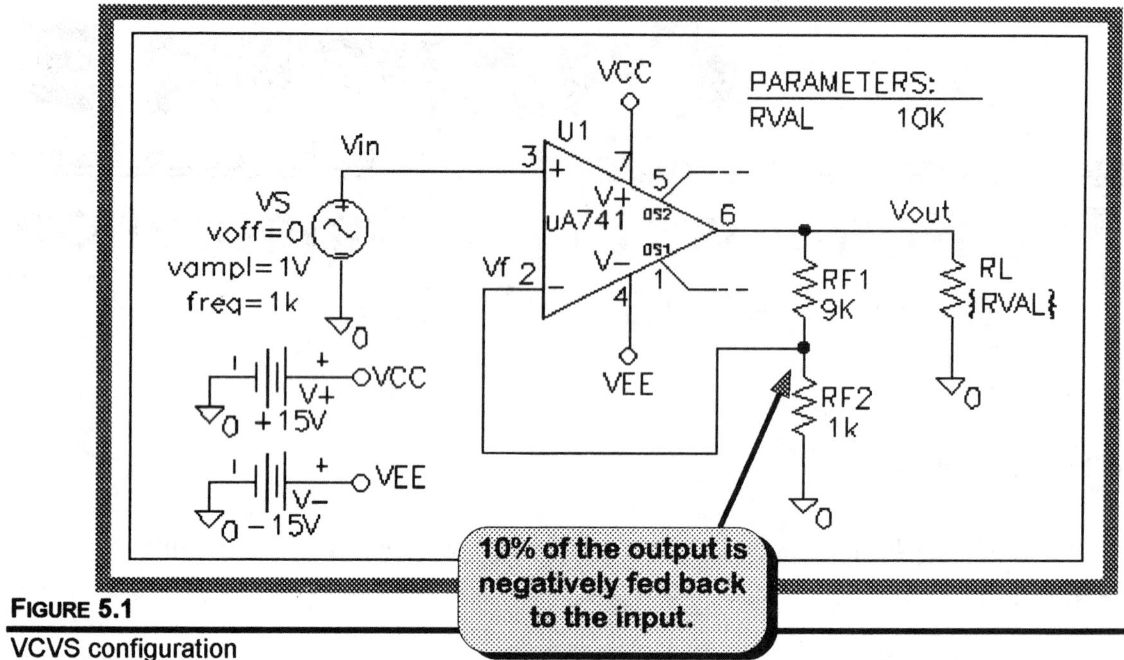

FIGURE 5.1
VCVS configuration

To calculate the closed-loop transfer function (A_{CL}) for the VCVS amplifier, we refer to the Thevenized version of Figure 5.2. We begin by writing the following equations:

$$V_{OUT} = A_{OL} \times (V_{IN} - V_F)$$

$$V_F = V_{OUT} \times \left(\frac{RF2}{(RF1 + RF2)} \right)$$

Solving these equations simultaneously yields:

$$A_{CL} = \frac{V_{OUT}}{V_{IN}} = \frac{1}{\frac{1}{A_{OL}} + \beta} = \frac{A_{OL}}{1 + A_{OL}\beta}$$

OL = open loop.
CL = closed loop.

where β = feedback ratio = RF2/(RF1 + RF2)

The ideal case

When performing theoretical calculations in the closed-loop (CL) case, a great simplification results if we assume the *ideal* case for the open-loop (OL) op amp specifications. *We will find that the errors that result from basing our calculations on the ideal case are very tiny.*

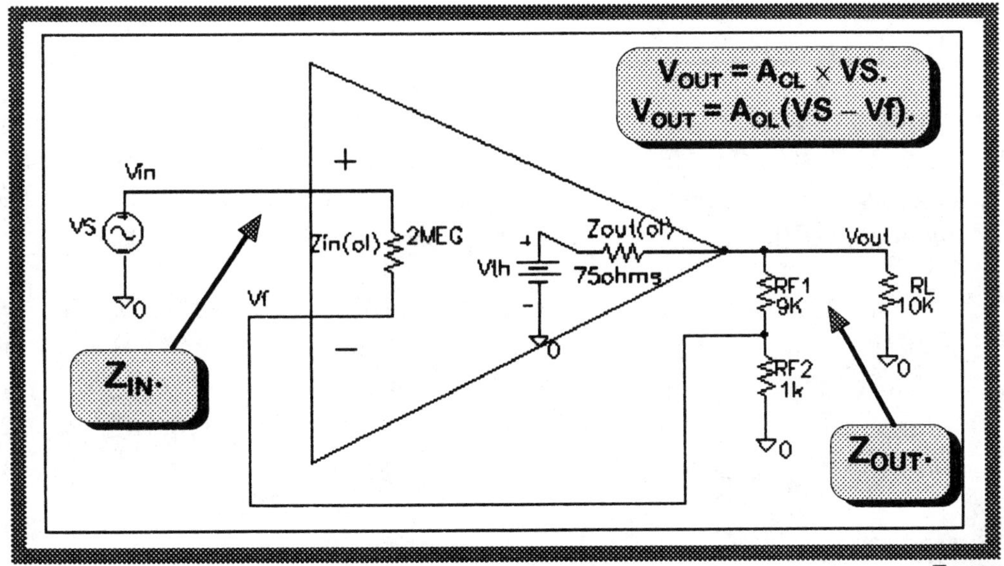

FIGURE 5.2
Thevenized VCVS circuit

The table below compares the 741 op amp published specifications for the "big three" parameters with their ideal counterparts.

Open-loop parameters

Item	Spec sheet	Ideal
A_{OL}	200,000	∞
Zin_{OL}	2MEG	∞
$Zout_{OL}$	75ohms	0

The ideal transfer function

As an example of the use of an ideal parameter, we substitute the ideal value for open-loop voltage gain (∞) into the equation developed earlier and generate the *ideal* closed-loop voltage gain (transfer function):

$$A_{CL} = \frac{1}{\frac{1}{A_{OL}} + \beta} = \frac{1}{\frac{1}{\infty} + \beta} = \frac{1}{\beta} = 1 + \frac{RF1}{RF2}$$

A perfect amplifier

When all the characteristics of the VCVS configuration are put together, we have a nearly perfect voltage amplifier. It gives us ease of design, DC or AC operation, high Z_{IN}, low Z_{OUT}, high bandwidth, and low distortion. It is called *non-inverting* because V_{IN} and V_{OUT} are in phase.

The shadow rule

During normal operation, negative feedback assures us that the inverting input will track (follow) the non-inverting input. This very useful analysis and troubleshooting aid will be known as the "shadow rule" (named after the famous vaudeville act "Me and My Shadow").

The shadow rule is a direct consequence of negative feedback. *In the ideal case*, the op amp will always generate an output voltage that will drive the input differential (error) voltage to zero. *Therefore, the voltage at the inverting input must track (shadow) the voltage at the non-inverting input.*

The VCIS configuration

The other major form of the non-inverting configuration is the VCIS (*voltage-controlled-current-source*) of Figure 5.3.

> ***A voltage-controlled-current-source is a device in which the output current is directly proportional to the input voltage, regardless of the load.***

As before, think of the VCIS configuration as a *voltage-to-current transducer*. There is a linear one-to-one correspondence between the input voltage and the output current—regardless of load value. Give me an input voltage, and I will give you the corresponding output current—the load (or output voltage) doesn't matter.

For this VCIS configuration, the transfer function is I_{OUT}/V_{IN}, which gives units of *transconductance*. Making use of the shadow rule, and assuming the ideal case, the closed-loop transfer function is calculated as follows:

$$V_{IN} = V_F \qquad I_{OUT} = \frac{VF}{RF}$$

$$\textbf{transconductance} = \frac{\mathbf{I_{OUT}}}{\mathbf{V_{IN}}} = \frac{\mathbf{1}}{\mathbf{RF}}$$

Because the load resistor (RL) is not referenced to ground, the VCIS configuration is seldom used. (It will appear in this book only as an advanced activity.)

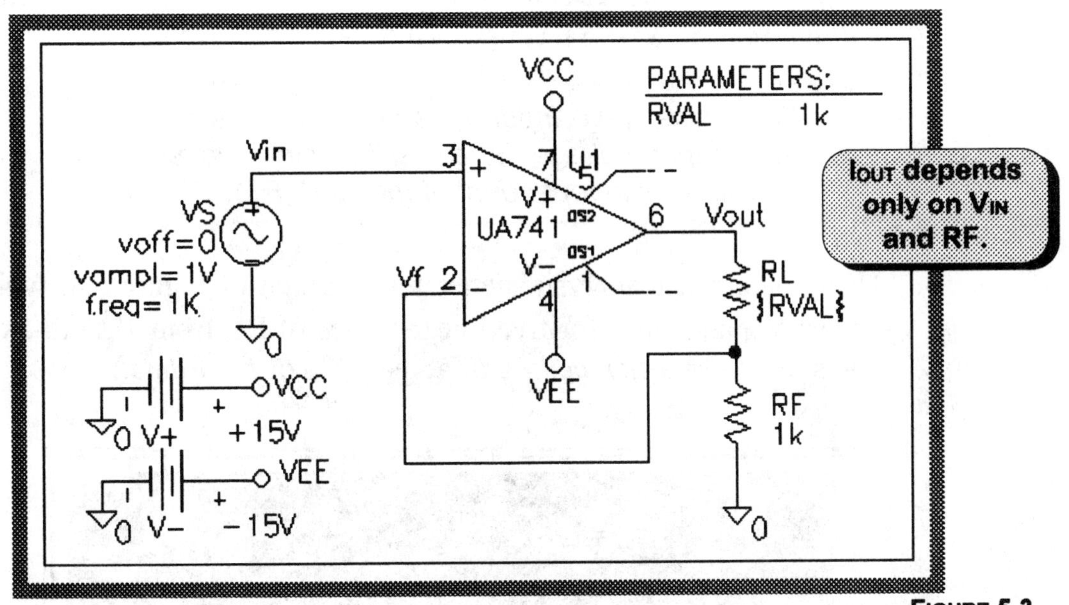

FIGURE 5.3
The VCIS non-inverting configuration

SIMULATION PRACTICE

VCVS

1. Draw the VCVS circuit of Figure 5.1, and set up the load for parametric analysis. (To draw the power supply connections, see *Schematics Note 5.1.*)

> **Schematics Note 5.1**
> ***How do I simplify power connections?***
>
> To simplify power connections (as shown in Figure 5.1), first move the DC power sources (VDC) to any convenient location on the schematic. Select part BUBBLE and position as desired. **DCLICKL** on each bubble to bring up the *Set Attribute Value* dialog box, enter desired bubble name (make sure the corresponding bubbles match), **OK**.

2. To prove that the circuit of Figure 5.1 is a *voltage-controlled-voltage-source* (VCVS), consider the following:

> *For a sine wave input voltage, a circuit is a VCVS if the output voltage is a perfect sine wave whose amplitude is independent of the load (RL).*

3. Using parametric analysis, generate the graph of Figure 5.4, which shows V_{IN} and V_{OUT} for five linear values of RL from 1kΩ to 5kΩ. (Be sure to use a *step ceiling* of at least 1% of *final time*.)

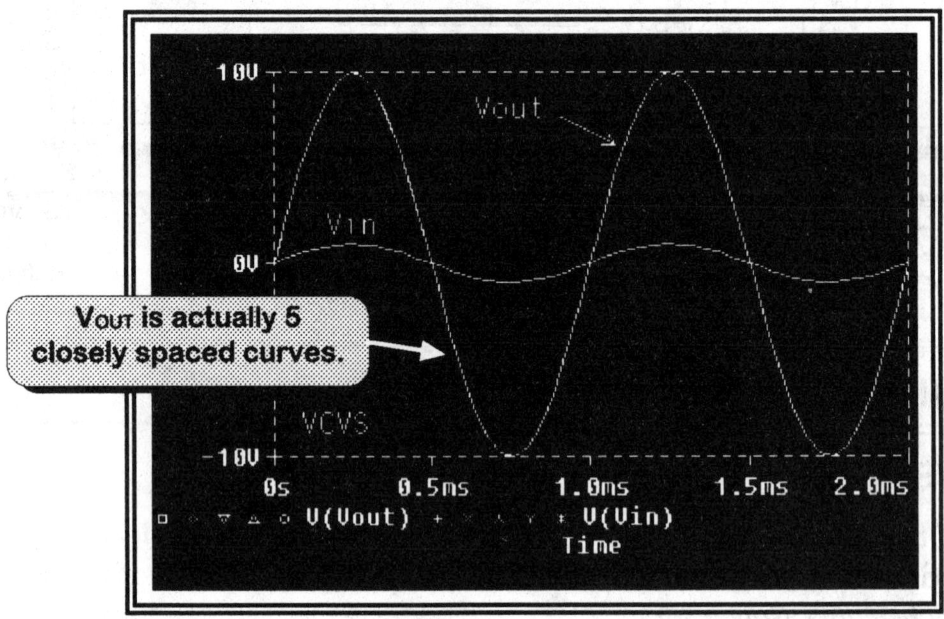

FIGURE 5.4
VCVS test results

4. Does the graph of Figure 5.4 prove that the circuit is indeed a near-perfect VCVS? (Does it appear that V_{OUT} is a sine wave whose amplitude is independent of RL?)

   Yes **No**

Zoom Area

5. Zoom in on any portion of the V_{OUT} curve and note that there are actually five closely spaced curves. Is it now fair to say that the VCVS circuit is a *very good* (but not *perfect*) VCVS?

   Yes **No**

Refit

Voltage gain (transfer function)

From this point on, disable the parametric analysis. (RL will automatically revert to 10k.)

6. Determine the voltage gain [A(CL)] for each case below.

   (a) The calculated case [assume A(OL) = 200,000].

   $$A_{CL} = \frac{1}{\frac{1}{A_{OL}} + \beta} = \_\_\_\_\_\_\_\_\_\_\_\_$$

   Reminder: $\beta = RF2/(RF1 + RF2) = .1$

   (b) Using PSpice.

   A(CL) (PSpice) = V_{OUT} / V_{IN} = ____________

Input impedance

7. Determine the input impedance [Z_{IN}(CL)] for each case below. [The equation shown below for Z_{IN}(CL) came from an analysis of Figure 5.2.]

   (a) The calculated case [assume Z_{IN}(OL) = 2MEG].

   $$Z_{IN}(CL)\ (real) = (1 + \beta A_{OL}) \times Z_{IN}(OL) = \_\_\_\_\_\_\_\_\_\_\_\_$$

   (b) Using PSpice. (Add *I(VS)* to Figure 5.4, and be sure to use peak-to-peak values in your calculations.)

   Z_{IN}(CL) (PSpice) = V_{IN} / I_{IN} = ____________

Output impedance

8. Determine the closed-loop output impedance [Z_{OUT}(CL)] for each case below.

   (a) The calculated case. [Assume Z_{OUT}(OL) = 75ohms.]

   $$Z_{OUT}(CL)\ (real) = \frac{Z_{OUT}(OL)}{1 + A_{OL}\beta} = \_\_\_\_\_\_\_\_\_\_\_\_$$

   (b) Using PSpice. (Apply a transient V_{IN} of 1V, measure V_{OUT} with high and low loads—such as 300Ω and 300MEG—and use algebra to determine Z_{OUT}.)

   Z_{OUT}(CL) (PSpice) = ____________

   Did you use a *step ceiling* of at least 1% of the *final time*?

9. To contrast the closed-loop big three characteristics, use previous results to complete the following table:

	A	Z_{IN}	Z_{OUT}
Calculated			
PSpice			
Ideal	10	∞	0

10. Based on the results of step 9, are we justified in using the ideal case for the big three closed-loop characteristics?

    Yes **No**

Bandwidth

11. Returning to Figure 5.1, assign attribute $AC = 1V$ to VS and set up the AC Sweep mode for 10 to 10MEGHz.

12. Determine the closed-loop bandwidth (BWCL) for each case below.

    (a) The calculated case. The bandwidth equals the frequency at which A_{CL} breaks (drops 3dB). As calculated below, this happens when A_{OL} drops to 27.65dB.

    $$\textbf{When } A_{CL} = 7.07 \text{ (70.7\% of 10)} = \frac{A_{OL}}{1 + A_{OL}(1/10)}$$

    $$A_{OL} = 24.129 \text{ (27.65dB)}$$

    From Figure 5.5, obtain the approximate frequency when A_{OL} is 27.65dB and fill in below. (This is the calculated closed-loop bandwidth!)

    BW_{CL} (calculated) = ___________

    (b) Using PSpice. Sweep from 1Hz to 10MEGHz, plot *DB(V(Vout)/V(Vin))* and generate the closed-loop curve of Figure 5.6.)

    BW_{CL} (PSpice) = ___________

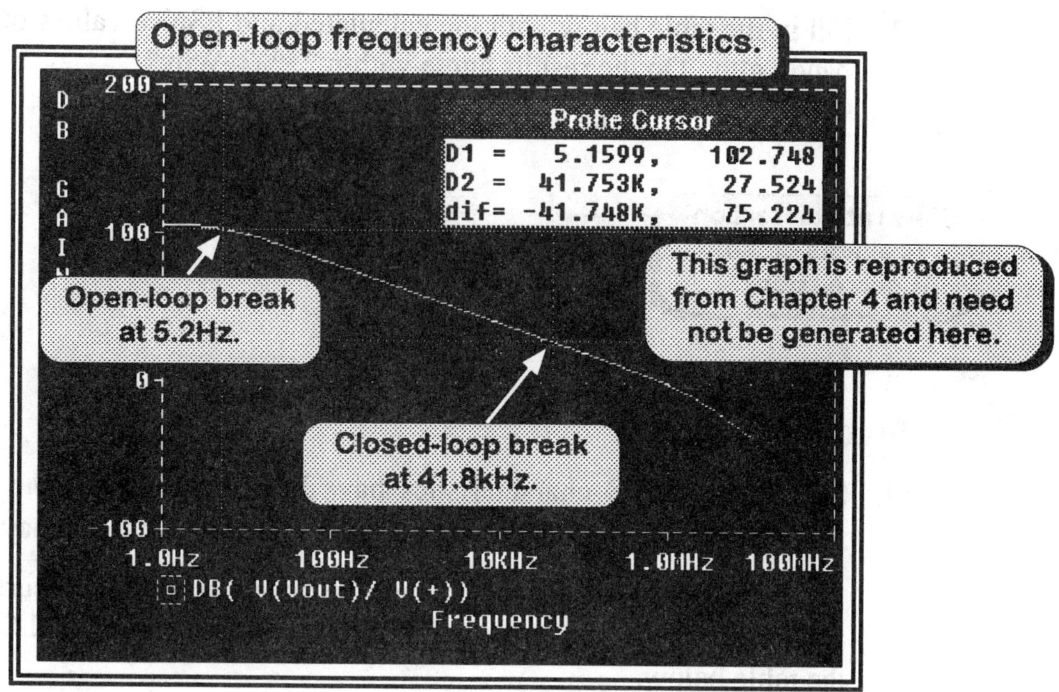

FIGURE 5.5
Open-loop Bode plot

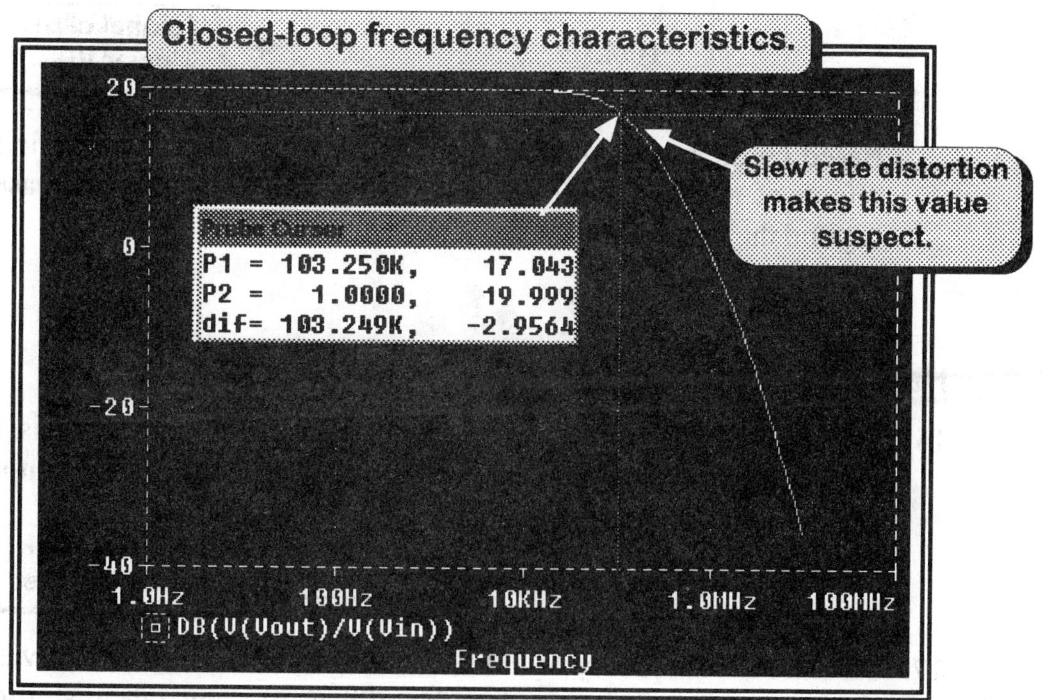

FIGURE 5.6
Closed-loop Bode plot

13. Fill in the table below with the calculated and PSpice values of the bandwidth. Do the results agree within 25%? Why?

    Yes **No**

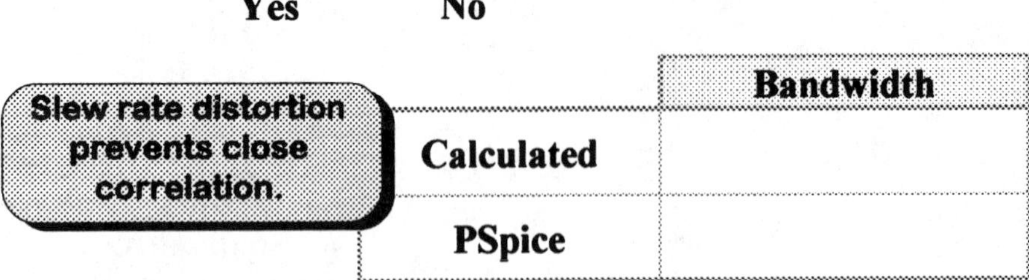

	Bandwidth
Calculated	
PSpice	

Frequency effects

14. So far, all VCVS characteristics have been measured at the low 1kHz frequency (below the ≈50kHz calculated break frequency).

    Pick one characteristic at random (such as Z_{IN}), measure its value at 100kHz (above the calculated break frequency), and fill in the table below.

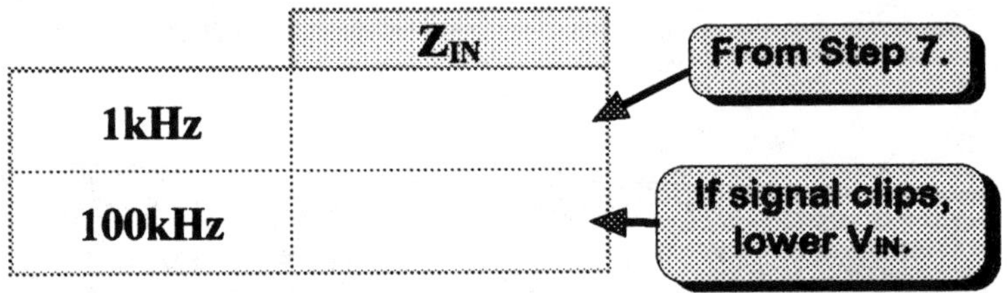

	Z_{IN}
1kHz	
100kHz	

15. Based on the results of Step 14, is it fair to say that the beneficial effects of negative feedback disappear as the frequency increases beyond the breakpoint?

    Yes **No**

Harmonic distortion

16. Returning to the circuit of Figure 5.1, use transient analysis to determine the total percent harmonic distortion of the output signal. (Also view the transient waveform.)

    Total HD_{CL} (PSpice) = ____________

    Reminder: Look in the output file.

17. Increase the frequency to 10kHz, and again measure HD and view the transient waveform. Are the effects of slew rate distortion evident? (See Chapter 4, Step 19.)

    Yes **No**

The shadow rule

18. Plot graphs of V_{IN} and V_F. Does the inverting input *approximately* track the non-inverting input? (Zoom in on the overlapping curves.)

    Yes **No**

VCVS summary

19. Based on all the information presented so far, how would you define and describe the VCVS op amp configuration of Figure 5.1? (Why should its use be restricted to audio-frequency applications?)

Advanced activities

20. Determine how negative feedback effects SNR (signal-to-noise ratio.)

21. For the VCVS configuration, return to parametric analysis and generate the family of Bode plot curves (AC sweep) of Figure 5.7 by sweeping RF1 over a range of values (such as 1k, 9k, and 99k).

    (a) For all curves, does the ideal formula for midband gain (1 + RF1/RF2) approximately hold?

    Yes **No**

    (b) For all curves, is the gain/bandwidth product *approximately* a constant? If so, what is this value?

    $A_{CL} \times BW_{CL}$ = ____________

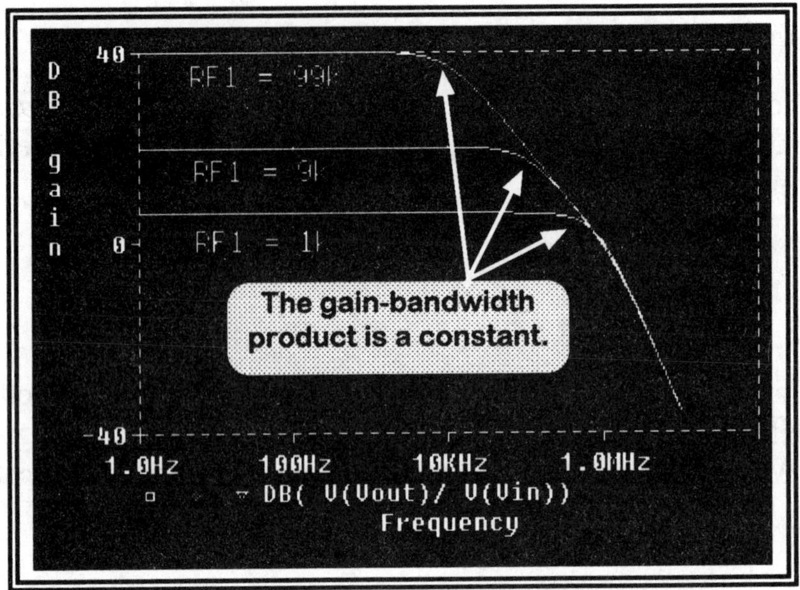

FIGURE 5.7
Family of Bode plot curves

22. Perform a complete analysis of the VCIS (*voltage-controlled-current-source*) of Figure 5.3. Determine any or all of the following and compare to the VCVS of Figure 5.1:

    (a) Graph of I_{OUT}(CL) versus V_{IN}(CL) for various values of RL from 1kΩ to 10kΩ. (Is the circuit a VCIS?)

    (b) The transfer function (transconductance).

    (c) Z_{IN}(CL), Z_{OUT}(CL), HD(CL), and BW(CL).

23. Compare the results of Step 22 with the ideal characteristics of a VCIS. Would we be justified in using the ideal case for most applications?

EXERCISES

- Investigate the output properties of the *buffered* VCVS configuration of Figure 5.8.
- Investigate the properties of the ideal diode of Figure 5.9.
- Perform a worst case analysis on the amplifier of Figure 5.1. Which resistor is most critical?

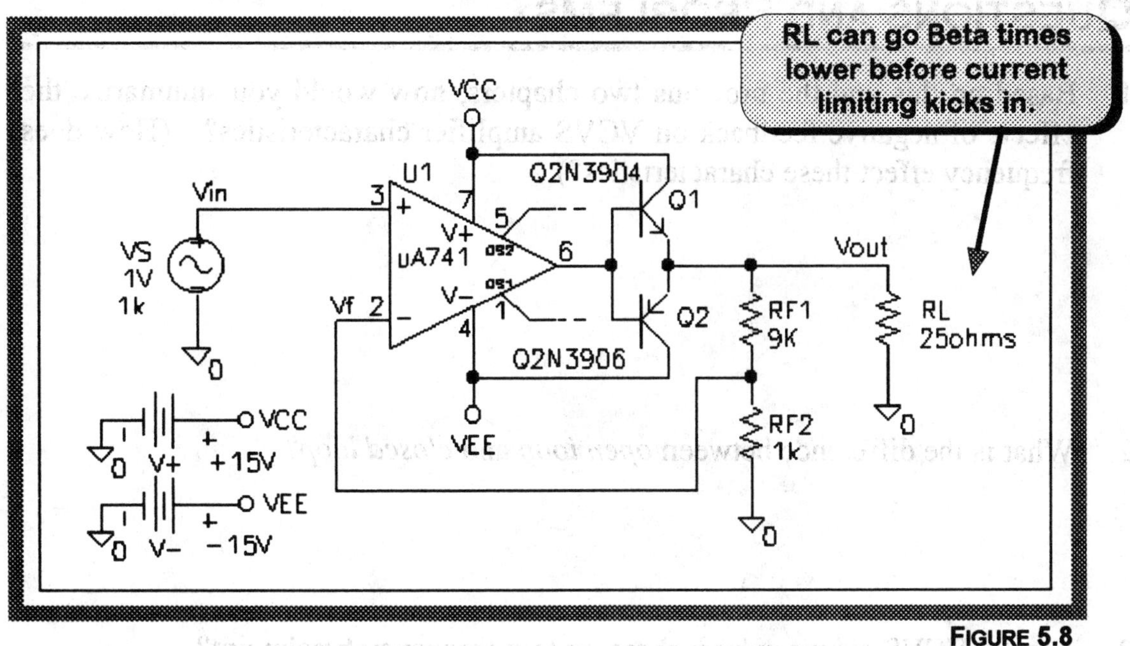

FIGURE 5.8
Buffered VCVS

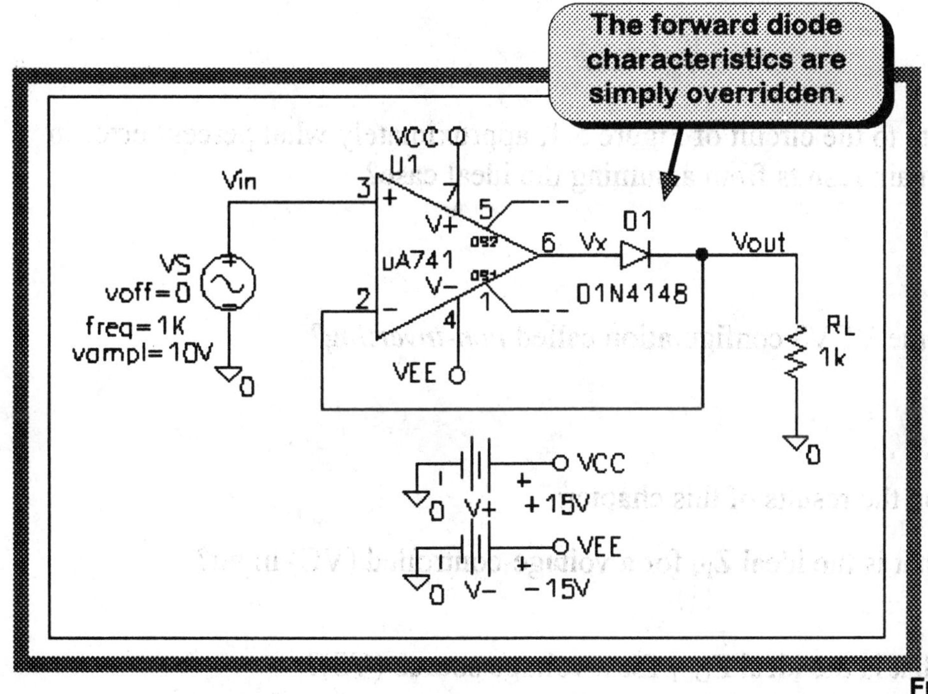

FIGURE 5.9
Ideal diode

QUESTIONS AND PROBLEMS

1. Based on this and the previous two chapters, how would you summarize the effects of negative feedback on VCVS amplifier characteristics? (How does frequency effect these characteristics?)

2. What is the difference between *open loop* and *closed loop*?

3. For the VCVS circuit, why is there no low-frequency breakpoint?

4. Referring to the circuit of Figure 5.1, approximately what percent error in voltage gain results from assuming the ideal case?

5. Why is the VCVS configuration called *non-inverting*?

6. Based on the results of this chapter:

   (a) What is the ideal Z_{IN} for a voltage-controlled (VC) input?

   (b) What is the ideal Z_{OUT} for a voltage source (VS)?

   (c) What is the ideal Z_{OUT} for a current source (IS)?

CHAPTER 6

The Inverting Configuration
ICVS and ICIS

OBJECTIVE

- To determine the characteristics of the ICVS and ICIS op amp configurations.

DISCUSSION

The four basic closed-loop configurations are listed below. The first two were the topic of the previous chapter; the last two are the subject of this chapter.

- **VCVS (voltage-controlled-voltage-source)**
- **VCIS (voltage-controlled-current-source)**
- **ICVS (current-controlled-voltage-source)**
- **ICIS (current-controlled-current-source)**

Because of the lessons we learned with the first two configurations, we can more quickly cover the remaining two. For example, we already know the following:

- The ideal case can be used for most applications with very little error.
- Ideal transfer functions are easy to develop when using the shadow rule.
- A voltage-controlled (VC) input has an ideal Z_{IN} of infinity, and a current-controlled (IC) input has an ideal Z_{IN} of zero.
- A voltage-source (VS) has an ideal Z_{OUT} of 0, and a current-source (IS) has an ideal Z_{OUT} of infinity.

The ICVS configuration

The most versatile negative feedback configuration for an op amp is the ICVS (current-controlled-voltage-source) of Figure 6.1.

> ***A current-controlled-voltage-source is a device in which the output voltage is directly proportional to the input current, regardless of the load.***

Think of the ICVS configuration as a *current-to-voltage transducer*, with a linear one-to-one correspondence between the input current and the output voltage—regardless of load value. Give me an input current, and I will give you the corresponding output voltage—the load doesn't matter.

Virtual ground

The great versatility of the ICVS configuration is closely related to the creation of the *virtual ground* at the inverting input. (A *virtual ground* is at zero volts, but it is not directly connected to ground.)

To see why a virtual ground is created, remember the *shadow rule*: The inverting input will equal (shadow) the non-inverting input. Since the non-inverting input is at zero volts (grounded), the inverting input must also be driven to zero volts.

The ICVS transfer function

For the ICVS configuration of Figure 6.1, the input is current and the output is voltage, and therefore the transfer function is its *transresistance*. In the ideal case, the transfer function depends only on the single feedback resistor (RF). (This time, it will be the student's responsibility to determine the transresistance equation.)

A handy current source

Because the inverting input is at zero volts (a *virtual ground*), we can substitute a voltage-source/resistor combination for the true current source (as shown by Figure 6.2). This configuration is called *inverting* because Vout is 180° out of phase with V_{IN}.

If the input is taken as VS, the configuration of Figure 6.2 is often called a voltage amplifier (although it does not have the desirable high Z_{IN} characteristics of the VCVS configuration of the last chapter).

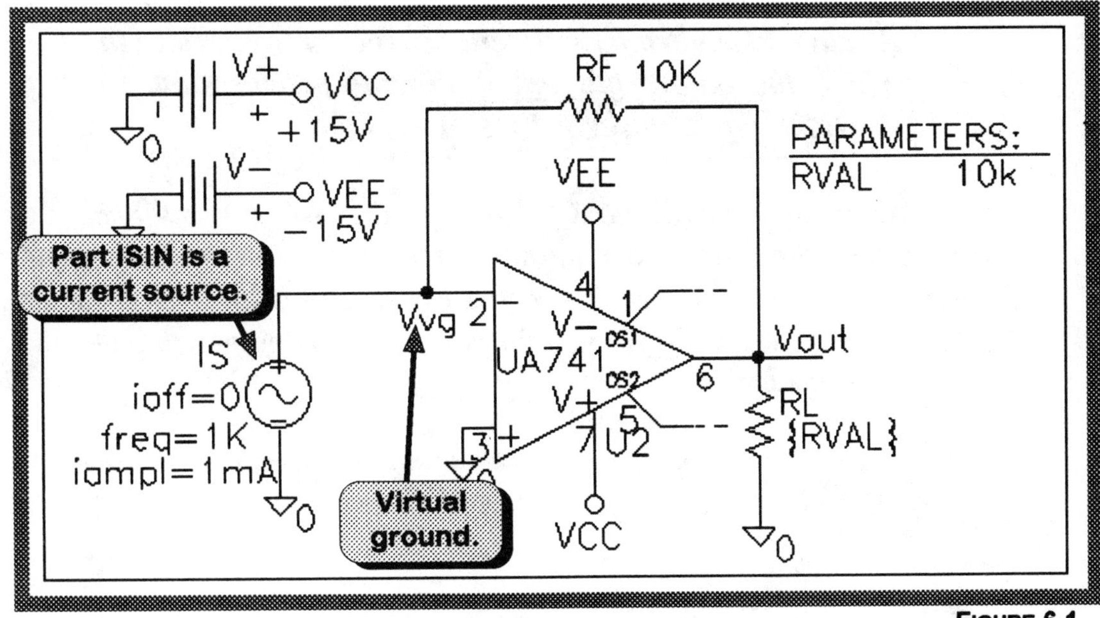

FIGURE 6.1
ICVS configuration using true current source (IS)

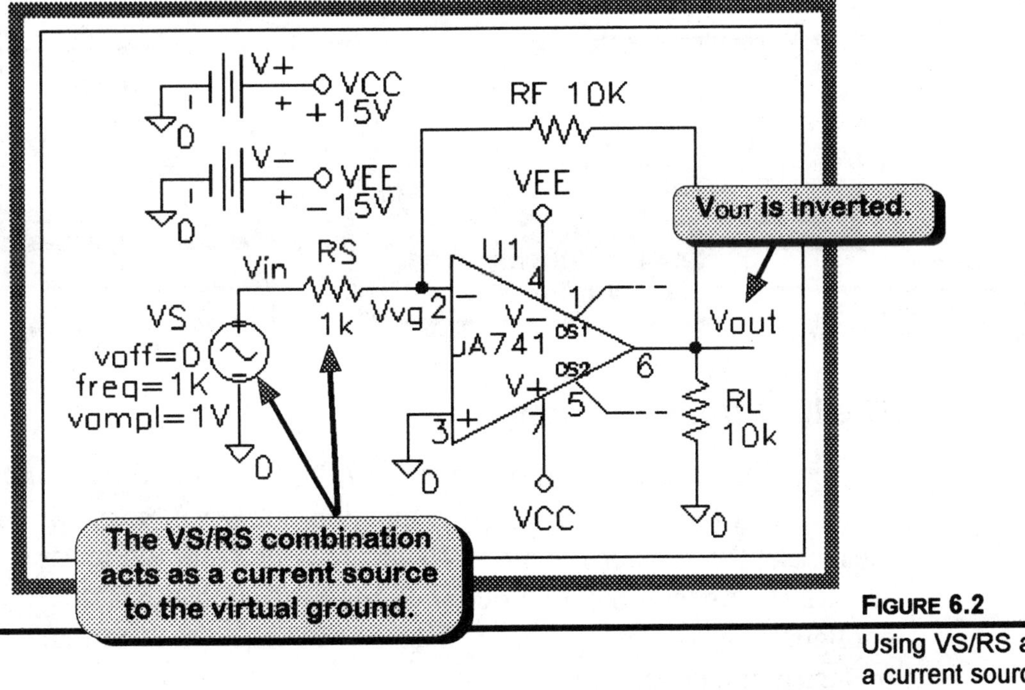

FIGURE 6.2
Using VS/RS as a current source

The ICIS configuration

The other major form of the inverting configuration is the ICIS (*current-controlled-current-source*) of Figure 6.3.

> ***A current-controlled-current-source is a device in which the output current is directly proportional to the input current, regardless of the load.***

As before, think of the ICIS configuration as a *current-to-current transducer*, with a linear one-to-one correspondence between the input current and the output current—*regardless of load value.* Give me an input current, and I will give you the corresponding output current—the load doesn't matter.

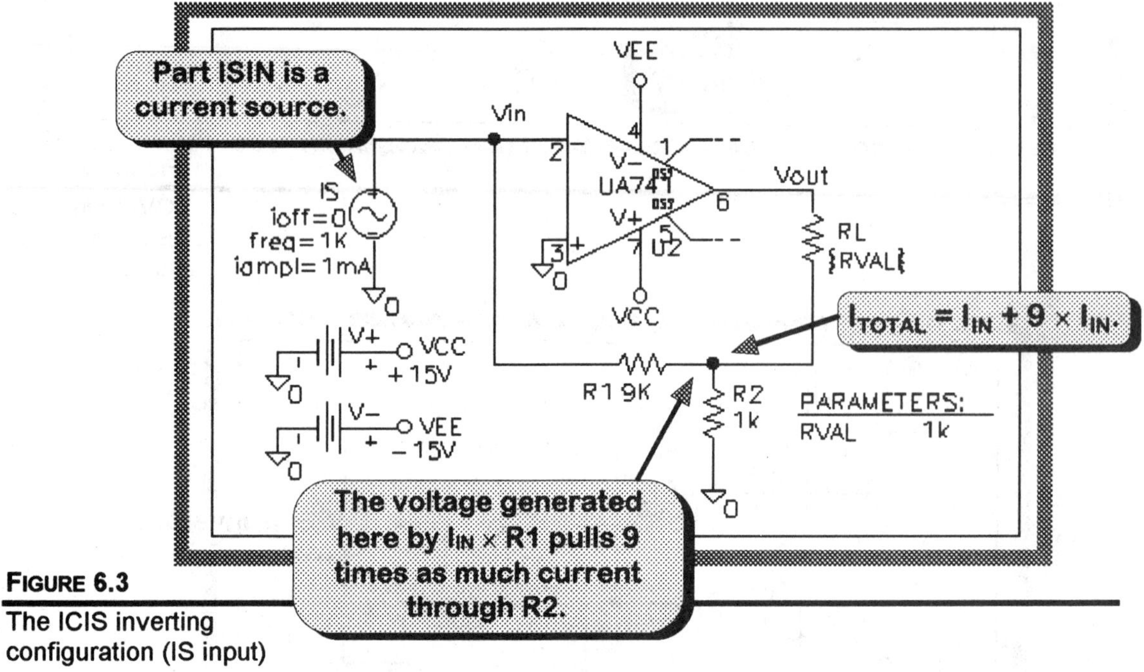

FIGURE 6.3
The ICIS inverting configuration (IS input)

The ICIS transfer function

For the ICIS configuration of Figure 6.3, both the input and output are current, and therefore the transfer function is *current gain* (β). In the ideal case, the transfer function depends only on feedback resistors (RF1 and RF2). As with the VCIS configuration of the last chapter, the ICIS configuration is seldom used because voltage (not current) is the preferred output variable.

SIMULATION PRACTICE

The ICVS configuration

1. Draw the ICVS circuit shown in Figure 6.1.

2. Using the ideal case, derive an ICVS *transresistance* equation for V_{OUT} versus I_{IN} in terms of RF. (*Hint*: Use the shadow rule.)

   V_{OUT} = (   ) × I_{IN}

3. Using the equation you developed in step 2, calculate the transresistance gain for the ICVS circuit of Figure 6.1.

   Transresistance = V_{OUT}/I_{IN} = ________________

4. To prove that the circuit of Figure 6.1 is an ICVS, consider the following:

   > *For a sine wave input current, a circuit is an ICVS if the output voltage is a perfect sine wave whose amplitude is independent of the load (RL).*

5. Generate the transient graphs of Figure 6.4, which show I_{IN} and V_{OUT} for various values of RL from 1kΩ to 5kΩ. (Be sure to set a *step ceiling* of at least *final time*/100.)

6. Based on the waveforms of Figure 6.4, determine the transfer function value (transresistance) as accurately as you can.

   Transresistance (PSpice) = ________________

   Is this experimental value close to the calculated value of step 3?

   Yes   **No**

7. Zoom in on any portion of V_{OUT} and note the five closely spaced curves. Is it fair to say that the circuit of Figure 6.1 is a very good (but not perfect) ICVS with a *transresistance* of 10kΩ?

   Yes   **No**

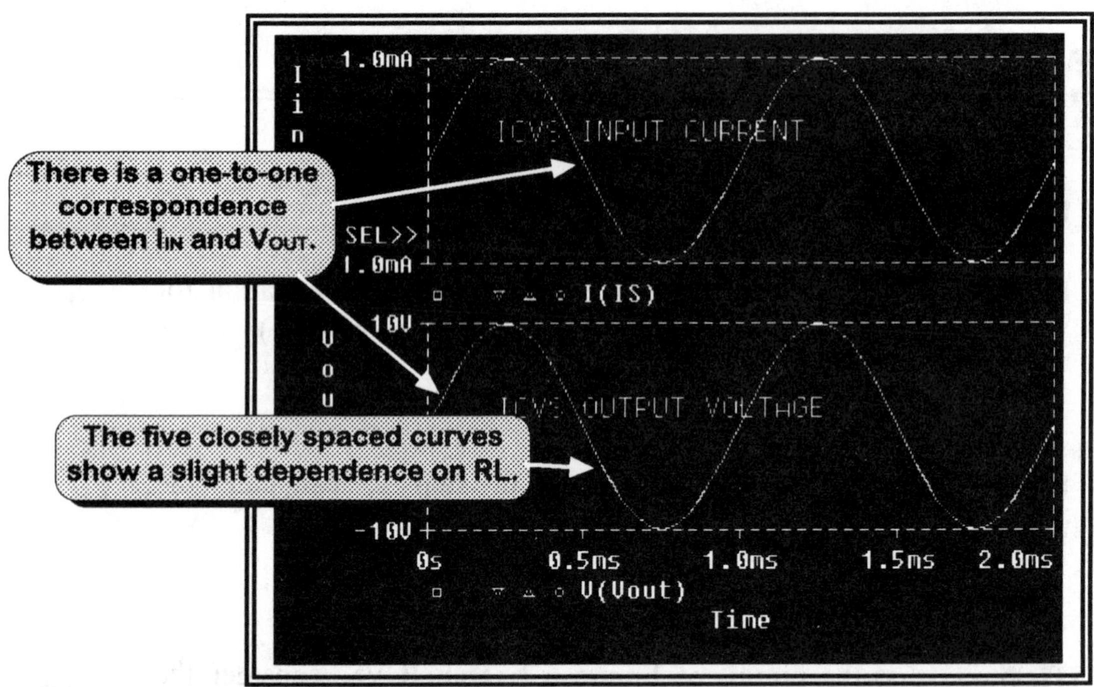

FIGURE 6.4

Plot of V_{out} versus I_{in}

8. If the input of the ICVS circuit of Figure 6.1 is a virtual ground and the output is a voltage source, *ideally* what would you expect the circuit's Z_{IN} and Z_{OUT} to be?

Z_{IN} (ideal) = __________ **Z_{OUT} (ideal) =** __________

9. Measure Z_{IN} and Z_{OUT} using PSpice. (From this step on, disable the parametric analysis and note that RL becomes 10k.)

Reminder: $Z_{IN} = V_{VG}/I(IS)$. To measure Z_{OUT}, determine V_{OUT} with and without a load. (Watch for signal clipping.)

Z_{IN} (PSpice) = __________ **Z_{OUT} (PSpice) =** __________

10. Comparing steps 8 and 9, are we justified in assuming ideal characteristics for the op amp?

Yes **No**

11. Is the virtual ground at *approximately* zero volts (less than 20mV)?

Yes **No**

The inverting voltage amplifier

12. Modify your circuit to generate the amplifier configuration of Figure 6.2, which uses a voltage-source/resistor combination for a current source.

13. Determine the circuit's voltage gain (V_{OUT}/V_{IN}) using:

    (a) The calculated ideal case.

    A(Ideal) = RF/RS = ________

    (b) Using PSpice (transient mode).

    A(PSpice) = ________

    (c) Are the ideal and experimental (PSpice) values approximately the same?

    Yes **No**

    (d) Are the input and output voltages (V_{IN} and V_{OUT}) 180° out of phase?

    Yes **No**

14. Using AC analysis, determine the bandwidth of the ICVS configuration of either Figure 6.1 or Figure 6.2. (Is the amplifier a DC amplifier?)

    Bandwidth = ________

Advanced activities

15. Determine one or more characteristics of the ICVS configuration (Figure 6.1 or Figure 6.2) before and after the break frequency. (Does the magic of negative feedback disappear after the break frequency?)

16. Perform a complete analysis of the ICIS (*current-controlled current source*) of Figure 6.3. Determine any or all of the following and compare them to the ICVS of Figure 6.1.

    - I_{OUT} versus I_{IN} for various RL
    - Z_{IN} and Z_{OUT}
    - Bandwidth (BW)
    - The transfer function (β)
    - Harmonic distortion (HD)

17. Compare the results of step 16 with the ideal characteristics of an ICIS. Would we be justified in using the ideal case for most applications?

18. Show how to use the ICVS configuration to convert the input current from the collector of a transistor (a current source) to output voltage.

19. As we learned in Chapter 4 (Step 19), slew rate distortion involving the 741 op amp sets in at about 8kHz for a voltage swing of 10V.

    For the popular ICVS version of Figure 6.2, view the output waveform at 10V and determine the HD (harmonic distortion) at 5kHz and 10kHz. Repeat for an output voltage amplitude of 5V. Summarize your conclusions. (Be sure to properly adjust the *center frequency* in the Transient dialog box.)

EXERCISES

- Draw the *summing* amplifier of Figure 6.5 and show that the output voltage waveform is the algebraic sum of the two input voltage waveforms.

> The 3k resistor (R8) is used to counter the effects of the *input bias current*. The problem arises when this current passes through unequal resistance in the inverting and non-inverting input circuits—resulting in an unwanted differential voltage input. By making the resistances approximately equal (3k ≈ 10k||10k||10k), this differential voltage is near zero.

- Determine the transfer function of the *difference amplifier* of Figure 6.6 and test your equation with PSpice.

- For the *logarithmic amplifier* of Figure 6.7, what is the relationship between V_{OUT} and V_{IN}? (Hint: Is the plot of V_{OUT} versus LOG(V_{IN}) a straight line?)

- Predict the output of the ideal clamper of Figure 6.8. Is it a positive or negative clamper? (Change V_{REF} to +5V and predict V_{OUT}.)

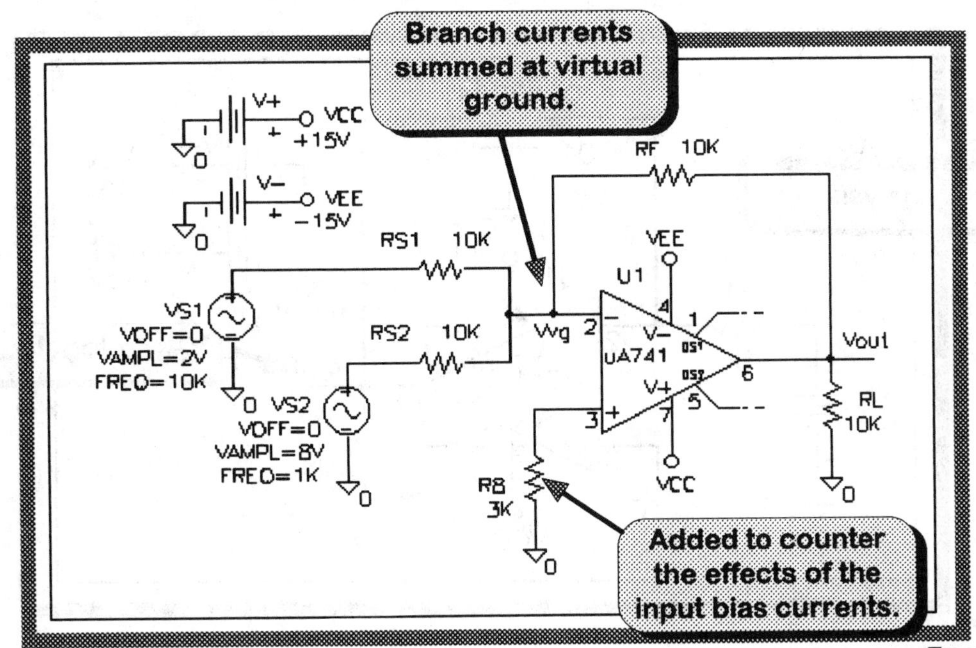

FIGURE 6.5
The summing amplifier

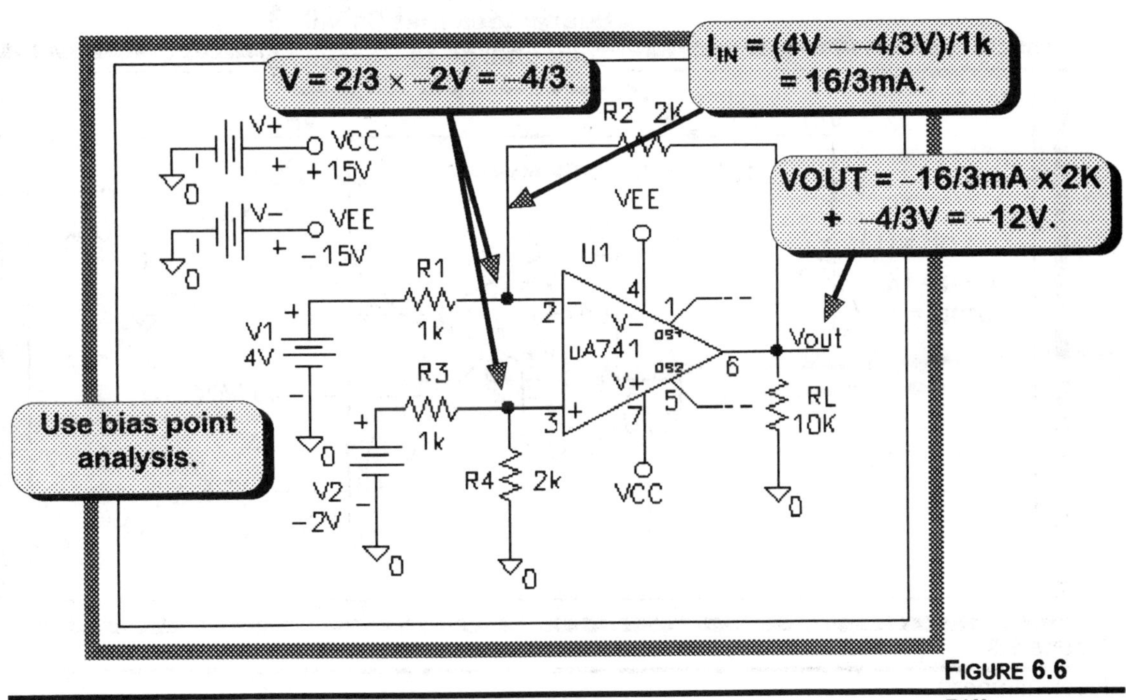

FIGURE 6.6
Difference amplifier

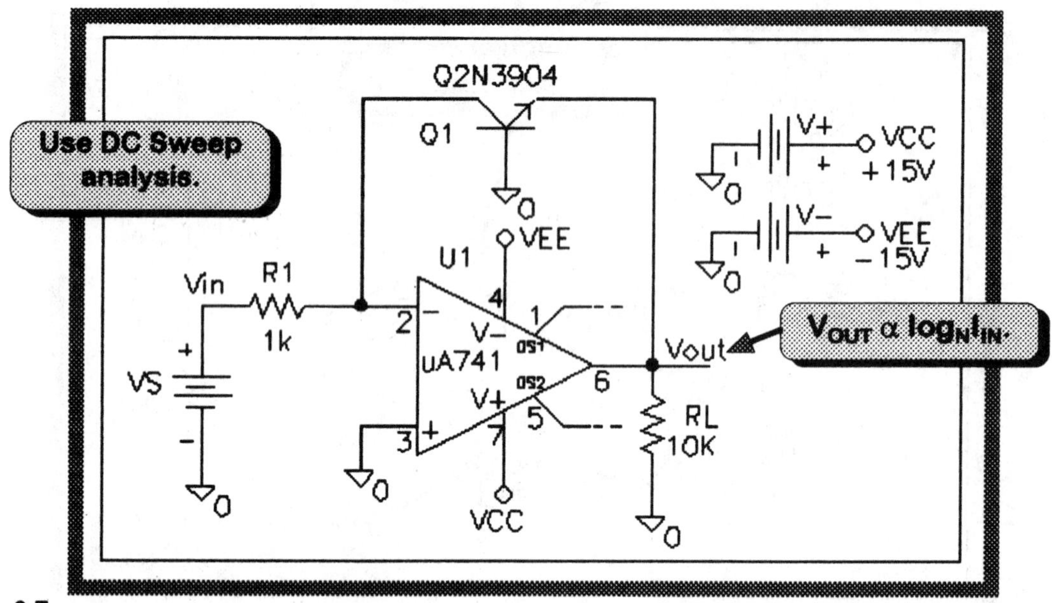

FIGURE 6.7
Logarithmic amplifier

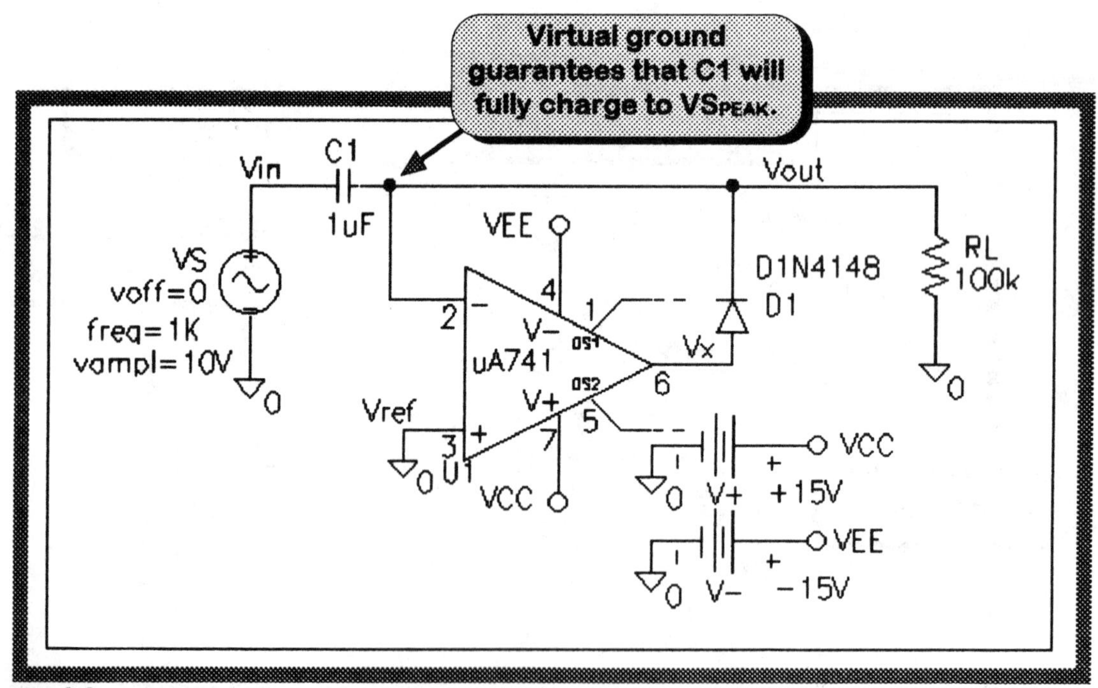

FIGURE 6.8
Ideal clamper

QUESTIONS AND PROBLEMS

1. What is the difference between *ground* and *virtual ground*?

2. When used as a *voltage amplifier*, why is the VCVS configuration of Chapter 5 considered superior to the ICVS configuration (Figure 6.2) of this chapter?

3. A certain ICVS amplifier has a gain/bandwidth product of 10MEG. If the gain is 80, what is the bandwidth?

4. Regarding the summing amplifier of Figure 6.5:

   (a) Could the circuit be used by an analog computer to perform the *add* process?

   (b) How does the summing amplifier eliminate *crosstalk* (V_{S1}'s circuit affecting V_{S2}'s circuit and vice versa)?

5. Place type VCVS, VCIS, ICVS, or ICIS after each of the following:

Z_{IN}	Z_{OUT}	Type
0	0	
0	infinity	
infinity	0	
infinity	infinity	

CHAPTER 7

Op Amp Integrator/Differentiator

Calculus

OBJECTIVES

- To review the fundamentals of calculus.
- To perform integration and differentiation using both RC and op amp circuits.

DISCUSSION

We are well aware of the relationships between distance (D), velocity (V), and time (t):

$$\mathbf{D = V \times t} \qquad\qquad \mathbf{V = D/t}$$

However, the well-known *multiply* and *divide* operators (× and /) only work if velocity is a *constant*. If velocity is a *variable* (a changing function of time), we must turn to the techniques of *calculus*, developed in the sixteenth century by Isaac Newton. We must *integrate* and *differentiate.*

$$D(t) = \int_0^t v(t)dt \qquad\qquad V(t) = \frac{dD(t)}{dt}$$

Integration **Differentiation**

As shown in Figure 7.1, integration is performed graphically by taking the accumulated area under the curve, and differentiation is performed graphically by taking the slope of the curve.

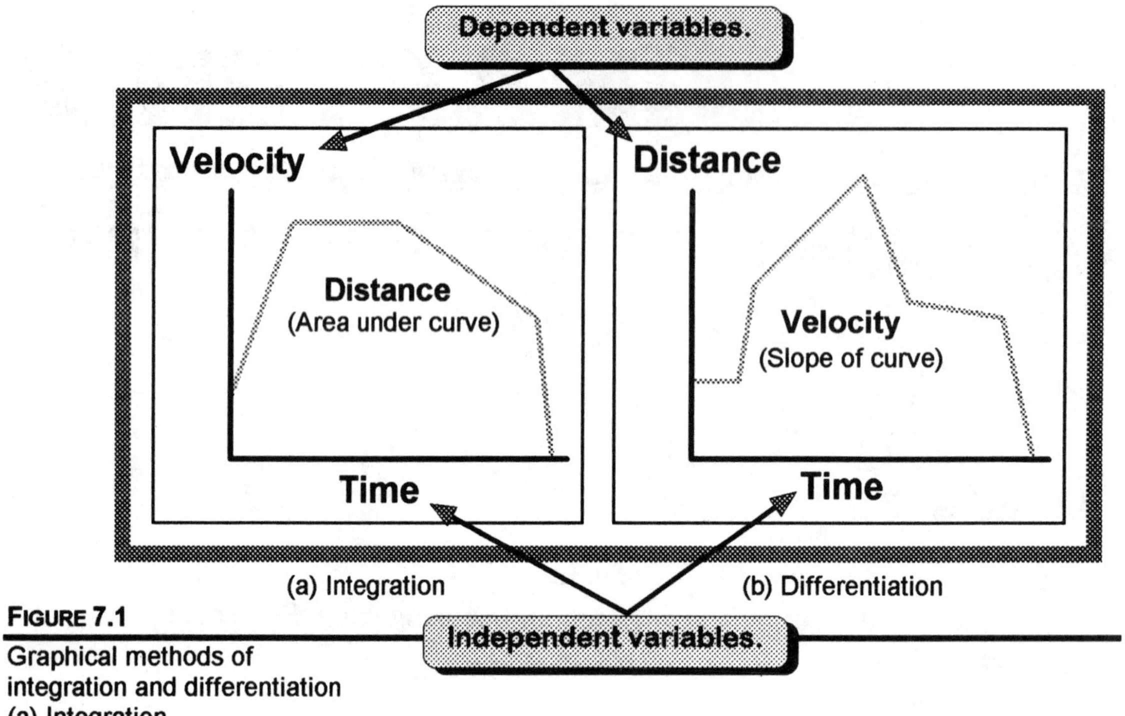

FIGURE 7.1
Graphical methods of integration and differentiation
(a) Integration
(b) Differentiation

Electronic integration and differentiation

To see how integration and differentiation can be performed electronically, we note the relationship between current and charge:

$$Q(t) = \int_0^t I(t)dt \qquad I(t) = \frac{dQ(t)}{dt}$$

However, it is more convenient if both input and output can be measured in voltage. Therefore, we use a resistor as a current-to-voltage transducer (V = IR), and we take advantage of the relationship between capacitor voltage and charge (Q = CV) to generate:

$$Vout(t) = \frac{1}{RC}\int_0^t Vin(t)dt \qquad Vout(t) = RC\frac{dVin(t)}{dt}$$

Based on these equations, we construct the simple RC-based circuits of Figure 7.2.

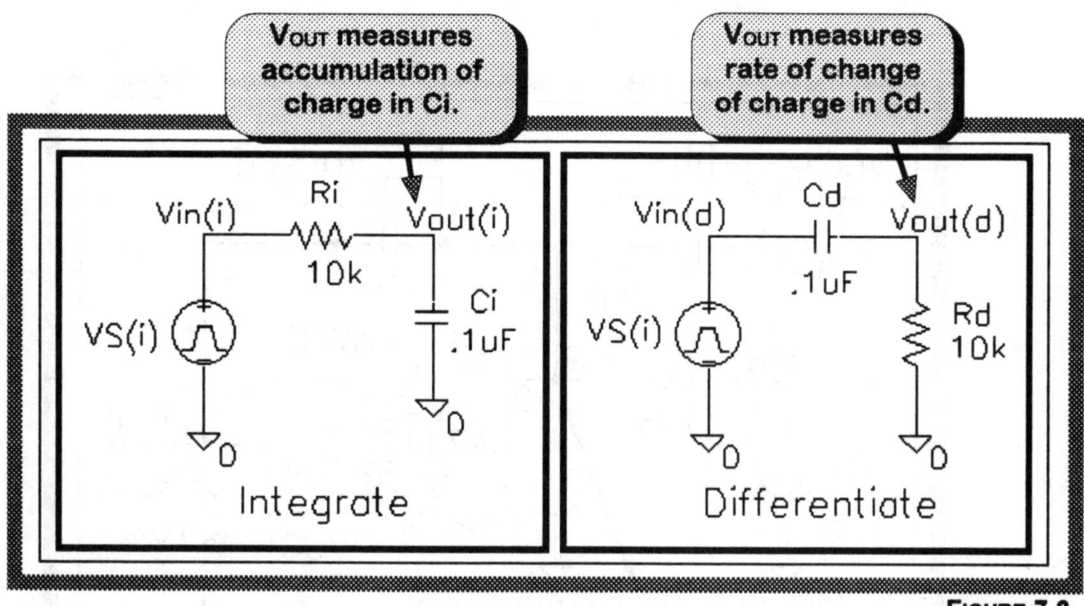

FIGURE 7.2
RC integrator and differentiator

Unfortunately the simple RC-based circuits of Figure 7.2 do not perform perfect mathematical integration and differentiation because the voltage in the capacitor feeds back and affects the current in the resistor. Therefore, to carry out mathematically correct integration and differentiation, we must uncouple the resistor from the capacitor by maintaining the junction at zero volts (a *virtual ground*).

Adding op amps to create the virtual grounds, we arrive at the circuits of Figures 7.3 and 7.4.

SIMULATION PRACTICE

Integrator

1. Draw the op amp integrator shown in Figure 7.3, and program VPULSE (VS) to generate the input waveform of Figure 7.5. The rise and fall times (td and tr) can be any small value, such as 10ns. (We initialize the capacitor to zero so the output waveforms will be predictable.)

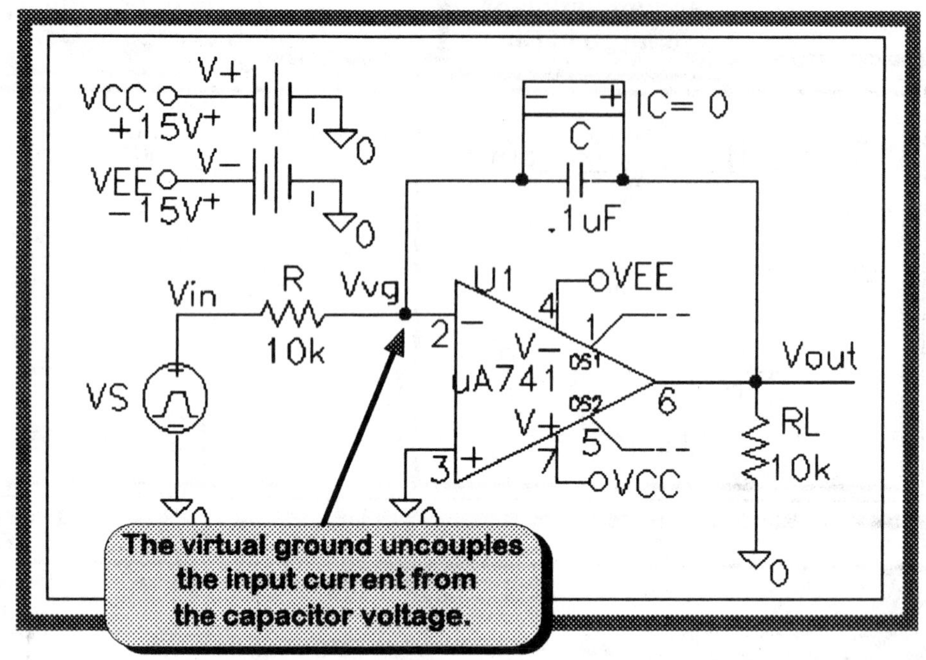

FIGURE 7.3

Op amp integrator

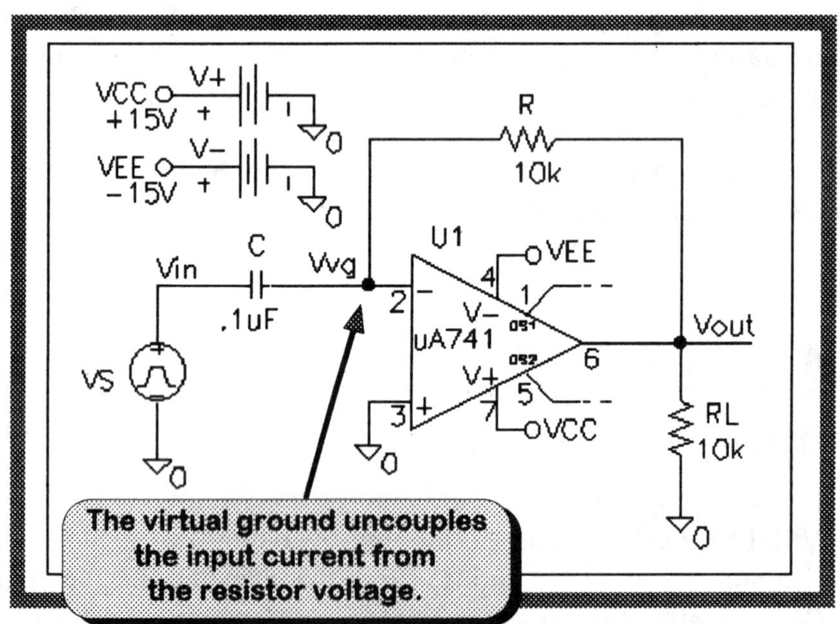

FIGURE 7.4

Op amp differentiator

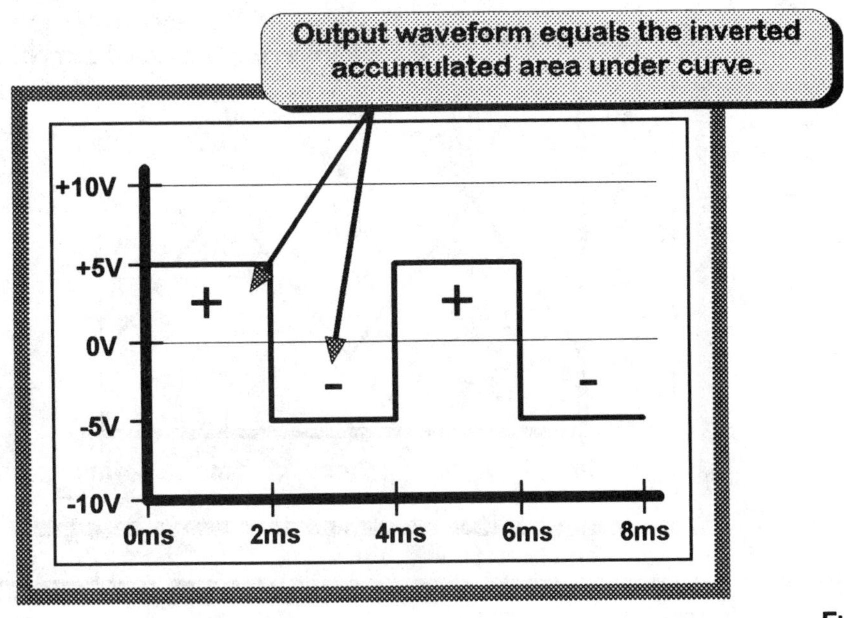

FIGURE 7.5
Integrator input waveforms

2. Predict the output waveform and sketch your answer on the graph of Figure 7.5. (Hint: The output amplitude profile starts at zero and is the inverted accumulated area under the curve of the input waveform.)

3. Generate the output waveform using PSpice and add your curve to Figure 7.5. Were your predictions correct?

   Yes **No**

Differentiator

4. Draw the op amp differentiator shown in Figure 7.4, and program VPULSE (VS) to generate the input waveform of Figure 7.6.

5. Predict the output waveform and sketch your answer on the differentiator graph of Figure 7.6. (Hint: The output waveform equals the inverted slope of the input.)

6. Generate the output waveform using PSpice. (Be prepared for an unexpected result.)

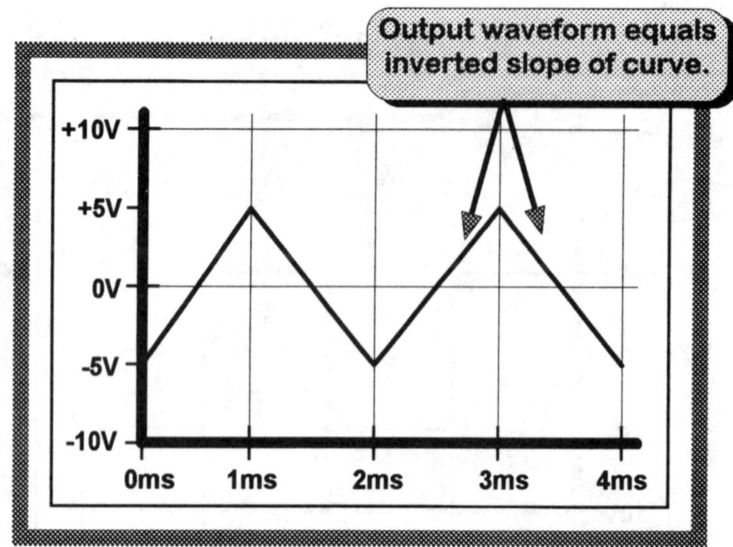

FIGURE 7.6
Differentiator input waveforms

7. It is clear that significant noise has been produced. This is because a differentiator amplifies noise [the derivative (slope) of a high-frequency signal can be greater than its amplitude].

   To solve the problem, add a small resistor to the input circuit (Figure 7.7) so the high-frequency noise will not "see" the capacitor.

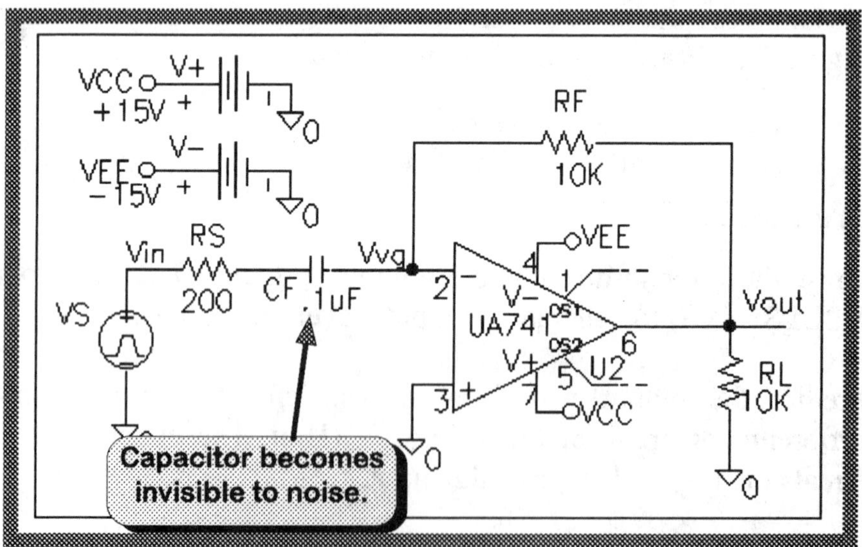

FIGURE 7.7
Differentiator with noise suppression

8. Again generate the output waveform using PSpice and add your curve to Figure 7.6. Has noise been greatly reduced, and is the new curve similar to your predictions?

Yes **No**

Advanced activities

9. For the input waveforms shown in Figure 7.8, predict the integrated and differentiated output waveforms and draw them on the graph.

10. For both the integrator of Figure 7.3 and the differentiator of Figure 7.7, change VS to VPWL and program the sources to match the waveforms of Figure 7.8. (See *Probe Note 7.1.*)

Using PSpice, generate both the integrator and differentiator waveforms. Did your predictions match the actual case?

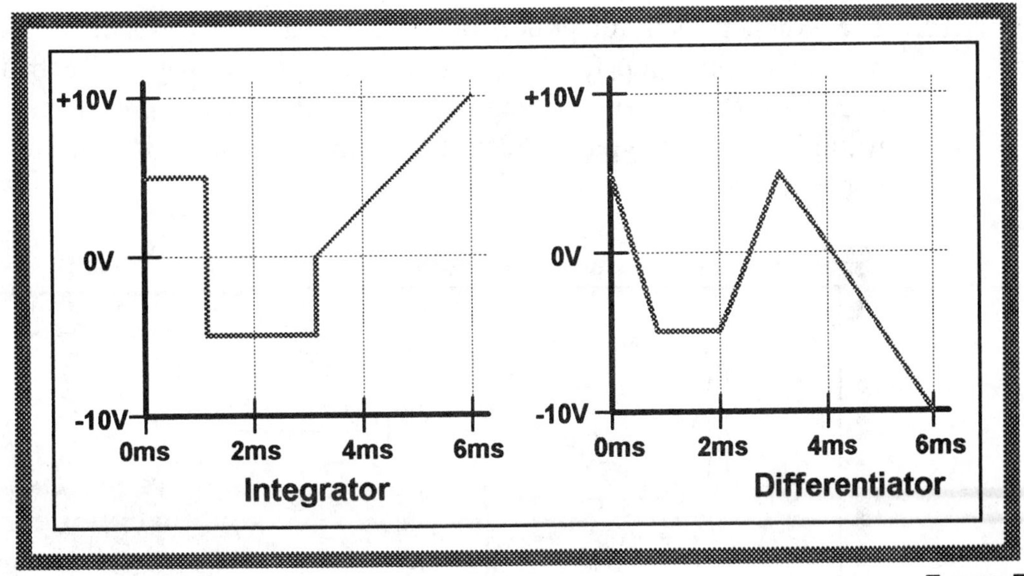

FIGURE 7.8
Input waveforms

11. Using various waveforms, compare the output of the simple RC integrator and differentiator of Figure 7.2 with the op amp counterparts of Figures 7.3 and 7.7. How do they differ?

Probe Note 7.1

How do I create custom waveforms out of voltage segments?

Place part VPWL (library *source.slb*) on your schematic. **DCLICKL** on the symbol to bring up the *Part Name* dialog box. Note the series of voltage and time pairs, where each pair represents a voltage/time node of the input waveform.

For example, to generate the integrator waveform of Figure 7.8, enter the following voltage/time pairs (be sure to **Save Attr** after each entry):

t1 = 0ms t2 = 1ms t3 = 1.01ms t4 = 3ms t5 = 3.01ms t6 = 4ms t7 = 6ms
v1 = +5V v2 = +5V v3 = −5V v4 = −5V v5 = 0V v6 = 0V v7 = +10V

For waveforms that have repetitive features, use source VPWL_ENH; for waveform values that are stored in a file, use VPWL_FILE.

12. The compensated integrator of Figure 7.9 uses resistor RF to prevent DC buildup (perhaps due to an offset voltage). Test the circuit over an extended range of time and over a range of pulse widths. Compare your results to the uncompensated integrator of Figure 7.3.

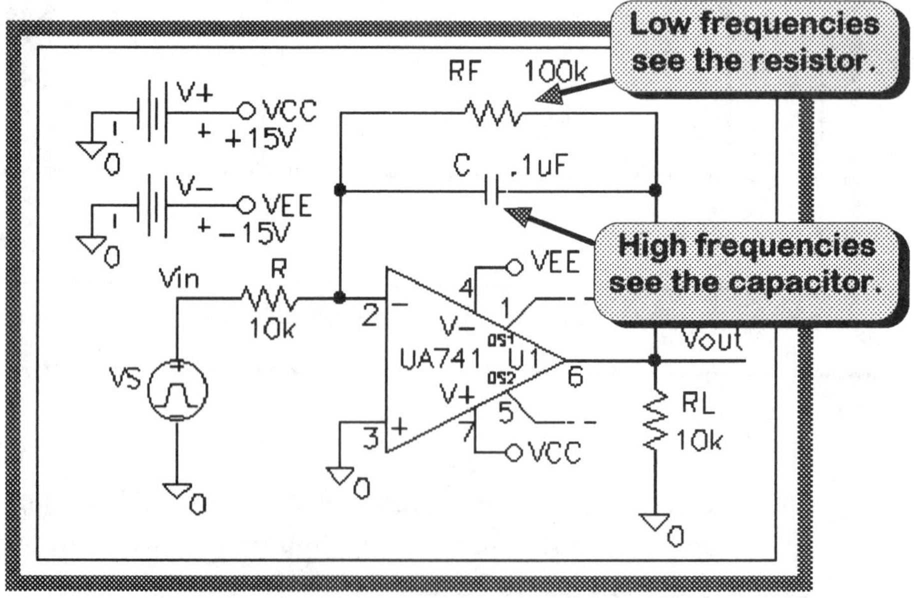

FIGURE 7.9
Compensated integrator

EXERCISES

- By cascading two differentiators, design a *double differentiator*. If the input is distance, what is the output? Also, test your design on various input waveforms.
- Making use of the circuit of Figure 7.10, test a moon-landing simulator that solves the equation below for an object of unity mass under control of thrust (*T*) and gravity (*G*).

  Using the transient mode, plot various trajectories. (Change the gravity value, and experiment with various VSIN, VPULSE, and VPWL for Vthrust.)

$$D(t) = \iint (T(t) - G(t))dt^2$$

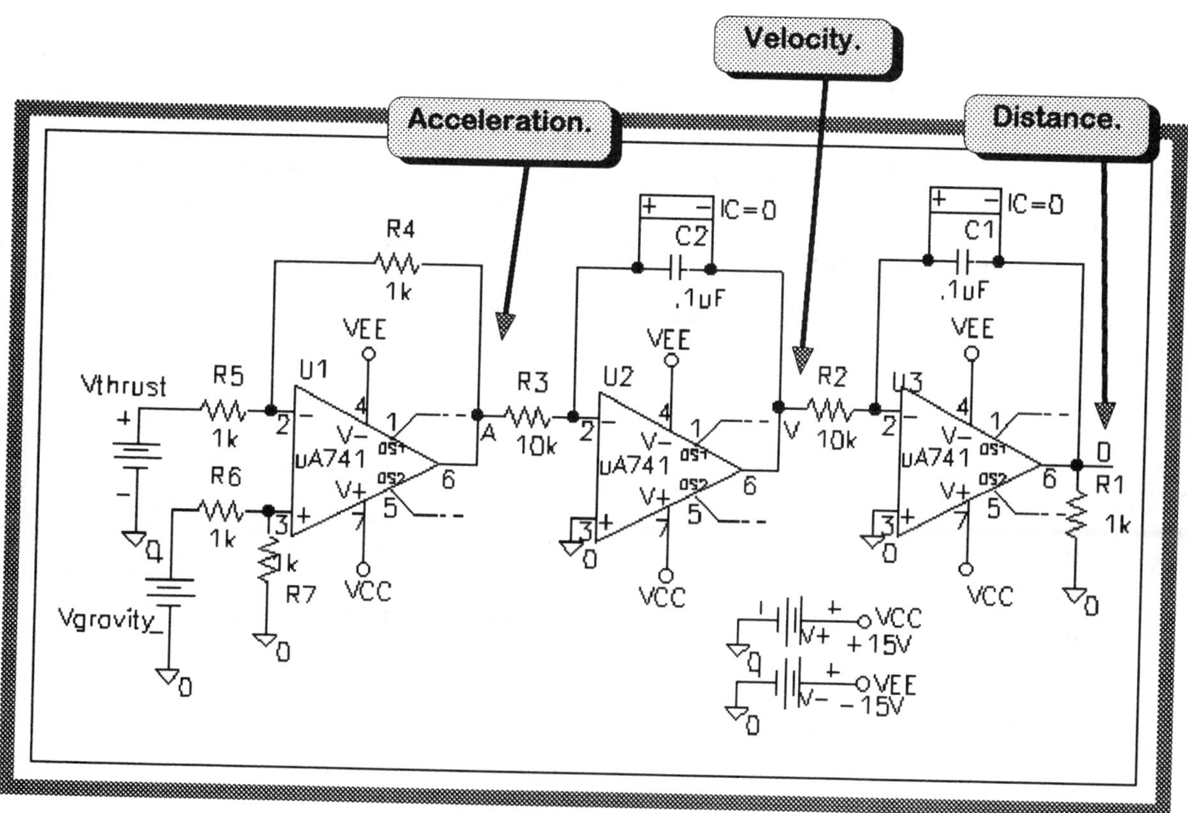

FIGURE 7.10
Moon-landing simulator

QUESTIONS AND PROBLEMS

1. What is the difference between *multiplication* and *integration*? Under what conditions does integration reduce to multiplication?

2. What is the difference between *division* and *differentiation*? Under what conditions does differentiation reduce to division?

3. When integration is depicted graphically, the output is the area under the curve of the input. Using this fact as a guide, how could you perform integration using an accurate scale and a pair of scissors?

4. Referring to the simple RC integrator of Figure 7.2, why is V_{OUT} equal to the integral of V_{IN} only when the capacitor is nearly empty of charge?

5. Why is noise amplified by a differentiator, but not an integrator?

6. Explain how to perform integration (over time) mechanically with the bucket and faucet of Figure 7.11.

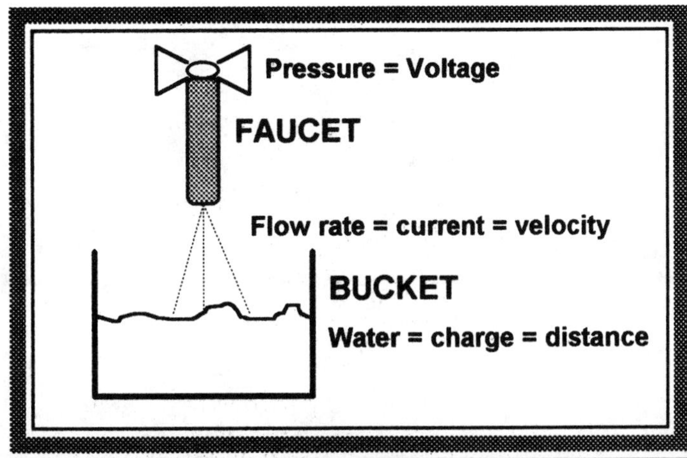

FIGURE 7.11 Mechanical integration

7. Mathematically, under what input conditions would the output of an integrated waveform go up when the input goes down?

8. Mathematically, under what input conditions would the output of a differentiated waveform be negative when the input was positive?

CHAPTER 8

Oscillators

Positive Feedback

OBJECTIVES

- To determine the conditions necessary for oscillation.
- To design and test sine and square wave oscillators.

DISCUSSION

Switching from negative to positive feedback requires only a tiny change in circuit configuration—but the switch results in an enormous change in operation.

For example, if our body temperature regulation mechanism changed from the stabilizing effects of negative feedback to the runaway effects of positive feedback, we would quickly succumb to a rapidly decreasing or increasing body temperature. When positive feedback is combined with negative feedback, the result is often oscillation.

Specifically, three conditions are necessary for oscillation:

1. **Closed-loop gain ≥ 1.**
2. **Positive feedback (loop phase shift = 360°).**
3. **Initial spark (noise).**

The two major classifications of oscillators are *sine* and *square*. Within each category are numerous varieties and types. This chapter will highlight several popular oscillator configurations.

SIMULATION PRACTICE

Sine wave oscillators

Phase-shift

1. Draw the *phase-shift* oscillator of Figure 8.1, a good choice for low-frequency applications.

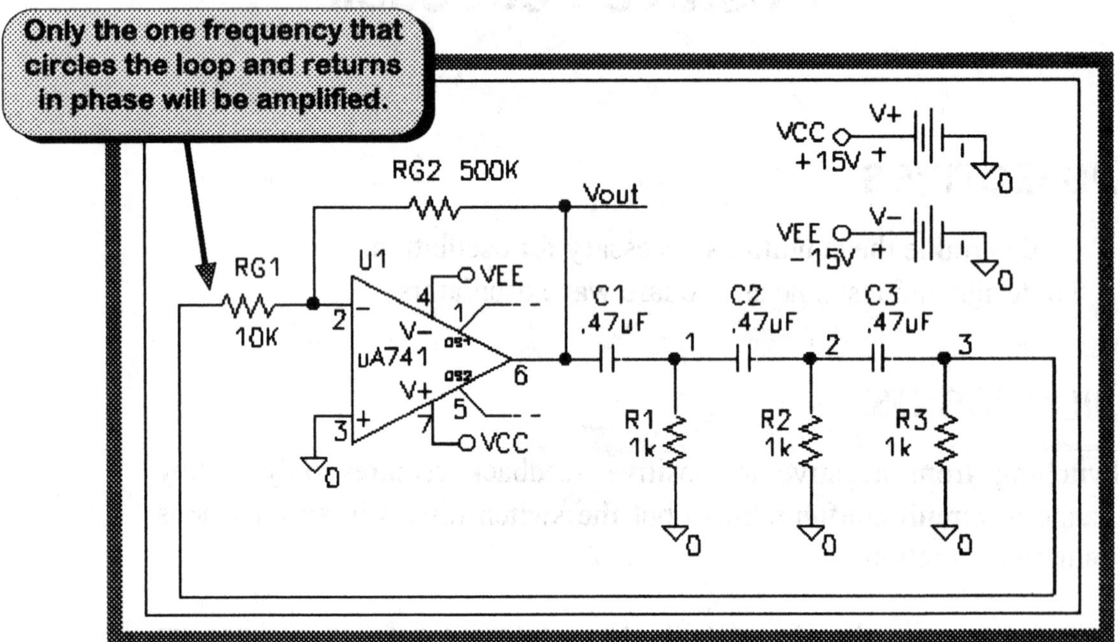

FIGURE 8.1
The phase-shift oscillator

2. To provide positive feedback, we must have 360° of phase shift. The op amp supplies 180°—so each RC network must supply 60°. To determine the frequency that results in a 60° phase shift, we note the following:

$$\tan 60° = 1.73 = X_C/R \qquad \text{where } X_C = 1/(2\pi fC)$$

Using this equation, determine the predicted frequency.

f = ___________

3. Analyze the circuit in the transient mode (0 to 150ms) and generate the graph of Figure 8.2.

   (a) Measure the oscillator frequency (1/period) and record the result below:

   Frequency of oscillation = ____________

   (b) Is the above frequency of oscillation *approximately* equal (within 50%) to the calculated value of step 2?

   Yes **No**

   (c) Why does the amplitude increase during the early cycles?

   (d) What limits the steady-state amplitude?

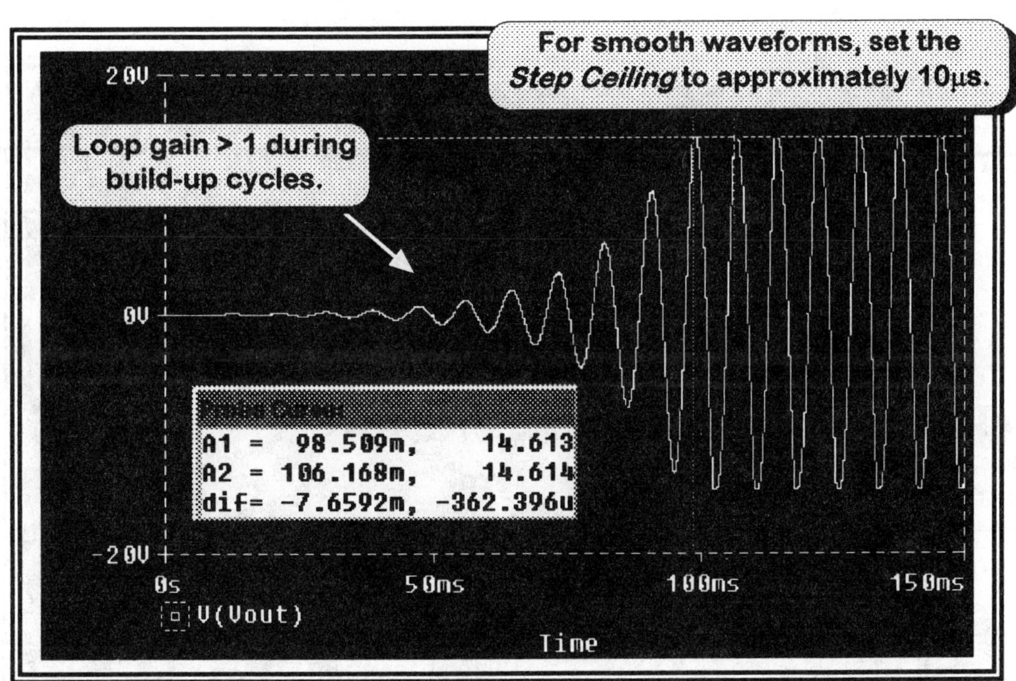

FIGURE 8.2
Phase shift oscillator output

4. Referring to Figure 8.1, display the waveforms at points 1, 2, and 3. (Be sure to label the wire segments 1, 2, and 3 as shown.)

   (a) Do the waveforms show the 60° phase shift? (<u>Suggestion</u>: Shift the X-axis to the 100ms-150ms range.)

   Yes **No**

   (b) Does the amplitude drop as we move from point 1 to point 3?

   Yes **No**

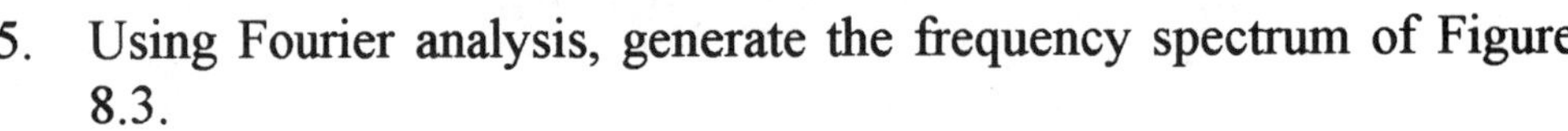

5. Using Fourier analysis, generate the frequency spectrum of Figure 8.3.

Fast Fourier transform

   (a) Is there a dominant fundamental frequency?

   Yes **No**

   (b) Does the fundamental frequency approximately equal the time-domain frequency measured in step 3?

   Yes **No**

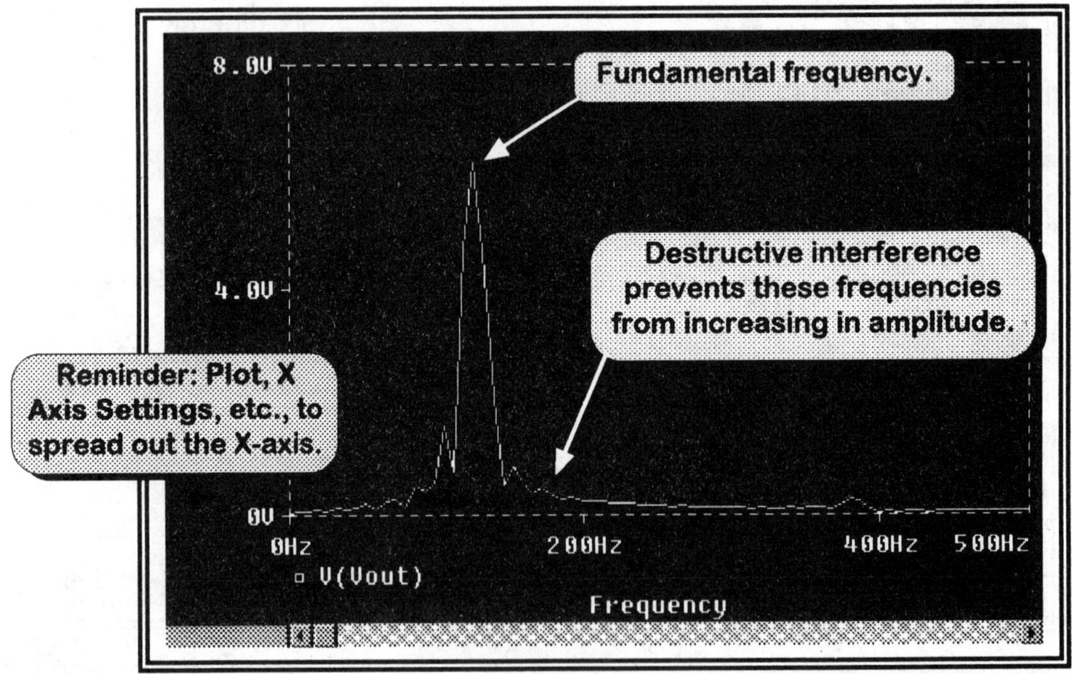

FIGURE 8.3
Phase-shift oscillator frequency spectrum

Colpitts

6. Draw the *Colpitts oscillator* of Figure 8.4, a good choice for higher frequencies.

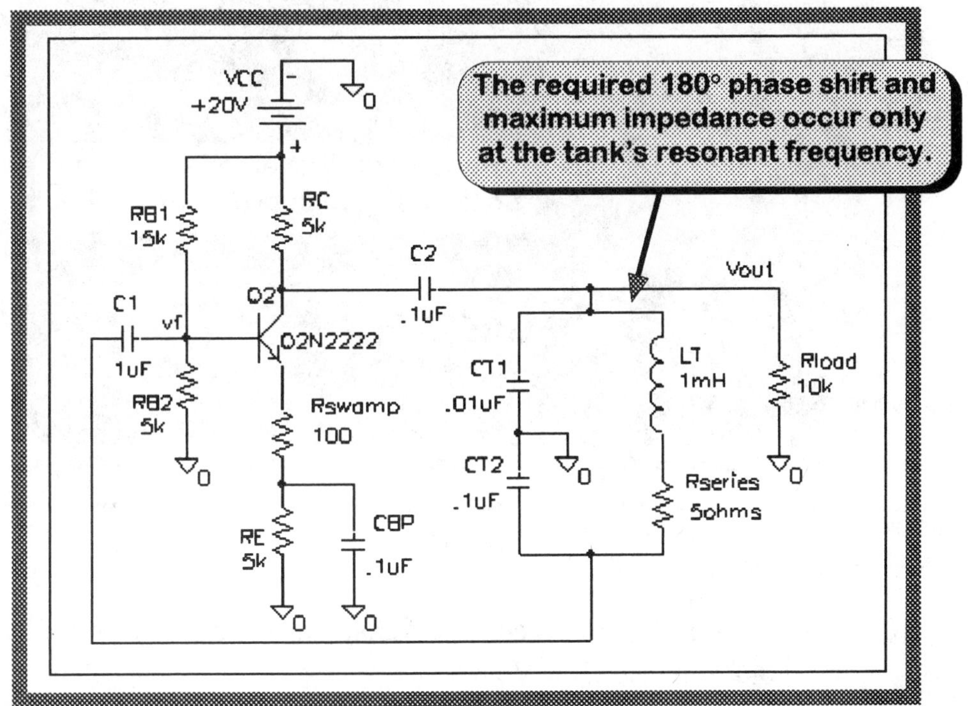

FIGURE 8.4
The Colpitts oscillator

7. Using the equation below, determine the resonant frequency of the tank circuit—which sets the oscillation frequency.

$$fr = \frac{1}{2\pi\sqrt{LC_T}} = \_\_\_\_\_\_\_\_ \qquad \text{where } C_T = \frac{CT1 \times CT2}{CT1 + CT2}$$

8. Perform a transient analysis to 500μs and display the output waveform of Figure 8.5.

9. Measure the oscillator frequency (1/period) and compare it to the calculated value of step 7. Are they similar?

f_r (PSpice) = __________

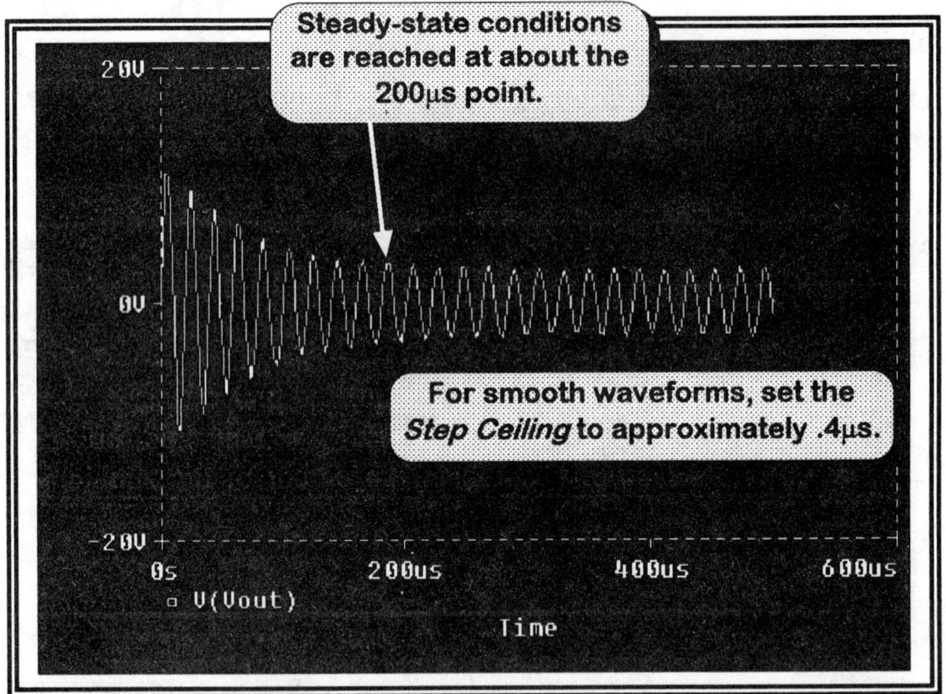

FIGURE 8.5

The Colpitts oscillator output waveforms

10. Add a plot of Vf (the feedback signal shown in Figure 8.4). Is Vf 180° out of phase with Vout?

    Yes **No**

11. Generate a Fourier spectrum for Vout of the Colpitts oscillator.

    (a) Is there a DC component? (Does the time-domain signal show a DC component?)

    Yes **No**

    (b) Is there slightly less distortion (a sharper fundamental frequency with a narrower bandwidth) than with the phase-shift oscillator?

    Yes **No**

    (c) Does the fundamental frequency approximately equal the calculated and measured values of steps 7 and 9?

    Yes **No**

Square wave

12. Draw the square-wave oscillator of Figure 8.6, which is good for audio frequencies.

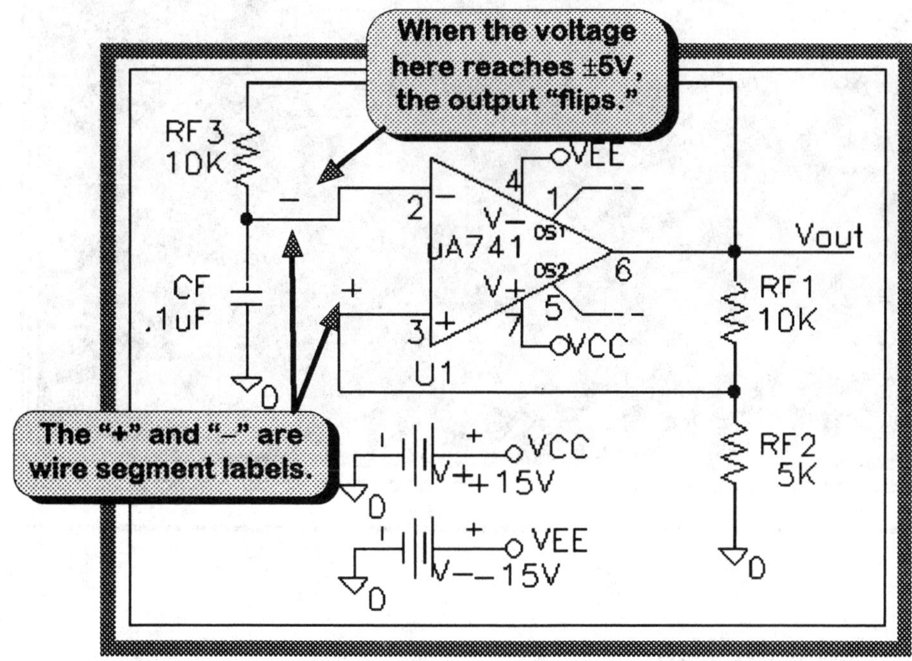

FIGURE 8.6
Square-wave oscillator

13. To predict the oscillation frequency we can reason as follows: The signal at the inverting input oscillates between –5V and +5V. Analysis of an RC pulsed waveform tells us that this requires approximately .7RC time constants. There are two RC swings per cycle. Therefore:

$$f = 1/(2 \times .7 \times RC) = 1/(2 \times .7 \times 10K \times .1\mu F) = \underline{714Hz}$$

14. Generate the transient output waveform using PSpice (0 to 4ms) and measure the oscillation frequency. How does the result compare to the predicted value of step 13?

f (measured) = ________

15. By adding waveforms at the V(+) and V(–) points, generate the graph of Figure 8.7. On the graph, identify all three waveforms (Vout, V(+), and V(–)).

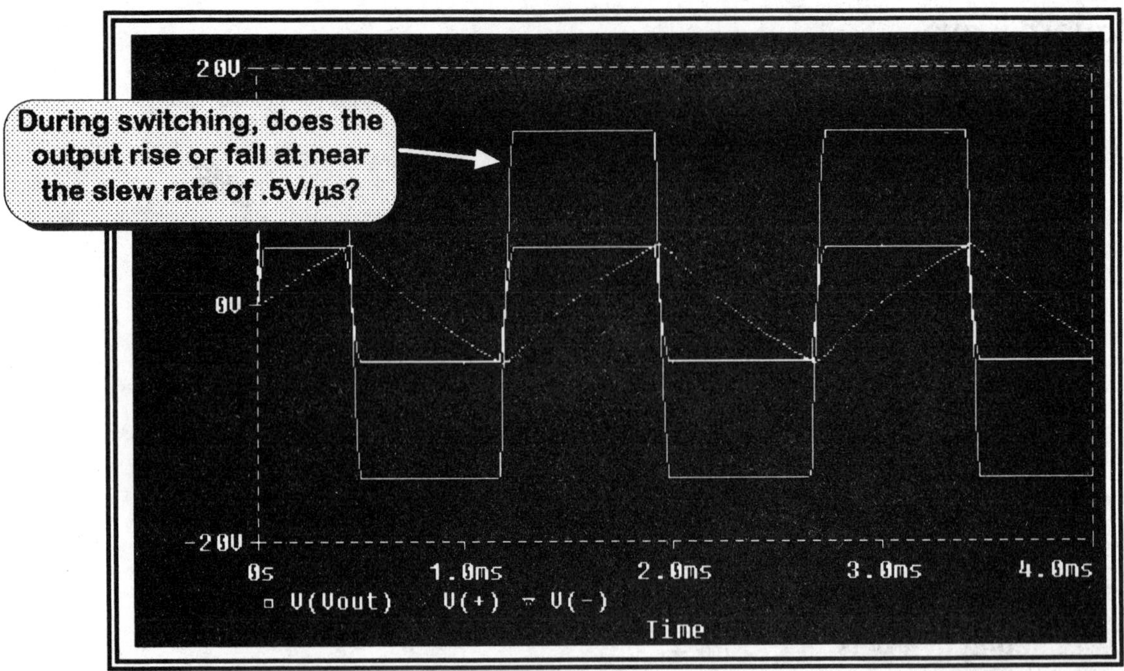

FIGURE 8.7
Square-wave waveforms

Advanced activities

16. By assigning 10% tolerances to the four tank circuit components of the Colpitts oscillator of Figure 8.4 (*Rseries*, *CT1*, *CT2*, and *LT*), generate the Gaussian frequency histogram of Figure 8.8.

- Use the following built-in goal function.

  Return the difference between the first and second X values at which the trace crosses the midpoint of its Y range with a positive slope (i.e., find the period of a time domain signal).

  ```
  Period(1) = x2-x1
  {
      1|Search forward level (50%, p) !1
        Search forward level (50%, p) !2;
  }
  ```

- To reduce computation time and to ensure steady-state analysis, perform 25 Monte Carlo runs (Gaussian distribution), and use *No Print Delay* to perform a transient analysis from 200µs to 250µs.

Performance analysis

- When the probe graph appears, switch to *Performance Analysis*, and enter *1/Period(V(vout))*.

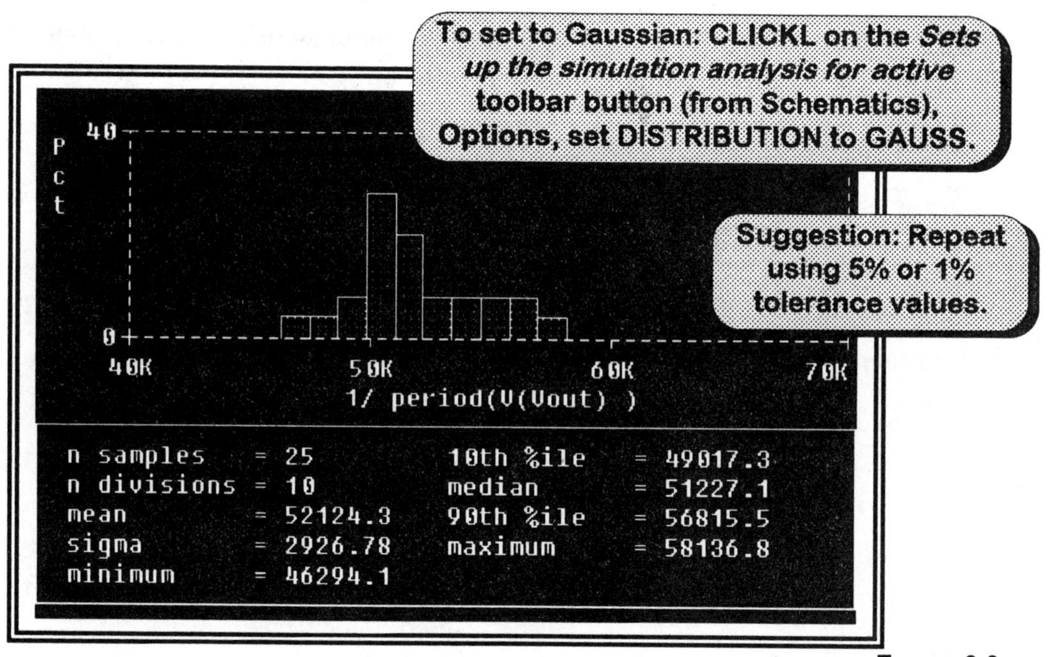

FIGURE 8.8
Frequency histogram of Colpitts oscillator

17. Looking at the results (Figure 8.8), what are the approximate odds that the frequency will lie between 45kHz and 55kHz? (Note: Your results may not necessarily be the same as Figure 8.8.)

EXERCISES

- By cascading a square-wave oscillator with an op amp integrator, design and test a triangle-wave generator.
- Design a Colpitts oscillator using an FET instead of a bipolar transistor.

QUESTIONS AND PROBLEMS

1. What are the three conditions necessary for oscillation?

2. During what period of oscillator operation is the closed-loop gain greater than 1? (Hint: How does the signal initially build up?)

3. For a real-world oscillator, what serves as the spark?

4. For the Colpitts oscillator of Figure 8.4, explain how the loop phase-shift is 360°.

CHAPTER 9

Filters
Passive and Active

OBJECTIVES

- To contrast passive and active filters.
- To compare low-pass, high-pass, and bandpass filters.

DISCUSSION

A *filter* is a device in which the gain is designed to be a special function of the frequency. The major classifications of filters are:

- Low-pass — passes low frequencies.
- High-pass — passes high frequencies.
- Bandpass — passes frequencies within a band.
- Band-stop (notch) — blocks frequencies within a band.

Any of these filters can be either passive or active. A passive filter uses only resistors, capacitors, and inductors. An active filter usually is implemented with an op amp that employs both positive and negative feedback. A *first-order* filter contains a single RC or RL network and rolls off at 20dB/dec; a *second-order* filter contains two RC or RL networks and rolls off at 40dB/dec.

The passive filter

A passive, first-order, low-pass filter is shown in Figure 9.1. To solve such a filter circuit, let's use complex numbers to determine its break frequency and rolloff.

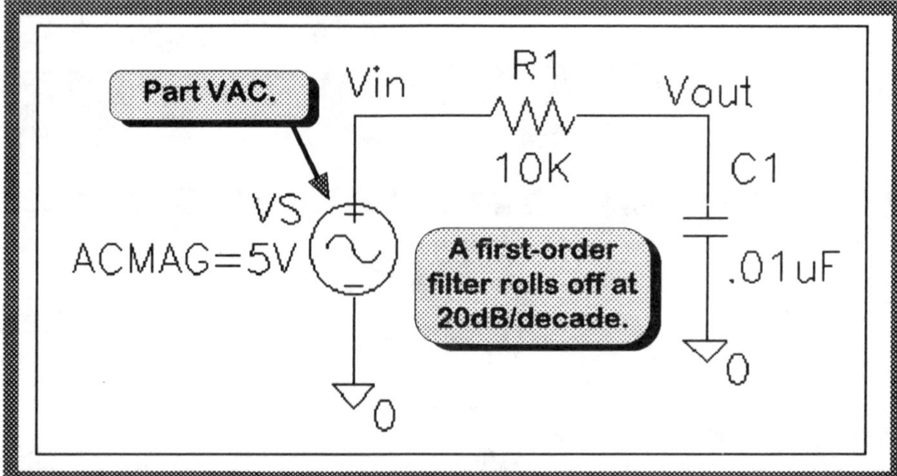

FIGURE 9.1
Passive, first-order, low-pass filter

We start with the filter's transfer function in rectangular form:

$$H(j\omega) = \frac{V_{OUT}}{V_{IN}}(j\omega) = \frac{-jX_C}{R - jX_C} = \frac{1}{1 + j\omega RC} \qquad \text{where } \omega = 2\pi f$$

Switching to polar form yields the magnitude and angle:

$$H = \frac{1}{\sqrt{1 + (\omega RC)^2}} \angle - \tan^{-1}(\omega RC)$$

By definition, the break (critical) frequency occurs when the angle equals 45° (or the magnitude drops by 3dB).

$$\tan(45°) = \omega RC = 1 \qquad \omega_C = \frac{1}{RC} \text{ or } f_C = \frac{1}{2\pi RC}$$

$$\text{Therefore, } f_B = \frac{1}{2\pi RC} = \frac{1}{2\pi \times 10k \times .01\mu F} = 1{,}591.5\text{Hz}$$

To determine rolloff, note that at high frequencies (well beyond the breakpoint):

$$\omega RC >> 1 \quad \text{So, the magnitude} \cong \frac{1}{\omega RC} = \frac{1}{2\pi fRC} \propto f^{-1}$$

Therefore, if **f** increases by a factor of ten (a decade), then:

$$\text{Rolloff} = 20\log_{10} 10^{-1} = -20\text{dB / decade}$$

An active filter

There are many variations of active filters. One of the most popular is the VCVS (*voltage-controlled-voltage-source*) *Sailen-Key network* configuration of Figure 9.2—popular because R and C network components are of equal value. Complex mathematical analysis yields parameters that are similar to the passive case: a break frequency of 1/2πRC (1,591.5Hz) and a rolloff of 40dB/decade.

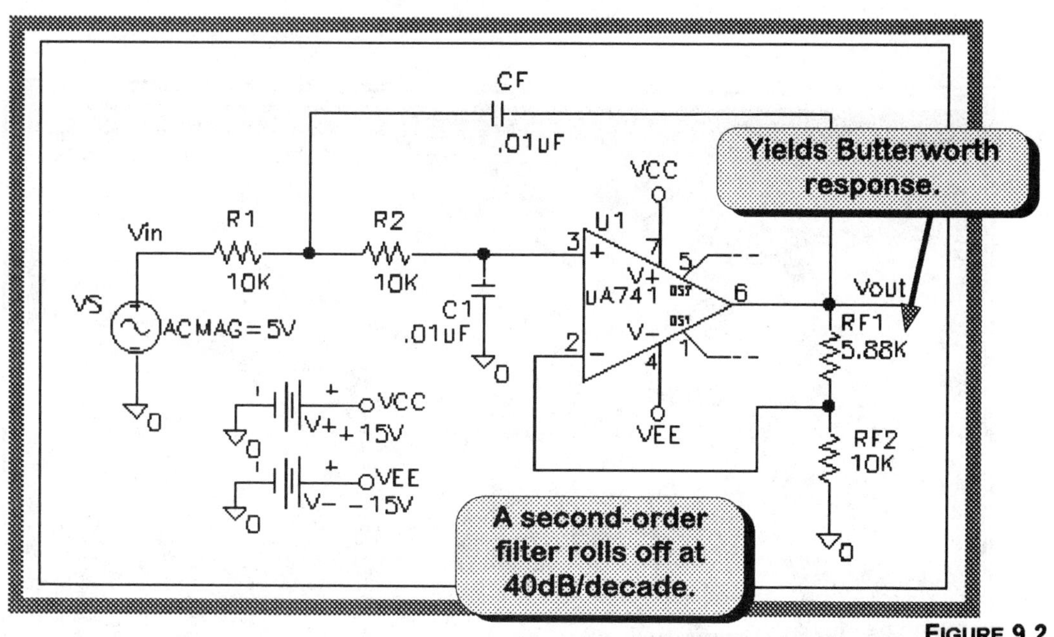

FIGURE 9.2
Active, second-order, low-pass filter

One advantage of using an active filter is that its characteristics depend on gain. For example, changing the gain of the circuit of Figure 9.2 changes its response, as shown in Table 9.1.

Response	RF1/RF2	Breakpoint	Gain
Bessel	0.288	1.27fc	1.27 (2.1dB)
Butterworth	0.588	1.00fc	1.588 (4.0dB)
1dB Chebyshev	0.955	0.863fc	1.955 (5.8dB)
2dB Chebyshev	1.105	0.852fc	2.105 (6.5dB)
3dB Chebyshev	1.233	0.841fc	2.233 (7.0dB)

TABLE 9.1
Major filter types

SIMULATION PRACTICE

1. Draw the first-order, passive, low-pass filter of Figure 9.1, and generate a Bode plot from 1Hz to 10MEGHz.

2. Repeat step 1 for the Butterworth active filter of Figure 9.2 and add (append) the results to the passive plot (**File**, **Append**, **CLICKL** on desired file, **OPEN**, **Do not skip sections**). The result is shown in Figure 9.3.

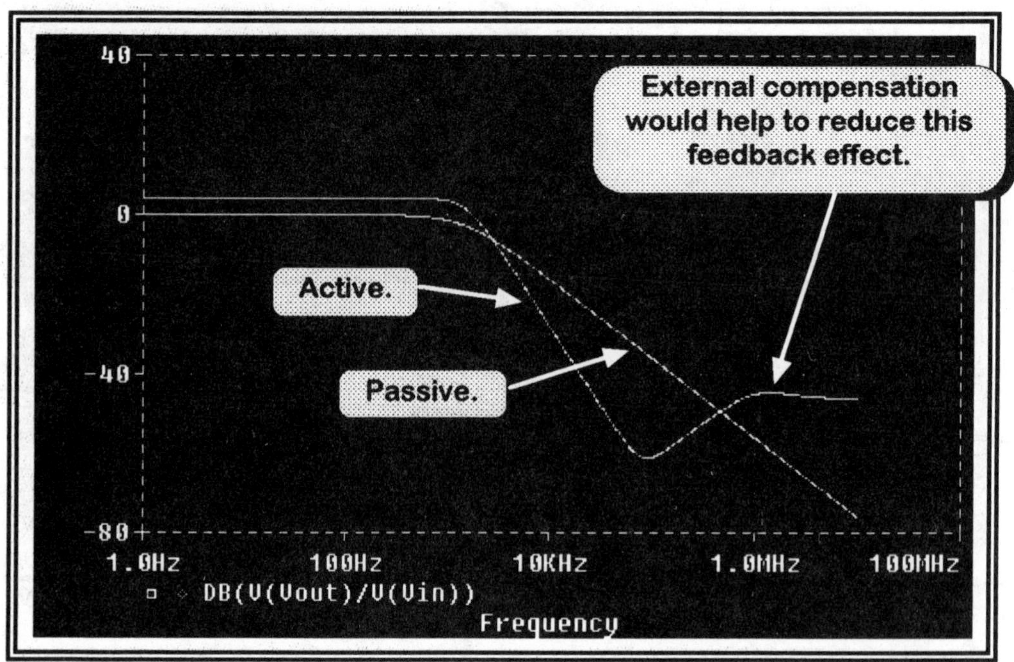

FIGURE 9.3
Bode plot

3. Looking at the curves:

   (a) Does the passive filter show breakpoint and rolloff values approximately equal to that calculated in the discussion (1,591.5Hz and –20dB/decade)?

   Yes **No**

   (b) Does the active filter show the predicted breakpoint and rolloff (1,591.5Hz and –40dB/decade)?

   Yes **No**

(c) Does the active filter show a small (≈4dB) gain?

Yes **No**

4. Sweep RF1/RF2 over the values shown in Table 9.1 and display curves for all five types of filters—as shown by Figure 9.4. (Suggestion: Keep RF2 constant and sweep only RF1.)

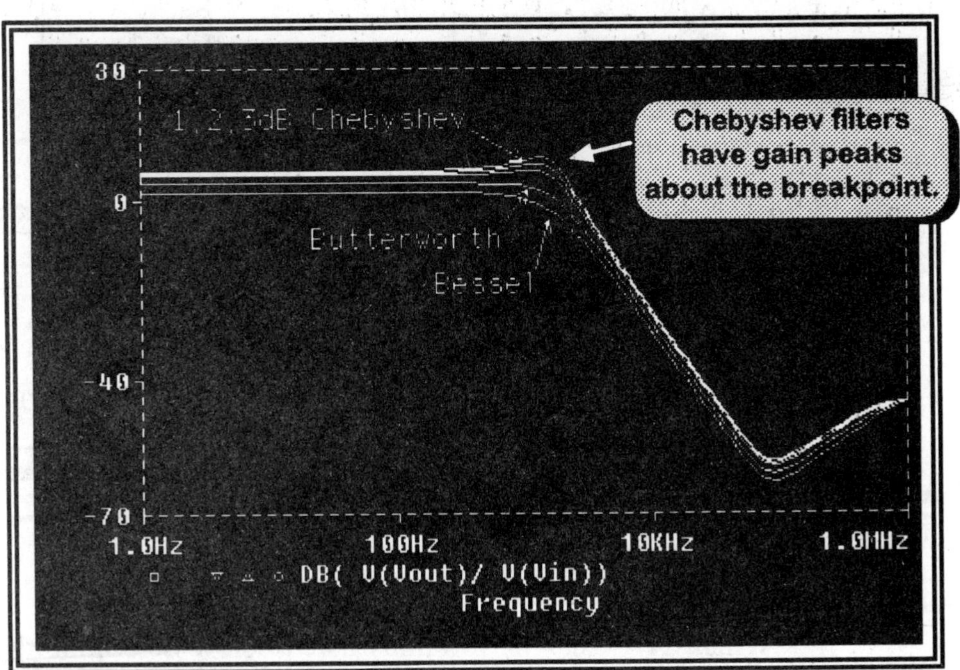

FIGURE 9.4
Bode plot

5. Based on Figure 9.4:

(a) Is the Butterworth filter the smoothest (maximally flat)?

Yes **No**

(b) Does the order of the Chebyshev filters (1dB, 2dB, and 3dB) equal its <u>increase</u> in gain about the breakpoint (that is, does $A_{DB}(peak) - A_{DB}(low\text{-}freq) = 1dB, 2dB, 3dB$)?

Yes **No**

Advanced activities

6. Draw the *multiple-feedback bandpass* filter of Figure 9.5.

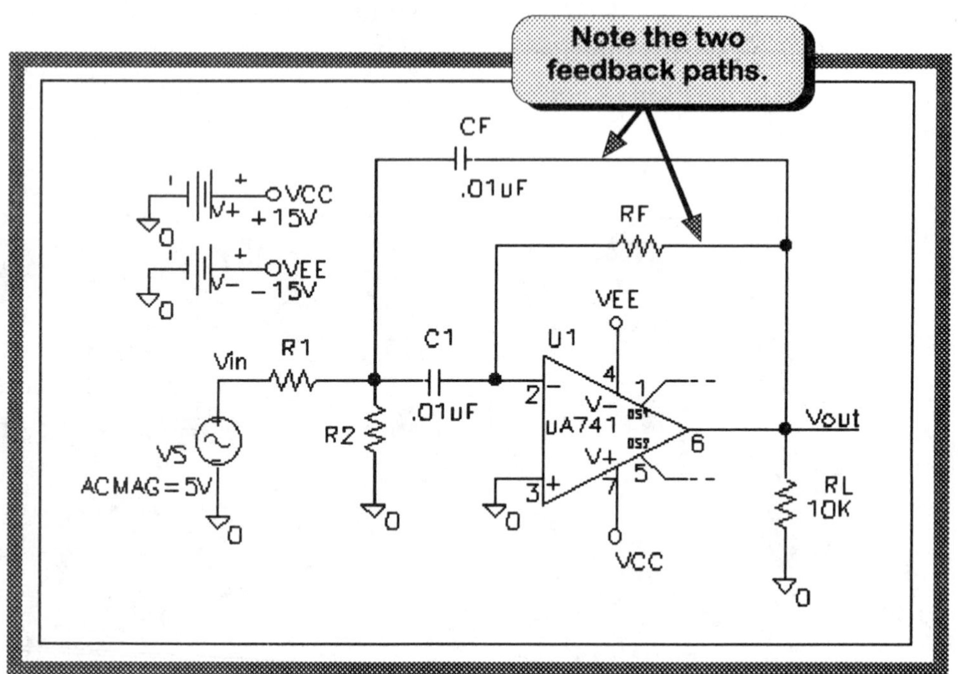

FIGURE 9.5
Bandpass filter

7. When C1 = CF, the formulas listed below determine the values of the resistors. Solve the equations for a center frequency (f_0) of 1kHz, a Q of 5, and a gain (A_0) of 1. (Assume C = .01μF, as shown for C1 and CF, and round off all Rs to generally available values.)

R1 $= Q/(2\pi f_0 C A_0) =$ __________

R2 $= Q/(2\pi f_0 C(2Q^2 - A_0)) =$ __________

RF = 2R1

Reminder: $Q = f_0/BW$.

8. Enter the calculated values into the circuit of Figure 9.5 and display the Bode plot of Figure 9.6. Do the displayed values of A_0, f_0, and Q *approximately* match the calculated values?

Yes **No**

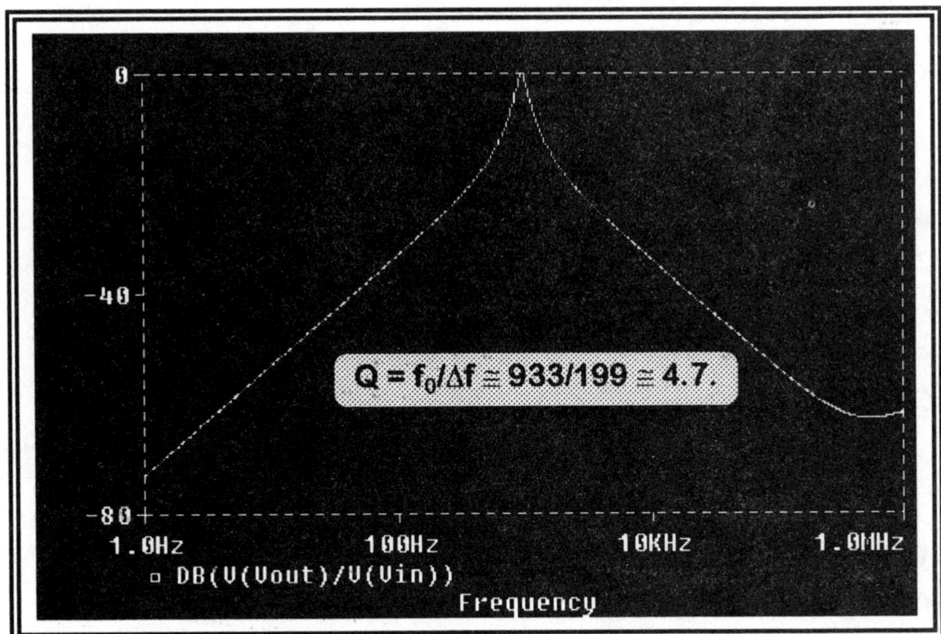

FIGURE 9.6

Bode plot

9. By reversing the Rs and Cs within the RC networks of Figures 9.1, 9.2, and 9.5, design and test passive and active high-pass and band-stop filters of your choice.

10. Substitute an inductor for the resistor in the passive low-pass filter of Figure 9.1. Set L to the value that gives the same breakpoint (when $X_L = R$). Is the filter still a low-pass filter, and is the rolloff still 20dB/decade?

EXERCISES

- Using the active bandpass filter of Figure 9.5, design a circuit to pass the note of A (880Hz) but reject by at least 10dB the next note below (G#, at 830.61) and the next note above (A#, at 932.33).

- Cascade the two second-order, low-pass filters of Figure 9.2 and determine their characteristics. (How does the breakpoint and rolloff compare to the single-stage configuration?)

QUESTIONS AND PROBLEMS

1. What are several advantages of using an active filter over a passive filter?

2. Why don't filters generally make use of inductors?

3. What order is the bandpass filter of Figure 9.5? (Hint: How many RC networks does the filter have?)

4. Show how a combined (cascaded) low-pass and high-pass filter can be used to construct a bandpass filter.

5. Given a summing amplifier, an inverter, and a bandpass filter, construct a band-stop filter.

CHAPTER 10

The Instrumentation Amplifier

Precision CMRR

OBJECTIVES

- To analyze the instrumentation amplifier.
- To compare the CMRR of an instrumentation amplifier with a single op amp amplifier.

DISCUSSION

Suppose we wish to design a biofeedback system. One approach is to attach two electrodes to the arm a slight distance apart. We then send the amplified differential voltage to a meter, and we close the loop by attempting to mentally control the voltage.

However, this approach confronts us with several technical problems. First of all, the biofeedback voltage is a very small *differential* signal, but the entire arm is immersed in a strong sea of electromagnetic interference. Second, the transducers attached to the arm have high output impedances and therefore supply only very small currents. For the feedback process to work, we must amplify the differential-mode information signal and reject the *common-mode* interference signal.

In purely technical terms, we require a high input-impedance differential amplifier of exceptionally high CMRR (common-mode rejection ratio). Also, to match the circuit to different individuals, it would be a real plus if we could easily and quickly control the gain. Such a circuit is called an *instrumentation amplifier* and is shown in Figure 10.1.

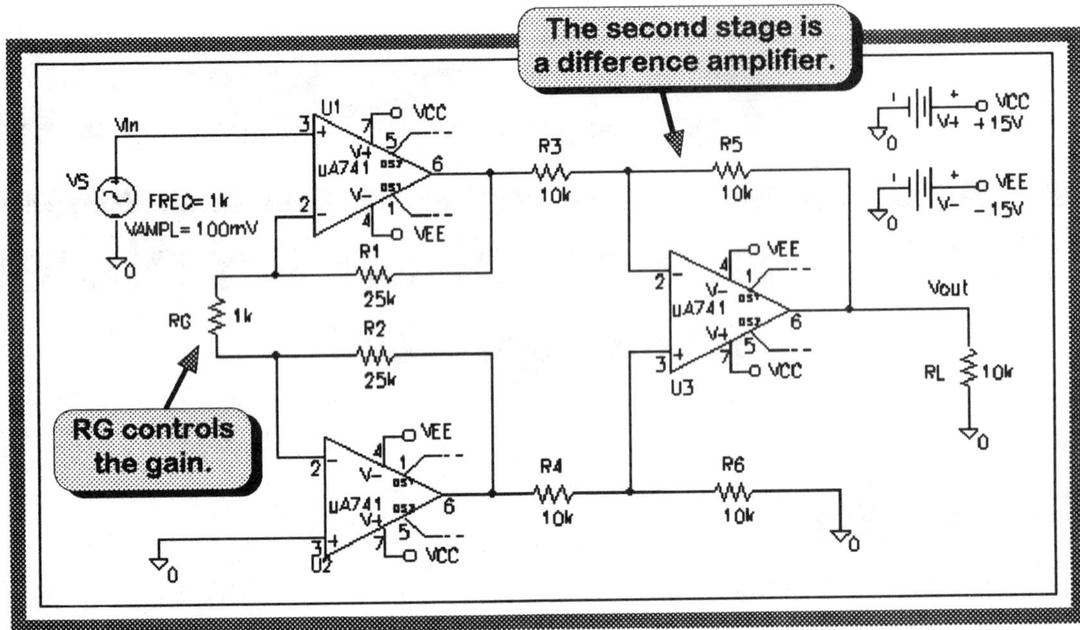

FIGURE 10.1
Passive, first-order, low-pass filter

Careful mathematical analysis shows that the differential mode (DM) voltage gain of the circuit is:

$$A_{DM} = \frac{2R1 + RG}{RG}$$

SIMULATION PRACTICE

1. Draw the circuit of Figure 10.1 and set the attributes as shown.
2. Using both calculations and PSpice, determine A(DM) when RG = 100Ω, 1kΩ, and 10kΩ, and place your answers in the table below.

Lower Vin to 10mV.

	Calculations	PSpice
A_{DM} at RG = 100Ω	________	________
A_{DM} at RG = 1kΩ	________	________
A_{DM} at RG = 10kΩ	________	________

3. With RG = 1kΩ, use PSpice to determine Z_{IN}(DM).

   Z_{IN}(DM) = ____________

4. Tie together the two inputs and determine the common-mode voltage gain A_{CM}. By combining A_{DM} and A_{CM}, determine the CMRR.

$$\text{CMRR} = \frac{A_{DM}}{A_{CM}} = \_\_\_\_\_\_\_\_\_\_\_\_$$

5. Based on the results of steps 3 and 4, is the Z_{IN}(DM) and CMRR much higher for an instrumentation amplifier than conventional, single, op amp amplifiers?

   Yes **No**

6. Generate a Bode plot of the instrumentation amplifier and report the bandwidth:

   Bandwidth = ____________

   Is the instrumentation amplifier a DC amplifier?

   Yes **No**

Advanced activities

7. Remove the ground from the right side of R6 (Figure 10.1) and substitute a +5V source. If this node were assigned a pin, what effect would various values of voltage have on Vout?

8. Generate a gain-bandwidth product family of curves for the instrumentation amplifier. Is the gain-bandwidth product a constant?

9. Set resistor tolerances and perform a worst case analysis. Is an instrumentation amplifier more stable than a conventional op amp amplifier?

EXERCISE

- Assuming a 1μV biofeedback signal and a .1V common-mode interference signal, test the effectiveness of the instrumentation amplifier as a biofeedback device. (Suggestion: Select different frequencies for the biofeedback and interference signals so they can be distinguished in the output signal.)

QUESTIONS AND PROBLEMS

1. What is the voltage gain range of the instrumentation amplifier of Figure 10.1?

2. What makes the instrumentation amplifier a DC amplifier?

3. For the amplifier of Figure 10.1, what is the voltage gain of the second stage (involving resistors R3 through R6 and op amp U3)?

4. Why do instrumentation amplifiers often interface transducers? (Hint: Typically, what is the Z_{OUT} of a transducer?)

5. By adding a Class B transistor buffer to the instrumentation amplifier of Figure 10.1, show how to reduce the amplifier's output impedance and increase its short-circuit output current. How low can we take RL before the output clips? (Hint: Place the Class B buffer inside the feedback loop of U3.)

PART II
Digital

In Part II we switch from analog to digital, where information is stored and transferred in two-state values rather than continuous levels.

To avoid excessive computations, special techniques are required to simulate digital circuits and to interface analog to digital.

CHAPTER 11

TTL Logic Gates
Analog/Digital Interfaces

OBJECTIVES

- To determine the specifications of the TTL logic family.
- To determine how PSpice interfaces analog and digital components.

DISCUSSION

The oldest, most widely used logic family is TTL (transistor-transistor-logic). In the ideal case, they are simply devices that carry out Boolean algebra. In the real world, however, they have threshold voltages, propagation delays, loading factors, and other considerations of which we should be aware. These specifications are the subject of this chapter.

TTL gates

Figure 11.1 shows that all six basic TTL gates are available to us when using the evaluation version of PSpice.

Analog-to-digital interfacing

In general, PSpice allows us to combine analog and digital devices in any way we wish. In most cases, it is not necessary for the designer to know the behind-the-scenes details concerning such mixed circuits. However, when displaying node values under *Probe*, more detailed knowledge of how analog and digital devices are combined may prove essential.

CMOS gates are not available in the evaluation version.

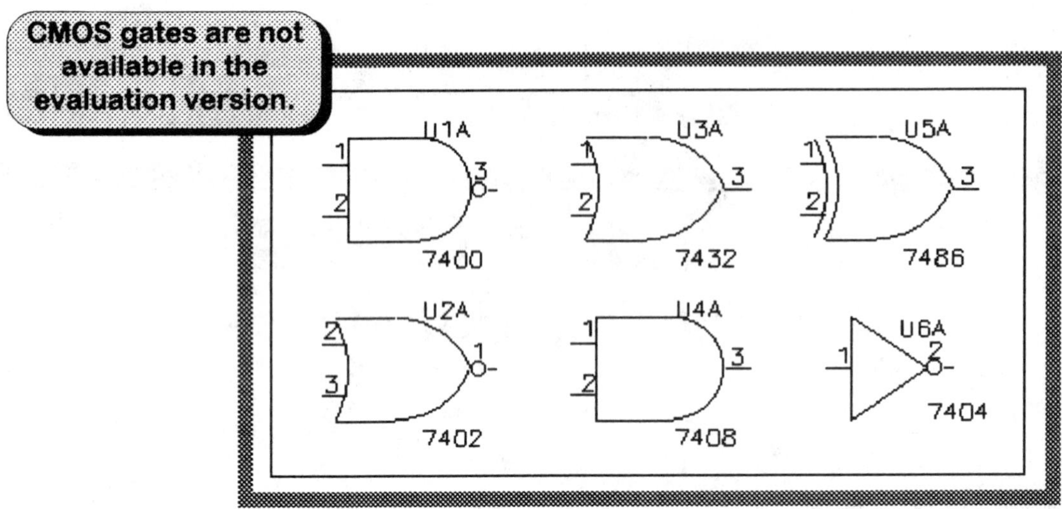

FIGURE 11.1
The basic TTL logic gates

Analog, digital, and interface nodes

PSpice evaluates three types of nodes: *analog*, *digital*, and *interface*. If all the devices connected to a node are analog, then the node is analog. If all the devices are digital, then the node is digital. If both analog and digital devices drive the node, then it is an interface node. For each type of node, certain *values* are assigned:

- For analog nodes, the values are voltages and currents.
- For digital nodes, the values are states, which are calculated from the input/output model for the device, the logic level of the node (0 or 1), and from the output strengths of the devices driving the node. Each strength is one of 64 ranges of impedance values between the default values of 2 and 20kΩ. To find the strength of a digital output, PSpice uses the logic level and the values of parameters DRVH (high-level driving resistance) and DRVL (low-level driving resistance) from the device's I/O models.
- For interface nodes, the values are both analog voltages/currents and digital states. This is because PSpice automatically inserts an AtoD or DtoA *interface subcircuit* (complete with power supply) at all interface nodes. These subcircuits handle the translation between analog voltages/currents and digital states.

Figure 11.2 shows a simple mixed analog/digital circuit. Although this is the way the circuit is drawn under *Schematics*, the interface nodes do not show the invisible circuitry automatically added by PSpice. For that we turn to Figure 11.3, where we have simply drawn in the interface blocks.

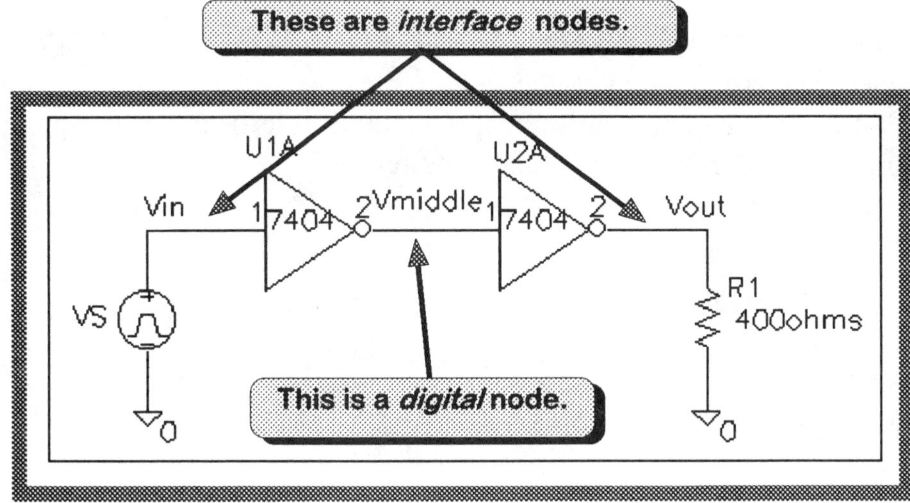

FIGURE 11.2
Mixed component circuit

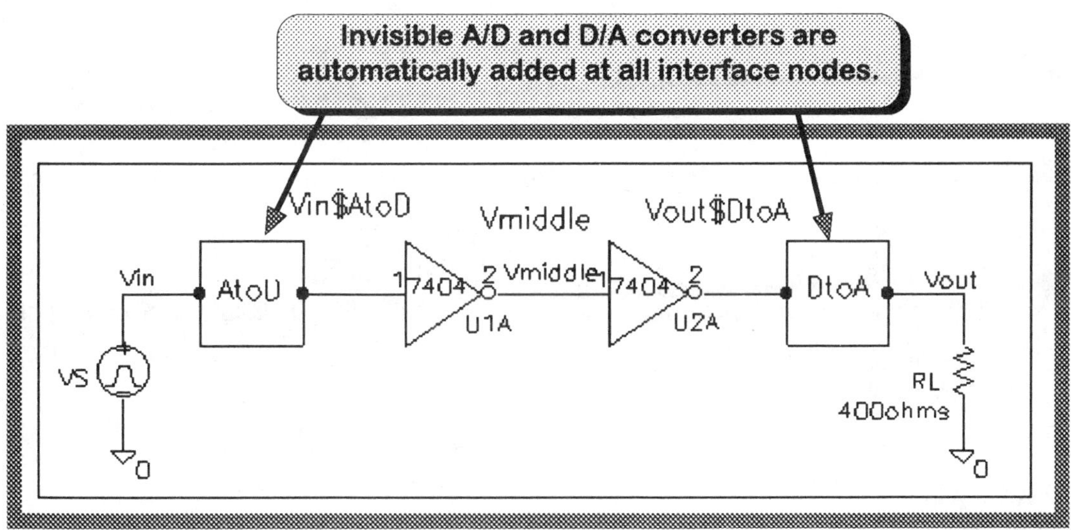

FIGURE 11.3
Circuit showing invisible nodes and interface devices

From Figure 11.3, we note the two interface circuits (AtoD and DtoA) created automatically by PSpice. With the wire segments labeled as shown, there are two analog voltage nodes [V(Vin) and V(Vout)] and three digital voltage nodes (Vin$AtoD, Vmiddle, and Vout$DtoA). As we will see, Probe displays digital values quite differently from analog values.

Although the interface circuits do not appear in the schematic, their characteristics are detailed in the output file—as listed below. (Note that PSpice also automatically creates a single interface power supply and ground for the AtoD and DtoA subcircuits.)

```
**** Generated AtoD and DtoA Interfaces ****

Analog/Digital interface for node Vin
Moving X_U1A.U1:IN1 from analog node Vin to new digital node Vin$AtoD
X$Vin_AtoD1 Vin Vin$AtoD $G_DPWR $G_DGND AtoD_STD
PARAMS: CAPACITANCE=   0

Analog/Digital interface for node Vout
Moving X_U2A.U1:OUT1 from analog node Vout to new digital node
Vout$DtoA
X$Vout_DtoA1 Vout$DtoA Vout $G_DPWR $G_DGND DtoA_STD
PARAMS: DRVH= 96.4   DRVL= 104   CAPACITANCE=   0

Analog/Digital interface power supply subcircuits
X$DIGIFPWR 0 DIGIFPWR

.END  ;(end of AtoD and DtoA interfaces)
```

In this chapter we will use several analog/digital circuits to study the basic characteristics and specifications of analog, digital, and interface nodes. Many of our results can be compared to the spec sheet values of Appendix D.

SIMULATION PRACTICE

1. Draw the test circuit of Figure 11.4 and set the attributes as shown.

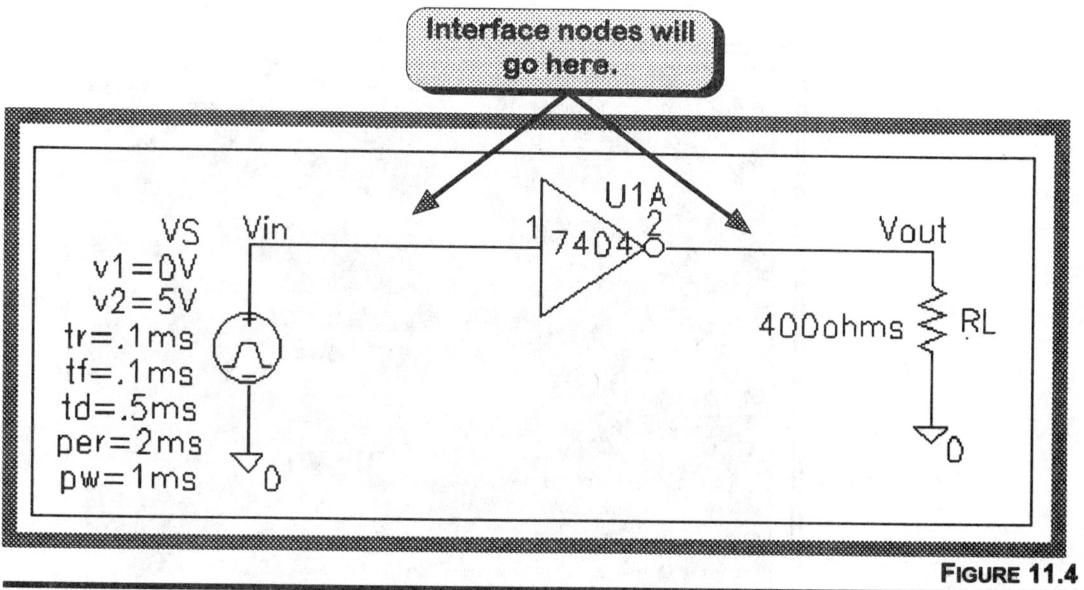

FIGURE 11.4
Analog/digital test circuit

2. Set the transient mode from 0 to 2ms (*step ceiling* of .02ms), run a simulation, generate the default Probe graph, and bring up the *Add Traces* dialog box.

   (a) Enable **Analog**, **Voltages**, and **Currents**. Note the following available analog traces:

   Time, V(Vin), V(Vout), V($G_DGND), V($G_DPWR), I(RL), I(VS), V(VS:–).

   (b) Disable **Analog** and enable **Digital**. Note the following available digital traces:

   Vin$AtoD, Vout$DtoA

   (c) For both the analog and digital cases, enable and disable **Alias Names** and note the additional (alternative) names of the analog nodes.

   (d) On the test circuit of Figure 11.4, sketch in an AtoD and DtoA converters (see Figure 11.3). From the list of (a) and (b), identify and label the analog and digital voltage nodes.

3. Using *analog* trace names, generate the input/output *analog* voltage and current curves of Figure 11.5.

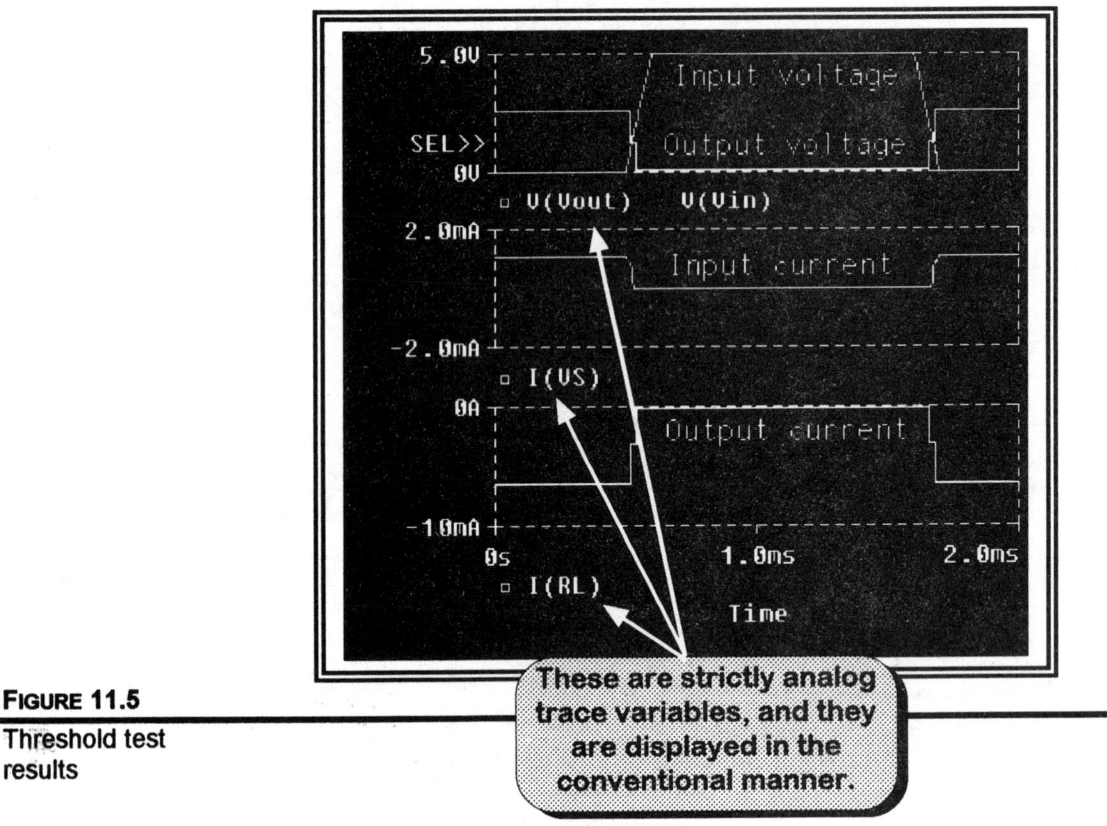

FIGURE 11.5
Threshold test results

In the following section, we will determine a number of TTL specifications. If time is short, be sure to choose those of greatest interest.

4. Based on the results of Figure 11.5, fill in the following voltages and currents in the tables provided and compare them to the 7404 specifications given:

   - **Input HIGH and LOW voltages (V_{IH} and V_{IL})** — *the minimum and maximum input voltages that are guaranteed to represent the input HIGH and input LOW states.* (What values of V_{IN} will give a V_{OUT} of .4V and 2.4V?)

	V_{IH}	V_{IL}
PSpice	________	________
Spec Sheet	2.0V (min)	0.8V (max)

- **Output HIGH and LOW voltages (V_{OH} and V_{OL})** — *minimum and maximum logic 0 and 1 output voltages.* (What is V_{OUT} when V_{IN} is 0V and +5V?)

	V_{OH}	V_{OL}
PSpice	______	______
Spec Sheet	2.4V (min)	0.4V (max)

- **Input HIGH and LOW currents (I_{IH} and I_{IL})** — *maximum input currents at the HIGH and LOW input states.* (What is I_{IN} when V_{IN} is 5V and 0V?)

	I_{IH}	I_{IL}
PSpice	______	______
Spec Sheet	40µA	1.6mA

- **Output HIGH and LOW currents (I_{OH} and I_{OL})** — *typical logic 1 and 0 output currents.* (What is I_{OUT} when V_{OUT} is maximum and minimum?)

	I_{OH}	I_{OL}
PSpice	______	______
Spec Sheet	16mA	400µA

Output short circuit current (I_{OS})

5. Short the output (reduce RL to .001ohms) and generate the output *analog* current plot of Figure 11.6. Based on the results, fill in the output short circuit current for the output HIGH state (V_{IN} = 0V) in the table below. (When done, return RL to 400Ω.)

	I_{OS}
PSpice	______
Spec Sheet	25mA

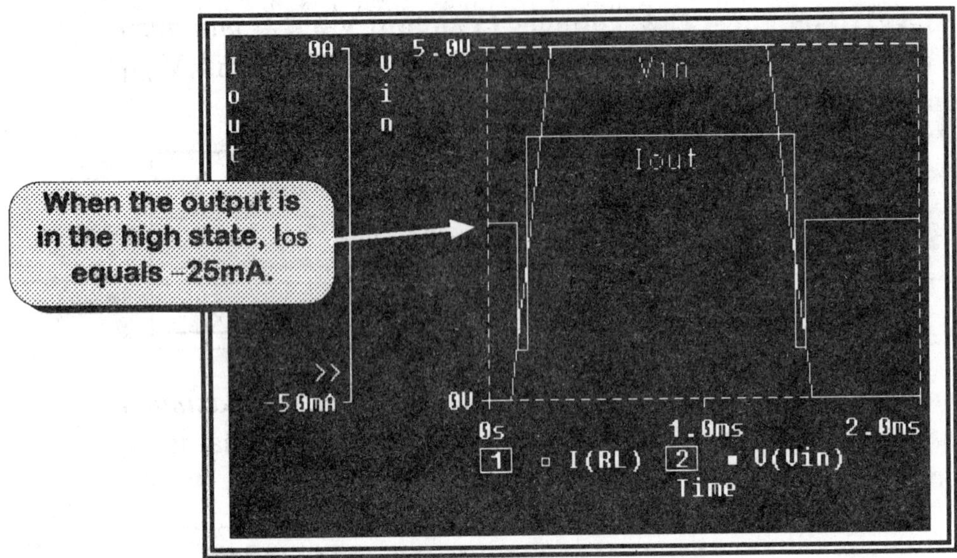

FIGURE 11.6
Output short circuit current

Propagation delay

6. Draw the modified test circuit of Figure 11.7 and set the attributes as shown.

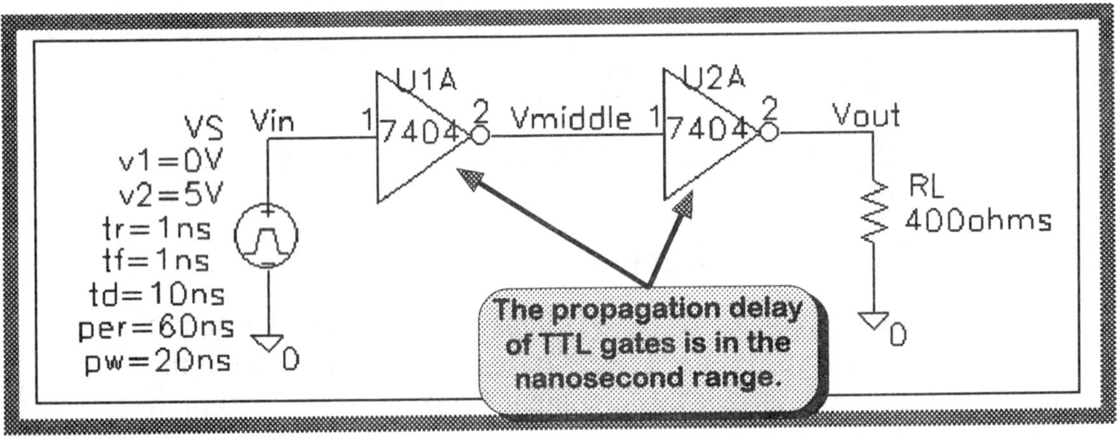

FIGURE 11.7
Propagation delay test circuit

7. Run PSpice and generate the input/output *analog* waveforms of Figure 11.8.

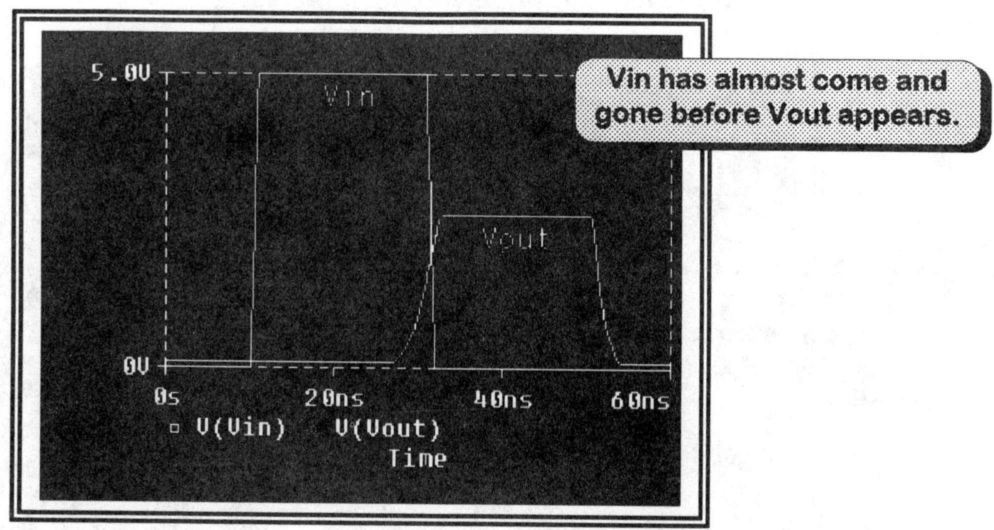

FIGURE 11.8
Propagation delay test results

8. Using any two corresponding points, fill in the value below. (Reminder: The signal has passed through two gates.)

	Propagation Delay
PSpice	________
Spec Sheet	15ns (typical)

Digital displays

9. To the propagation delay graph of Figure 11.8, add the three digital traces shown in Figure 11.9. Note that all digital signals are listed separately as logic state diagrams.

> Remember: To help select the proper variables, enable **Digital** and **Analog** in the *Add Traces* dialog box as necessary.

10. Enable the cursor and note the cursor window (as shown in Figure 11.10).

> Reminder: Use arrows to move cursor 1, and Shift/arrows to move cursor 2.

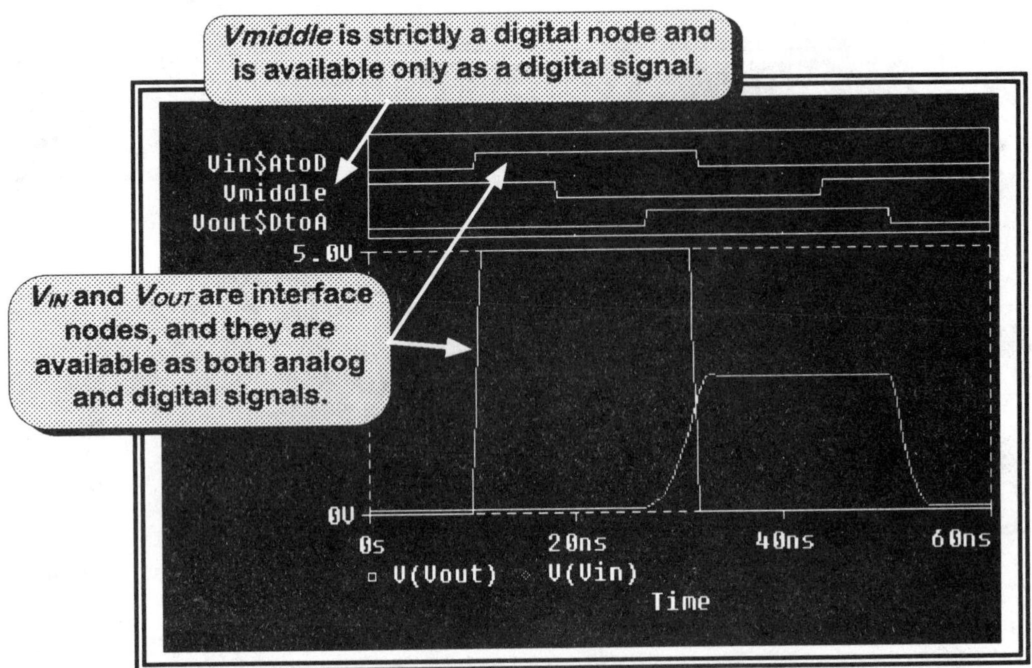

FIGURE 11.9

Adding a purely digital node voltage

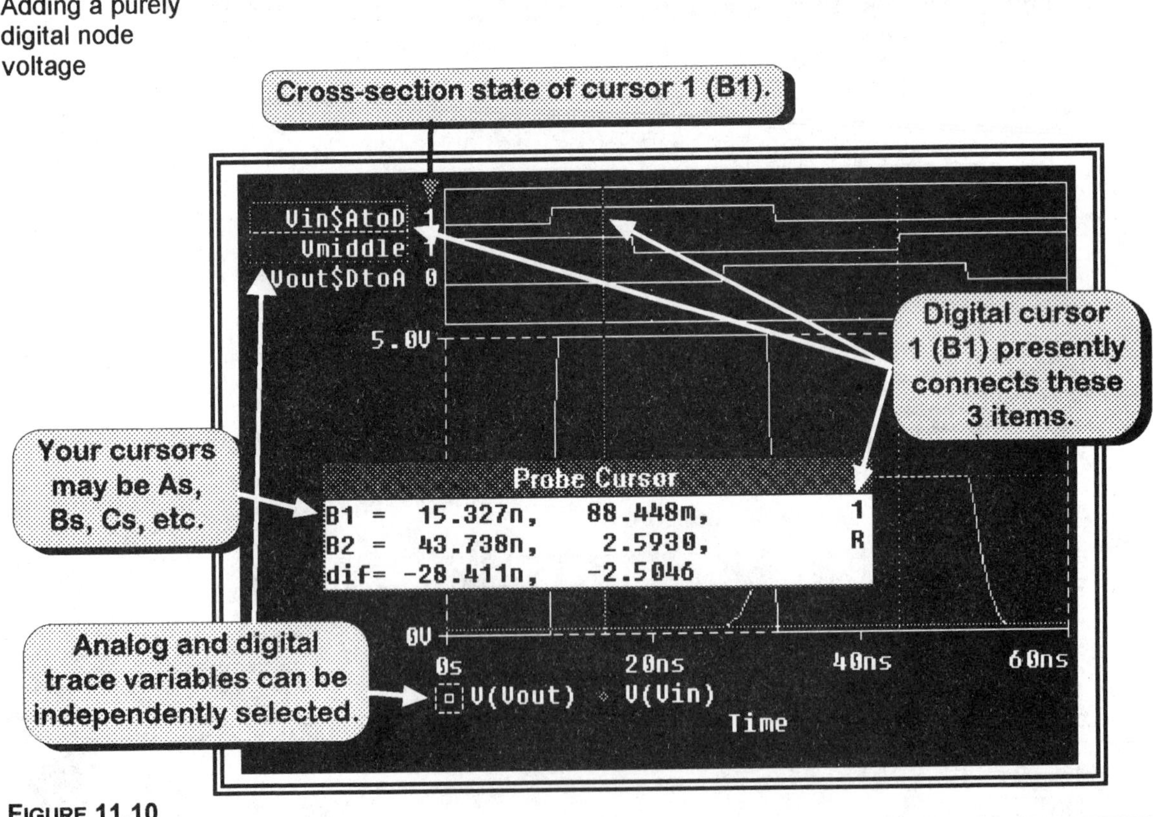

FIGURE 11.10

The cursor in mixed analog/digital displays

11. To gain experience in the use of cursors on mixed analog/digital waveforms, refer to Figure 11.10 and answer as many of the following as you can.

(a) Do the vertical axes of both cursors extend through both the analog and digital sections of the display?

Yes **No**

(b) Locate the cross-section values right after the digital trace variables (Figure 11.10). Move both cursors and note the values. Are the cross-section values based only on cursor 1?

Yes **No**

Does state R seem to indicate rising, and does state F seem to indicate falling?

Yes **No**

(c) Locate the cursor window. Does it have an analog section and a digital section?

Yes **No**

(d) Using **CLICKL** (for cursor 1) and **CLICKR** (for cursor 2), select various analog and digital trace variables. Can the analog and digital trace variables be independently selected?

Yes **No**

(e) Move the cursors to the locations shown in Figure 11.10. Within the cursor window, do the 1 and R correctly reflect the states of the digital waveforms?

Yes **No**

(f) Associate cursor 1 with analog trace *V(Vin)* and digital trace *Vin$AtoD*. Starting at 10ns, move the cursor to the right and note both the digital and analog readings. Did the R state fall approximately between analog values 1V and 2V?

Yes **No**

Does this approximately match the input HIGH and LOW voltages (V_{IH} and V_{IL}) of step 4?

Yes **No**

12. Select various digital and analog trace variables, move the cursors, and note the displays until you become familiar and comfortable with the overall cursor system.

Digital zoom techniques

13. Referring to *Probe Note 11.1,* place zoom bars as shown in Figure 11.11 and zoom in on the selected area to generate the waveform of Figure 11.12.

   (a) Are both the analog and digital sections expanded in unison?

      Yes **No**

   (b) Reactivate the cursor system. Do the parallelograms within the rise and fall times correspond to the R and F regions?

      Yes **No**

These parallelograms are also known as *ambiguity regions* and will be a major topic of Chapter 14.

Zoom Area

Probe Note 11.1

How do I expand (zoom in on) digital waveforms?

Refit

To expand a digital waveform: **CLICKL** on the *Zoom in on selected area of graph* toolbar button, move the mouse cursor arrow to either side of the desired zoom area, **CLICKLH** to create the first bar, drag the second bar (right or left) to the desired location (as shown in Figure 11.11), and release the left button. Note that both the analog and digital regions are expanded in unison.

To expand an analog waveform: **CLICKL** on the *Zoom in on selected area of graph* toolbar button, **CLICKLH** to drag a square around the area to be zoomed, and release the left button.

Also, we are always free to zoom in on a waveform by changing the X-axis range (**Plot**, **X-Axis**, etc.).

To return to the original waveforms: *Zoom to show all traces and labels* (Refit) toolbar button.

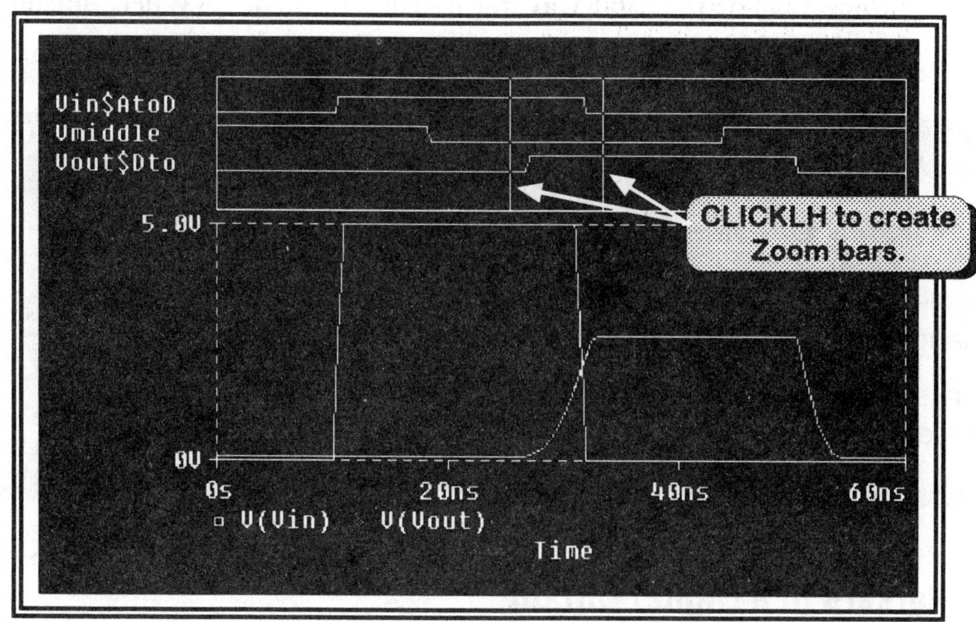

FIGURE 11.11
Placing zoom bars

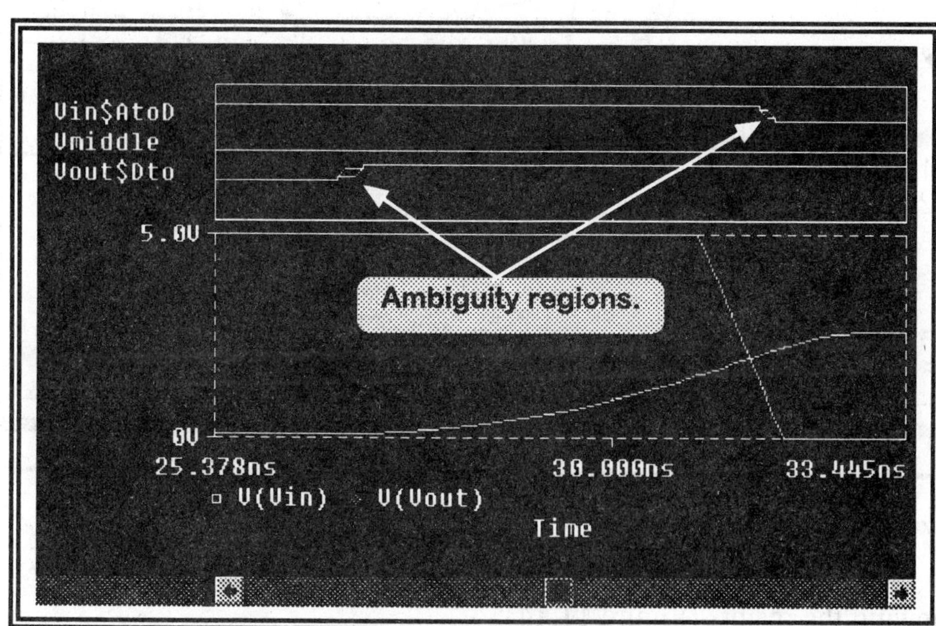

FIGURE 11.12
Expansion to zoom bars

14. Examine *Probe Note 11.2* and change the size of the digital display from 33% of the total waveform space to 50%. (When finished, return to 33%.)

Probe Note 11.2

How do I change the size of the digital display?

To change the size of the digital display in the Probe window: **Plot, Digital Size**, change the *Percentage of Plot to be Digital* from 33 to 50, **OK**.

Also, if the digital trace names are cut off, increase the default *Length of digital trace names.*

Markers in a Digital Circuit

15. Place markers on our test circuit as shown in Figure 11.13 and re-run PSpice to generate the waveforms of Figure 11.14.

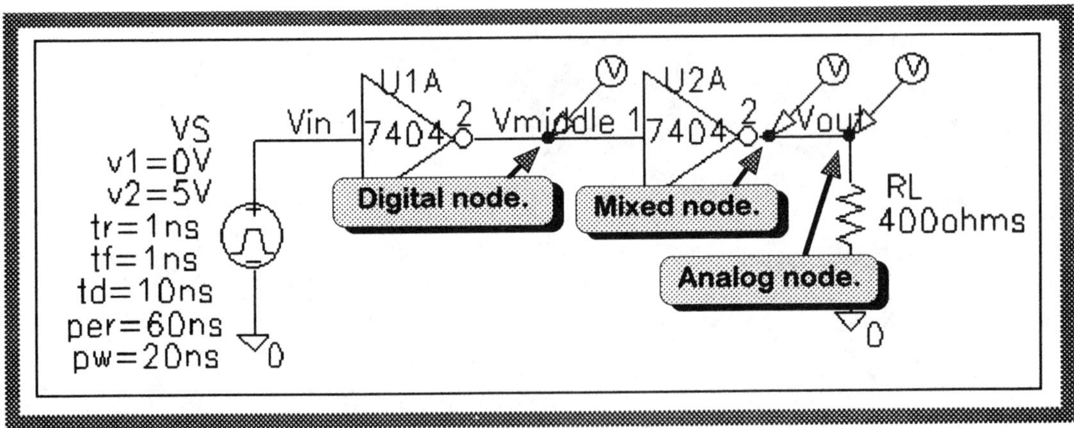

FIGURE 11.13
Placing analog and digital markers

16. Viewing the results (Figure 11.14):

    (a) Are digital waveforms generated only when markers are placed at digital nodes?

    Yes **No**

(b) Are analog waveforms generated when markers are placed at *either* analog or mixed nodes?

Yes **No**

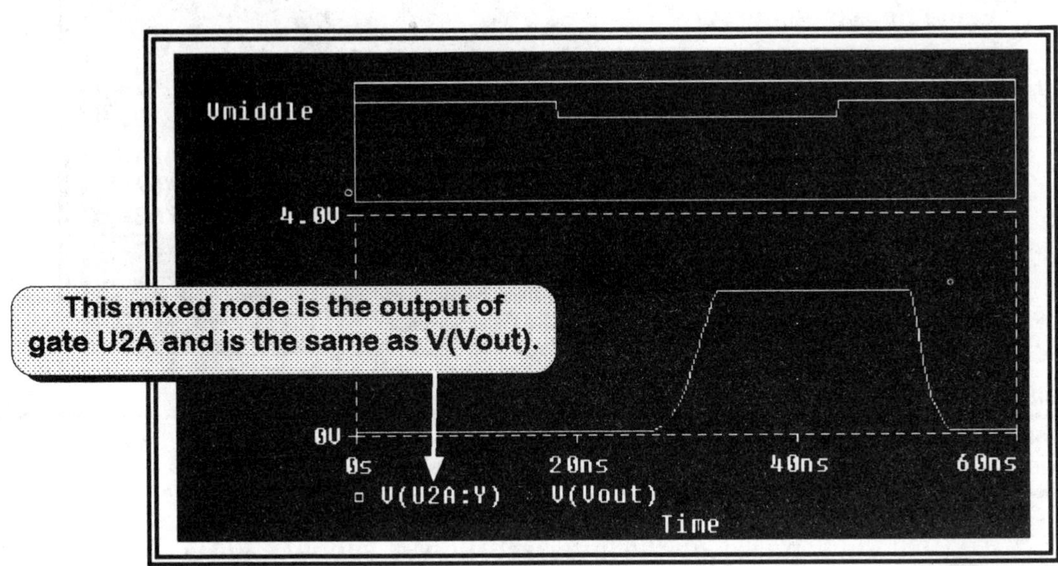

FIGURE 11.14
Marker waveforms

Advanced activities

17. From step 4, we find that the minimum allowed logic 1 output voltage level is 2.4V.

    Using the built-in goal function below, set up RL (of Figure 11.4) as a parametric variable and generate the graph of Figure 11.15. Based on this graph, what is the minimum allowed value of RL that will guarantee a logic 1 output voltage? Is the value below 400Ω?

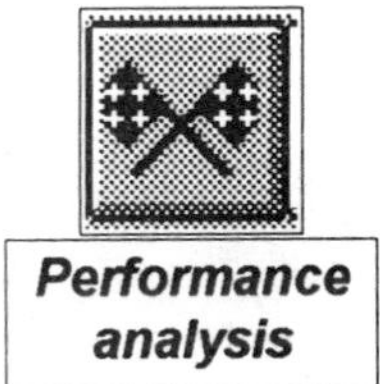

Performance analysis

```
Return the Maximum value of the trace.
Max(1) = y1
{
    1|Search forward max !1;
}
```

18. Remaining with Figure 11.4, perform a parametric sweep of temperature (perhaps from −50 to +50, in increments of 25) and display the waveforms. Does temperature influence the analog and digital waveforms in any significant manner?

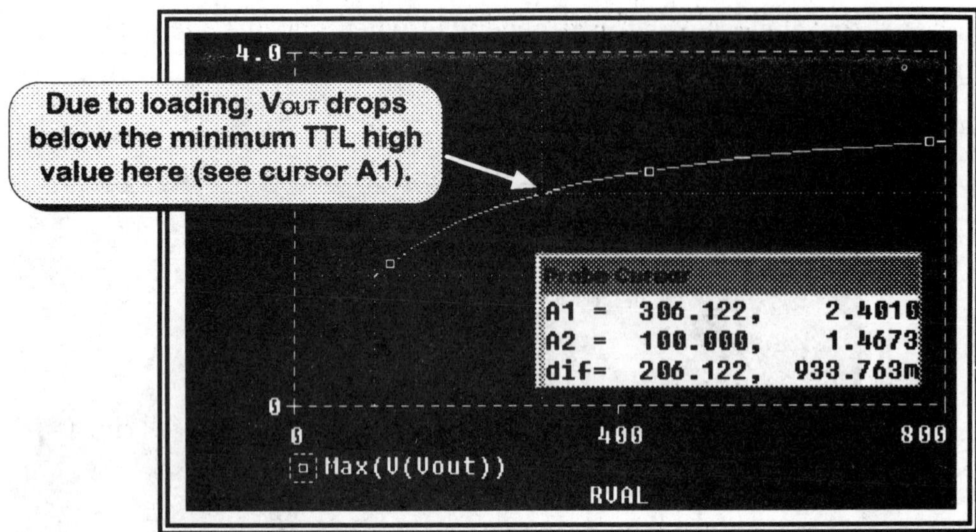

FIGURE 11.15

Performance analysis of maximum Vout versus RL

19. View traces of the interface power supply (V($G_DGND)) and V($G_DPWR)). Are the results as expected?

EXERCISES

- Add the necessary source and load, and test the *clock-edge detection* circuit of Figure 11.16.
- Using analog signals, generate the transfer functions of Figure 11.17 for the 7404 inverter and 7414 Schmitt trigger. (What is *hysteresis*?)
- Why is the circuit of Figure 11.18 called a *data selector*?
- Determine the frequency of oscillation of the circuit of Figure 11.19.
- Add the necessary sources and test the full adder of Figure 11.20. (Verify that A=1 + B=0 + C_{IN}=1 equals 0 with a carry out of 1.)
- Add the necessary sources and test the 4-bit even parity generator of Figure 11.21.
- What mathematical operation does the 2-bit circuit of Figure 11.22 perform? (Could the process be extended to any number of bits?)
- What is the purpose of the mystery circuit of Figure 11.23? (Under what conditions is the output high?)

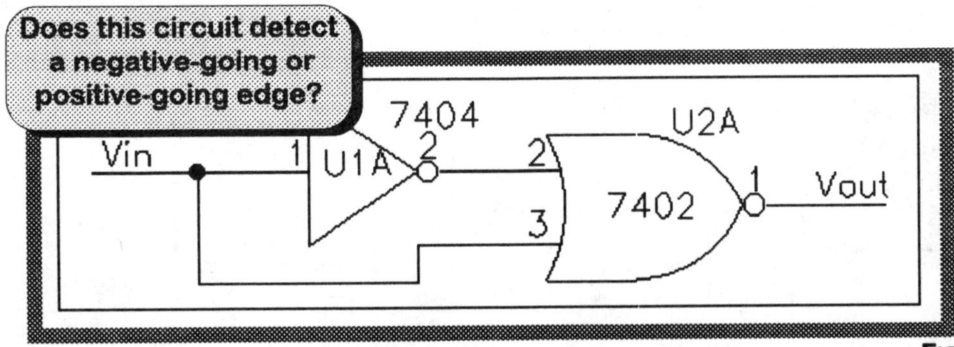

FIGURE 11.16

Clock-edge detection circuit

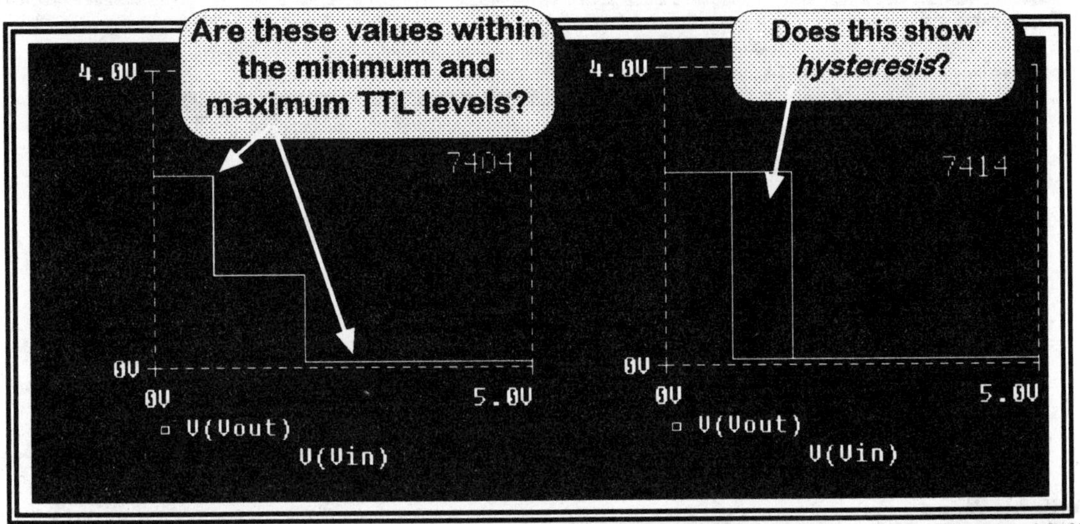

FIGURE 11.17

Transfer function comparison

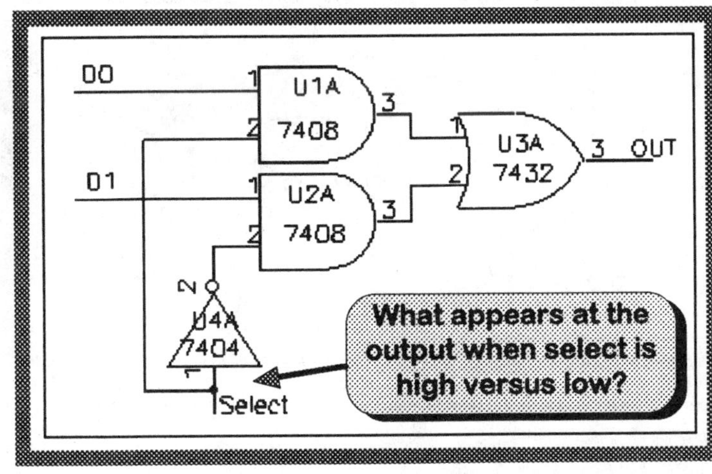

FIGURE 11.18

Data selector

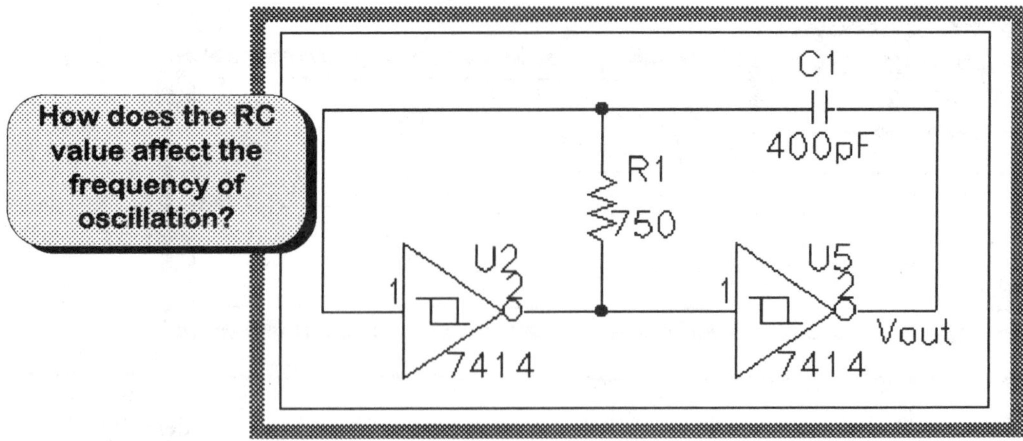

FIGURE 11.19

Oscillator circuit

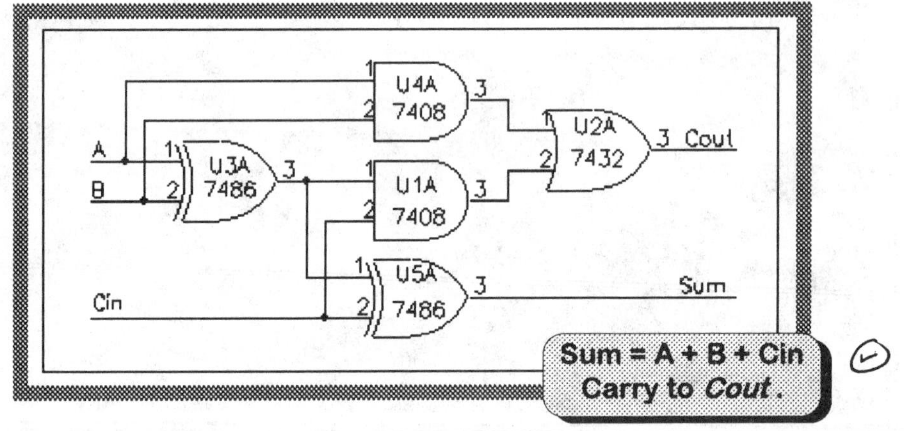

FIGURE 11.20

Full adder

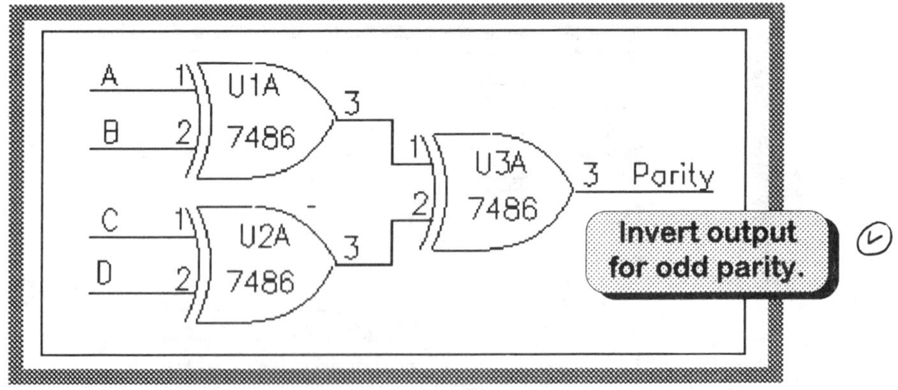

FIGURE 11.21

4-bit even parity generator

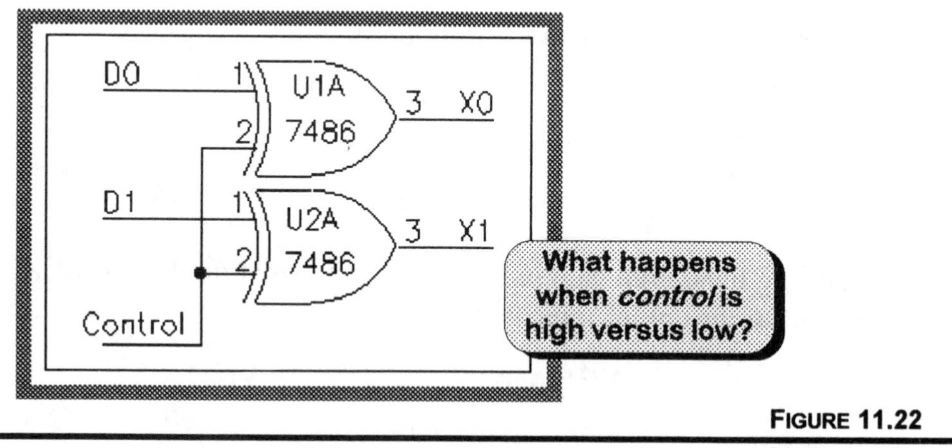

FIGURE 11.22
Math circuit

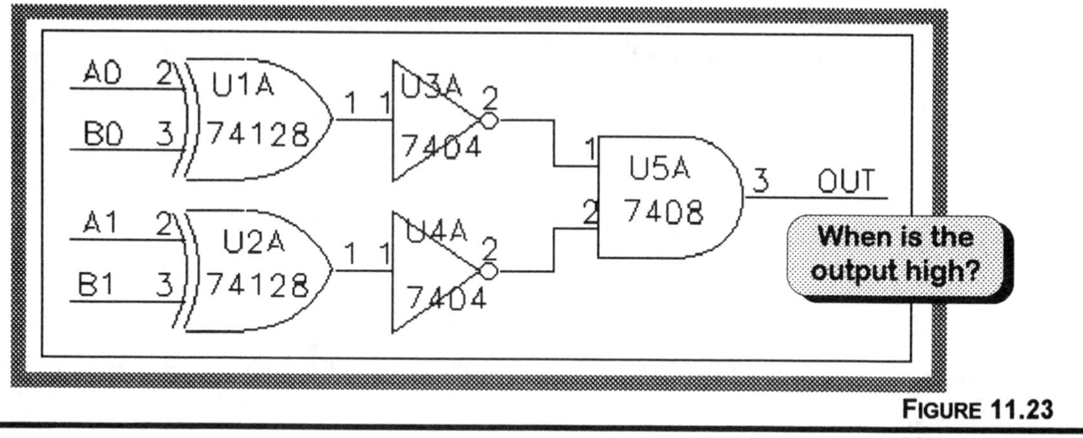

FIGURE 11.23
Mystery circuit

QUESTIONS AND PROBLEMS

1. What are the three possible types of nodes in a mixed analog/digital circuit?

2. Where does PSpice automatically place an AtoD interface circuit? A DtoA interface circuit?

3. Can we measure current at a purely digital node?

4. Regarding the digital gate indicators below, what do the terms "A" and "Y" (to the right of the colon) indicate?

   V(U1A:A) **V(U2A:Y)**

5. What is an *ambiguity region*?

6. Explain how the 7414 transfer function of Figure 11.17 shows the effects of *hysteresis*.

CHAPTER 12

Digital Stimulus
File Stimulus

OBJECTIVES

- To generate a variety of digital waveforms using the digital stimulus generator feature of PSpice.
- To generate digital waveforms from information stored in a file.

DISCUSSION

In modern computer systems, it is not unusual for the majority of components and interfaces to be purely digital. Quite often, only the initial input and final output are interfaced to analog devices. Furthermore, it is common for digital signals to appear in parallel—that is, in groups of 4, 8, 16, or 32.

For these reasons, PSpice added several important features to their simulation software. For routing parallel digital signals, they provided a *bus system*; for generating parallel signals, they provided the special *logic level* and *digital stimulus* devices of Figure 12.1.

The system bus

In a digital system, information is often stored and moved about in *groups* of bits. To simplify the diagramming of such circuits, schematics provides the system bus—a single heavy line drawn on the schematic screen and properly labeled. All bus-system wires are simply connected to this single bus line, and all connections are identified by labels.

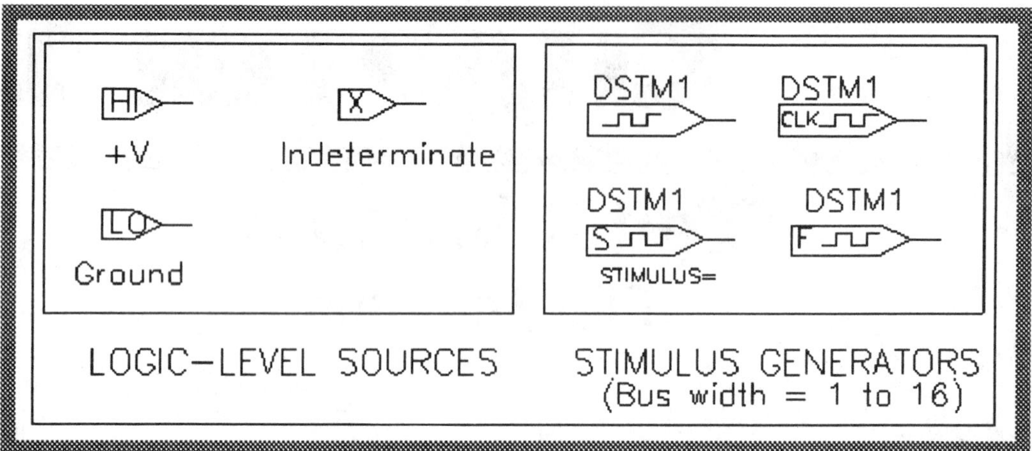

FIGURE 12.1
The logic-level and digital-stimulus devices

Logic-level devices

The logic-level devices are straightforward and easy to use. They simply apply a constant logic 1 (+5V for TTL) or logic 0 (0V for TTL) to any single node.

For example, the two-input NAND gate of Figure 12.2 uses the HIGH logic level (+5V) source to permanently enable one input and turn the NAND gate into an inverter.

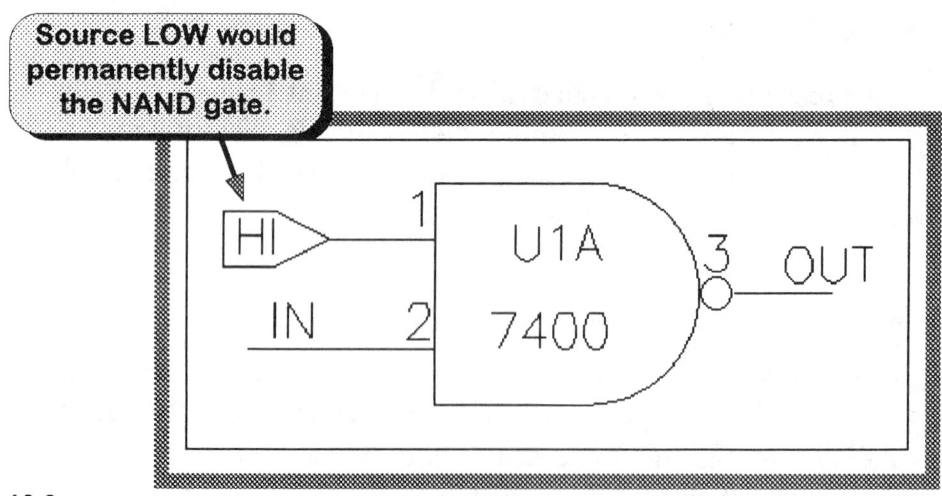

FIGURE 12.2
Use of the logic-level source

The digital stimulus devices

The digital stimulus generators are divided into four categories:

- The *general stimulus* device (upper left of the *STIMULUS GENERATORS* section of Figure 12.1) comes in four versions (STIM1, STIM4, STIM8, and STIM16) for bus widths of 1, 4, 8, and 16 lines. They are custom programmed for any desired waveform set.
- The *clock stimulus* device (upper right) is easy to program and generates a continuous clock signal.
- The *stimulus editor* (lower left) is the digital counterpart to analog's VSTIM, and provides a stimulus editor window to create a variety of waveforms that can be easily modified and selected.
- The *file stimulus (FileStim)* device (lower right) is used when the number of stimulus commands is very large. It obtains the stream of digital signals from a file in a hard or floppy disk.

As always, the best way to learn the use of such sophisticated devices is by way of example.

SIMULATION PRACTICE

Example 1 — Using the clock stimulus generator.

For our first example, we will use the clock stimulus generator to produce a continuous clock waveform.

1. Draw the all-digital test circuit of Figure 12.3 and set the attributes as shown.
   - The digital clock generator is part *DigClock* from library *source.lib*.
   - To set the attribute values, **DCLICKL** on the clock device. The attribute values shown will generate a 1MEGHz clock with a duty cycle of 25% and an initial delay of 2μs.
2. Simulate the circuit and generate the input/output waveforms of Figure 12.4. Were the results as expected?

   Yes **No**

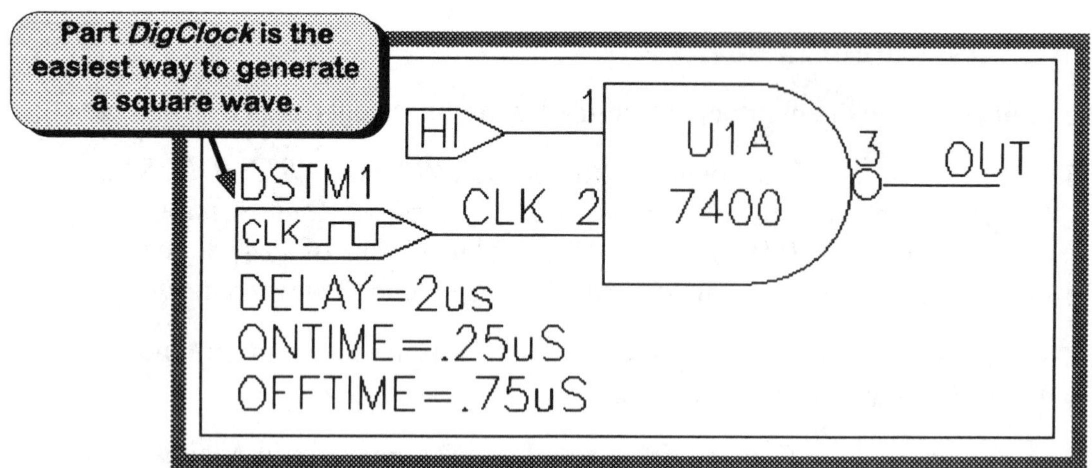

FIGURE 12.3

Example 1 circuit using digital clock

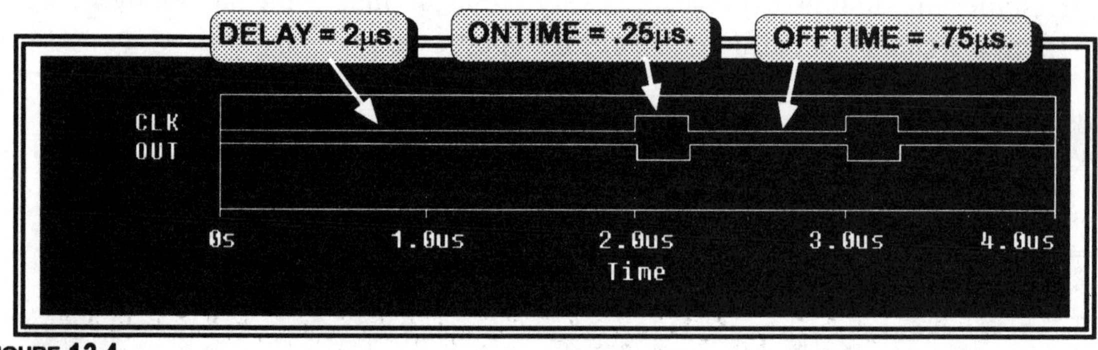

FIGURE 12.4

Example 1 digital waveforms

Example 2 — Using the stimulus editor.

For our second example, we will use the stimulus editor to help us custom design a desired waveform. (Note: The only waveform type available with the evaluation version stimulus editor is the simple clock waveform.)

3. By replacing part *HI* with part *DigStim*, modify the test circuit of Example 1 as shown in Figure 12.5. [To document the customized waveform, **DCLICKL** on attribute *Stimulus=* to bring up the *Set Attribute Value* dialog box, enter any desired name (such as MYSIGNAL), **OK**.]

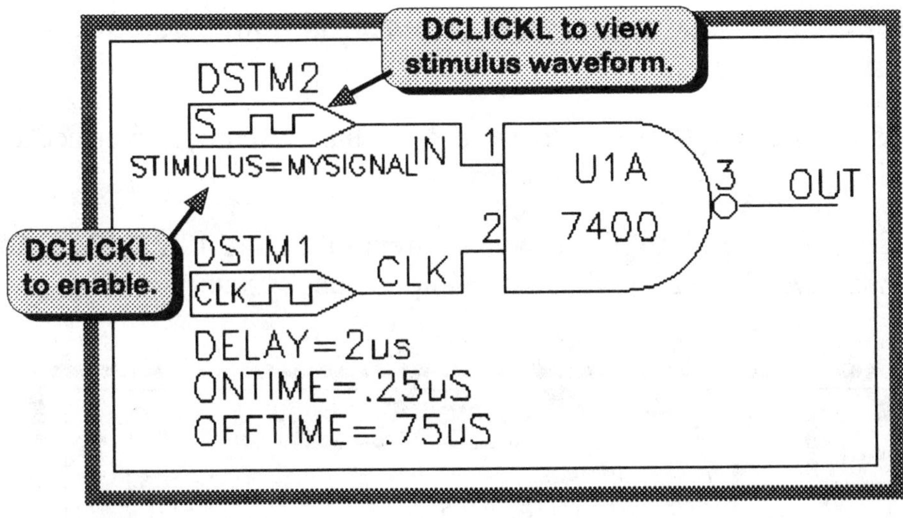

FIGURE 12.5
Example 2 circuit using stimulus editor

4. To edit a waveform, **DCLICKL** on the stimulus editor device and fill in the *Stimulus Attributes* dialog box as follows:

Frequency = 100k	Duty cycle = .5
Initial value = 0	Initial delay = 0

Apply to view the waveform, **OK,** close *Stimulus Editor* window.

5. Simulate the circuit and generate the input/output waveforms of Figure 12.6. Were the results as expected?

Yes **No**

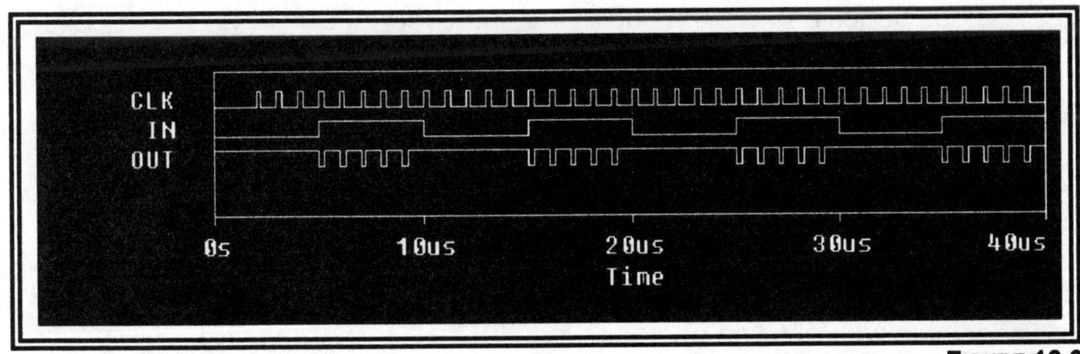

FIGURE 12.6
Example 2 digital waveforms

Example 3 — Using the 4-bit general stimulus generator to drive a system bus.

In this third example, we will use a 4-bit bus system to interface a stimulus generator with a small array of gates.

6. Using device STIM4, draw the test circuit of Figure 12.7. (Refer to *Schematics Note 12.1* when setting up the bus system.)

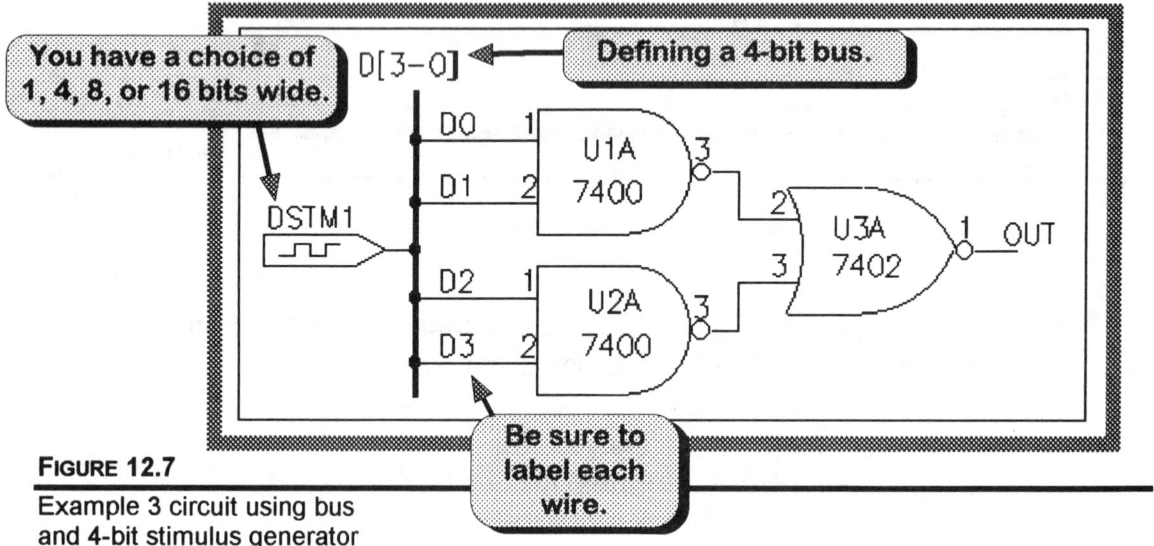

FIGURE 12.7
Example 3 circuit using bus and 4-bit stimulus generator

Schematics Note 12.1
How do I add a system bus?

To add a system bus: **CLICKL** on the *Draws a new bus* toolbar button to create a bus pencil and draw a bus in the same manner as a wire.

Next, label the bus: **DCLICKL** on any part of the bus to bring up the *Set Attribute Value* dialog box, enter the name of the bus and the numerical bus elements in square brackets. For example, the following name specifies a 4-bit data bus (D0, D1, D2, and D3).

D[3-0]

Next, make all connections to the bus using **Wire** segments, and label all wire segments with allowed names (such as D0, D1, D2, and D3).

Finally, a bus can connect to another bus if one is a subset of the other.

7. To program DSTM1, **DCLICKL** on the symbol, and fill in according to Table 12.1. (Suggestion: Display the values of the four commands—as shown by Figure 12.8.)

ITEM	VALUE	DESCRIPTION
TIMESTEP=	<leave blank>	Used only if timing specified by clock cycles
COMMAND1=	0s 0000	At 0s, binary D3-D0 = 0000
COMMAND2=	REPEAT 32 TIMES	Repeat 32 times
COMMAND3=	+1μs INCR by 0001	1μs later, increment the output by 1
COMMAND4=	ENDREPEAT	End of repeat loop
WIDTH=	4	Drives a 4-bit wide bus
FORMAT=	1111	Output signals specified in binary
DIG_PWR=	$G_DPWR	Default +5V power supply
DIG_GND=	$G_DGND	Default 0V ground
IO_MODEL=	IO_STM	Default IO model
IO_LEVEL=	0	Default interface subcircuit

TABLE 12.1
Example 3 digital stimulus parameters

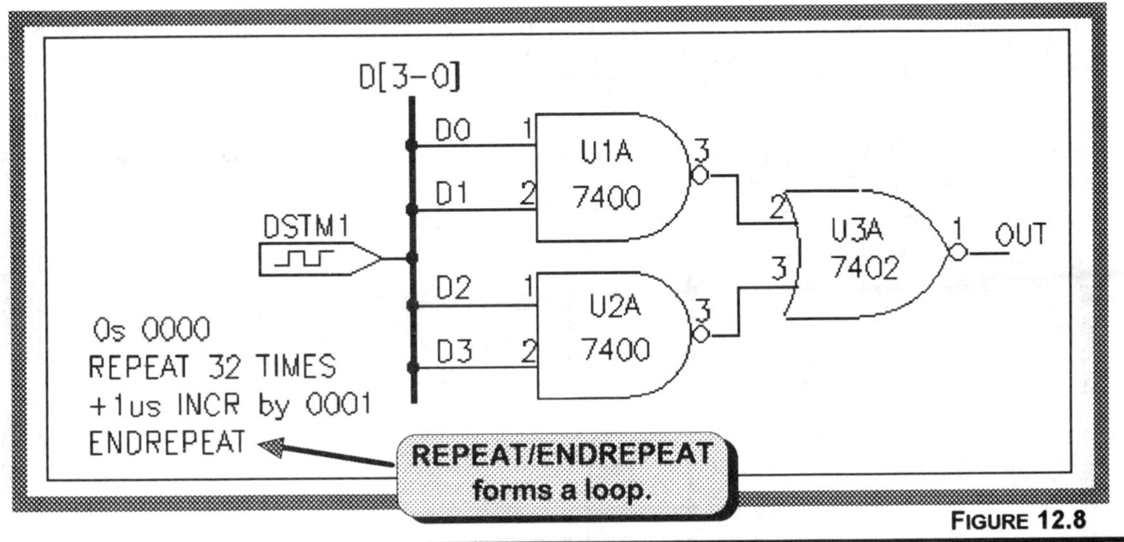

FIGURE 12.8
Example 3 circuit programmed for signal generation

8. Simulate the circuit and generate the input/output waveforms of Figure 12.9. Were the results as expected?

Yes **No**

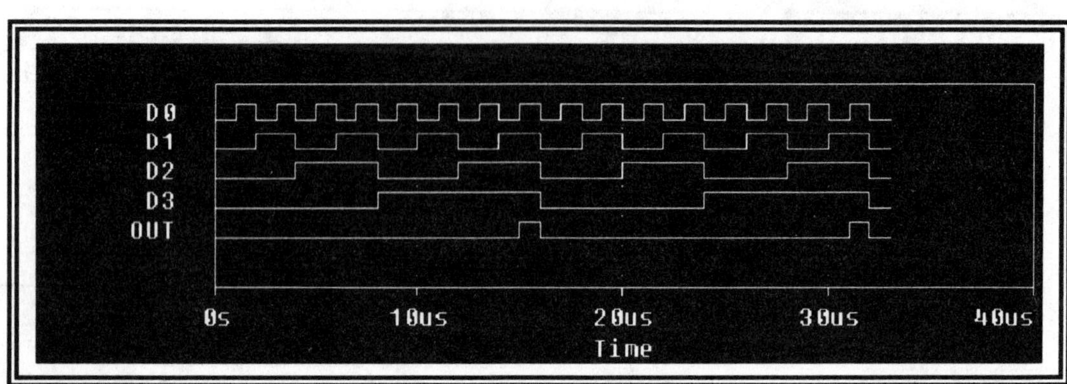

FIGURE 12.9
Example 3 digital waveforms

Example 4 — Generate a signal stream with File stimulus.

Our fourth example will generate the 4-bit digital signal stream from a file.

9. Using device *FileStim*, draw the test circuit of Figure 12.10, and set the attributes for DSTM1 as shown.

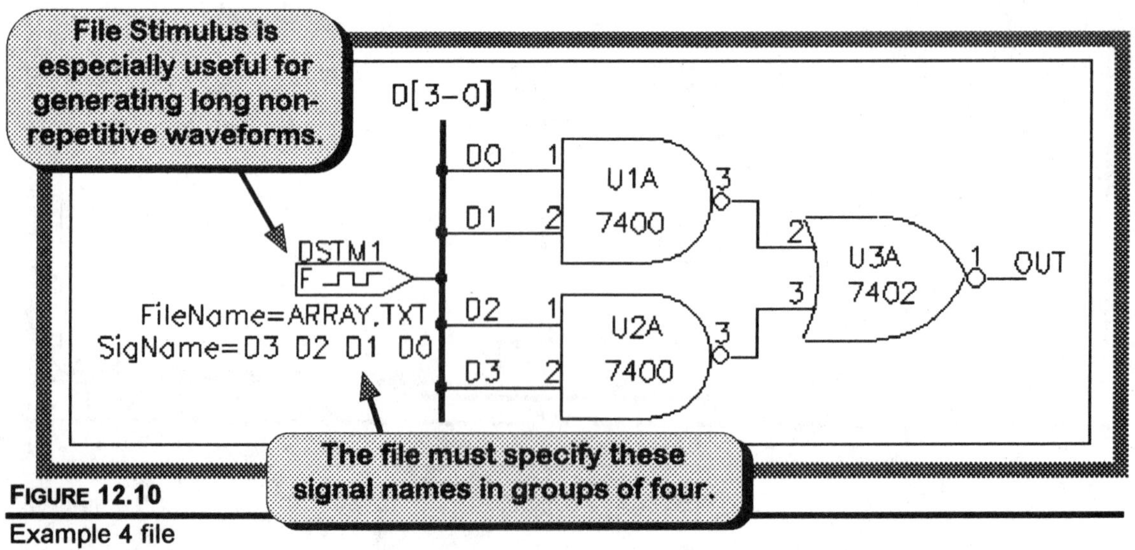

FIGURE 12.10
Example 4 file stimulus circuit

10. Using a text editor, open a file (ARRAY.TXT) in the current directory, enter the data below, and close the file.

> Hint: **DCLICKL** on *FileName=ARRAY.TXT* and Windows' text editor (notepad) will automatically open.

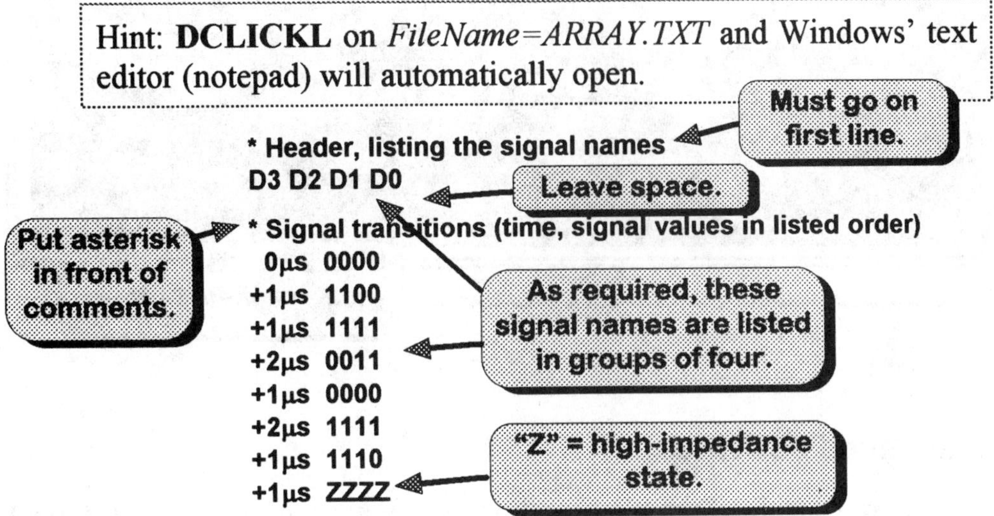

11. **DCLICKL** on the file stimulus device, bring up the Part Name dialog box, and enter the data of Table 12.2.

Name	Value	Comment
FileName=	ARRAY.TXT	File storing data
SigName=	D3 D2 D1 D0	Signal names
IO_MODEL=	IO_STM	Default value
IO_LEVEL=	0	Default value
ipin[PWR]=	$G_DPWR	Default value
ipin[GND]=	$G_DGND	Default value

TABLE 12.2
FSTM1 Part Name data

12. Run PSpice and generate the input/output waveforms of Figure 12.11.

13. Referring to Figure 12.11:

(a) Do the input signal data (D3, D2, D1, D0) specified by the file (ARRAY.TXT) match the corresponding waveforms?

Yes **No**

(b) Does the output signal (OUT) match your expectations?

Yes **No**

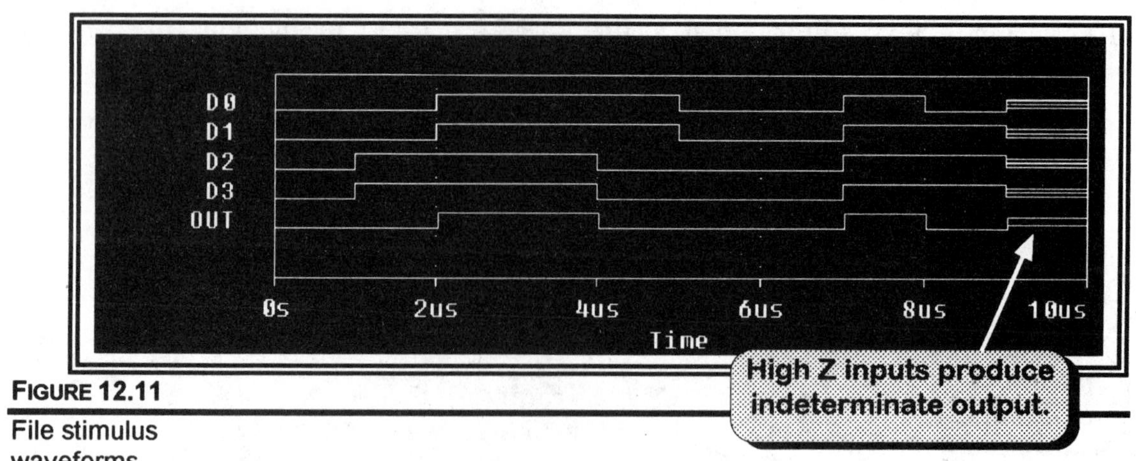

FIGURE 12.11
File stimulus waveforms

Advanced activities

14. Using the FileStim device shown in Figure 12.12, enter the data below into file SERIAL.TXT, and generate serial waveforms CLK and OUT. Identify each of the state values (0, 1, R, F, X, and Z) for both CLK and OUT. (Identify the one time period in which CLK and OUT differ, and explain the reason for the difference.)

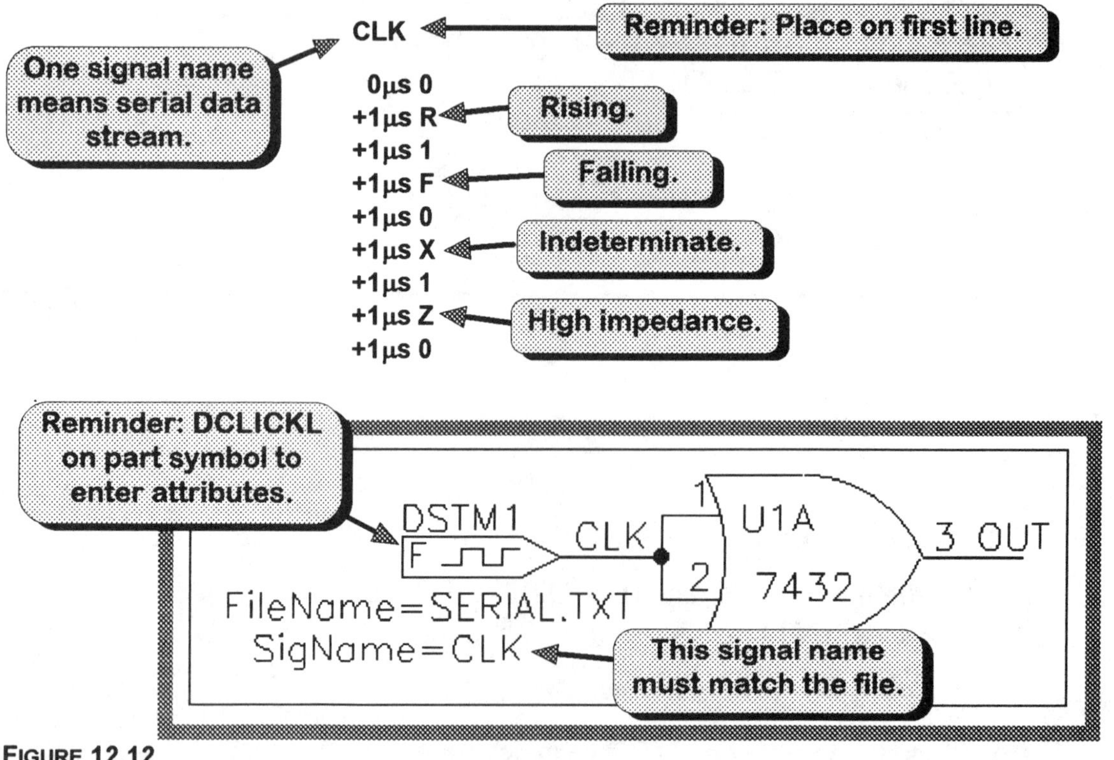

FIGURE 12.12
FileStim test circuit

Exercise

- The 7451 IC of Figure 12.13 consists of two AND and one OR gate internally connected. Using a bus and digital stimulus device of your choice, determine how the gates are arranged inside the IC block.

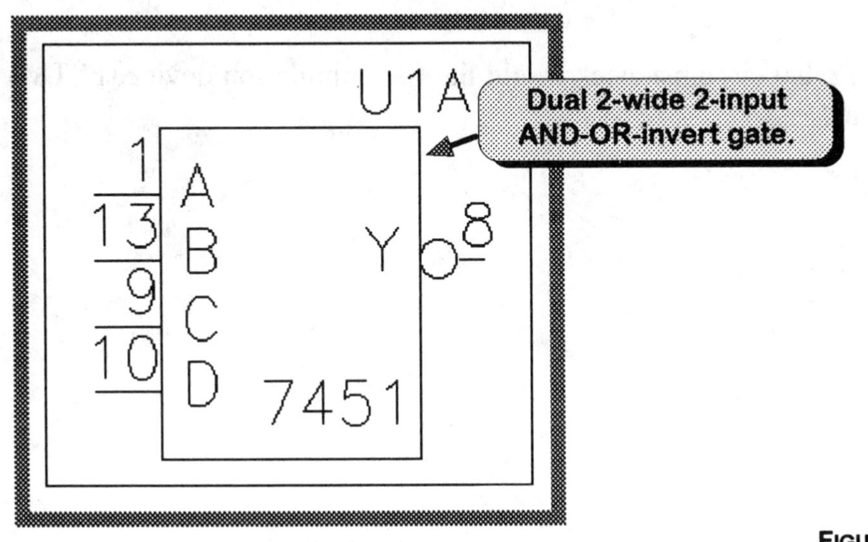

FIGURE 12.13
7451 AND/OR digital IC

QUESTIONS AND PROBLEMS

1. How does the use of a bus simplify digital schematics?

2. What is the digital stimulus symbol for each of the following?

   (a) Unknown

   (b) High Z

3. What is the purpose of the FORMAT parameter? (Hint: See Table 12.1.)

4. Under what circumstances would the file stimulation device (FSTM) be helpful?

CHAPTER 13

Flip-Flops

Edge- Versus Level-Triggered

OBJECTIVES

- To compare the level-active and edge-triggered D-latch.
- To demonstrate the characteristics of the JK flip-flop.

DISCUSSION

The flip-flop is the basic unit of data storage (memory). As shown by the circuit of Figure 13.1, a single data bit is stored by positive feedback latching action.

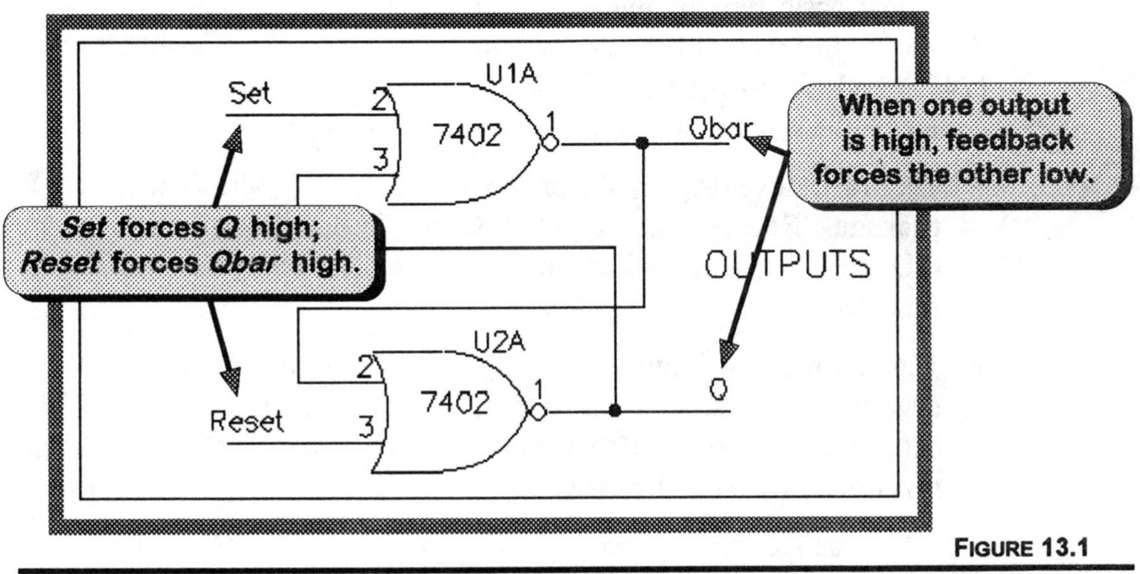

FIGURE 13.1
The SR (set/reset) flip-flop

Synchronous versus asynchronous

The set (S) and reset (R) inputs to the SR latch of Figure 13.1 are *asynchronous* (unclocked) because they can occur at any time. By *gating* the *set/reset* inputs, as shown by Figure 13.2, we create a *synchronous* (clocked) flip-flop.

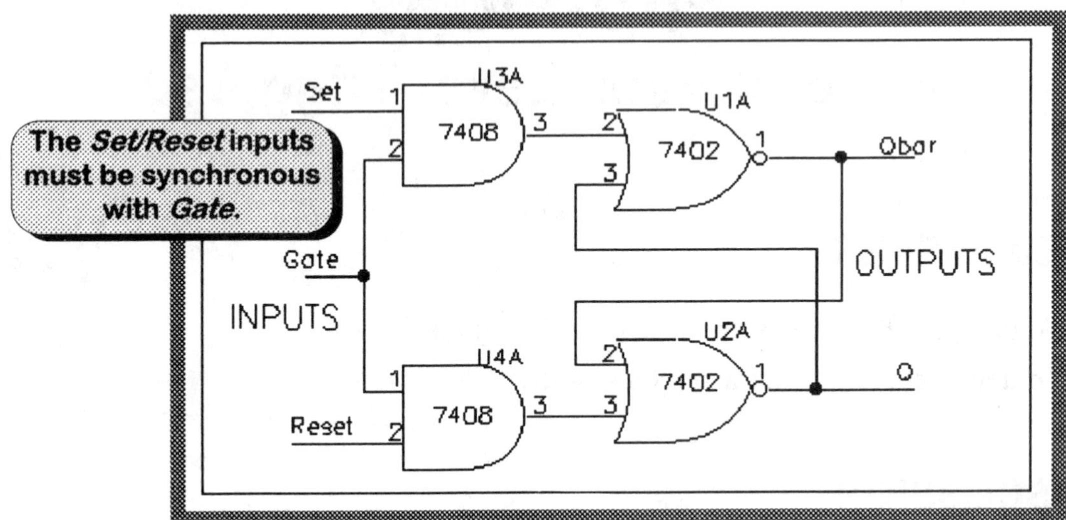

FIGURE 13.2
Gating the input

The D-latch

The most basic type of integrated circuit (IC) flip-flop is the D-latch. As shown in Figure 13.3, it comes in two major forms: *level-triggered* (7475) and *edge-triggered* (7474). The function table of Table 13.1 compares the synchronous inputs of the 7474 and 7475.

- The 7475 level-triggered version offers two flip-flops in a single package. Because the outputs (1Q and 2Q) follow the data inputs (1D and 2D) only when the clock (C) is high, the inputs are synchronous.
- The synchronous input to the 7474 is similar to the 7475, except it is edge-triggered rather than level-triggered. That is, with the 7474, data on the D input is transferred to the Q output on the *low-to-high transition* of the clock pulse.

The active low *preset* (PRE) and *clear* (CLR) signals are asynchronous because they can be activated at any time. They override the clock input and directly set the Q output to 1 and 0.

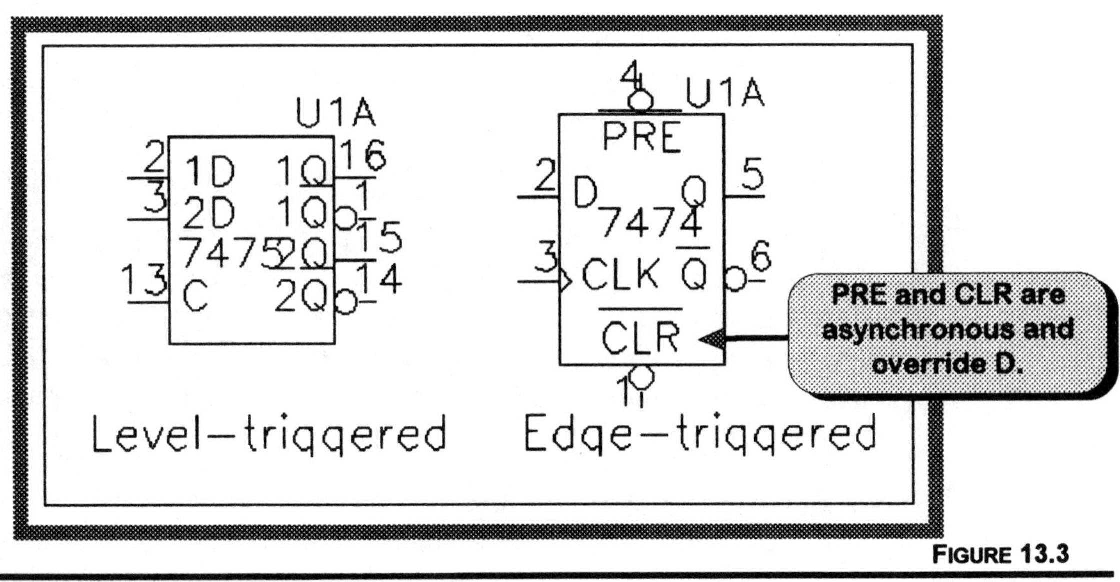

FIGURE 13.3
The D-latches

"X" means "don't care."

C	D	Q	Qbar	Comments
1	0	0	1	Q = D = 0
1	1	1	0	Q = D = 1
0	X	Q	Qbar	Data latched

(a) 7475

C	D	Q	Qbar	Comments
0	X	Q	Qbar	No change
1	X	Q	Qbar	No change
↑	1	1	0	Set
↑	0	0	1	Reset

(b) 7474

"↑" means "low-to-high transition."

TABLE 13.1
The D-latch function tables
(a) 7475
(b) 7474

The JK flip-flop

The JK flip-flop is the most versatile of the three basic types (SR, D, and JK). First and foremost, it offers a new mode of operation: a *toggle* mode, in which the outputs *change* state.

As shown by Figure 13.4, the JK flip-flop comes in three major versions:

- Positive pulse-triggered (7473, 7476, and 74107): These older flip-flops are known as *master-slave* devices because they consist of two latches: a master latch to receive data while the input clock is HIGH, and a slave latch to receive and output data from the master when the clock goes LOW.
- Positive edge-triggered (74109, 74LS109): With these newer devices, the flip-flop latches and outputs JK data upon the positive-going LOW-to-HIGH clock transition. Transitions of the JK inputs before or after the active clock are ignored.
- Negative edge-triggered (74LS73, 74LS76, 74LS107, 74LS112): The flip-flop latches and outputs JK data upon the negative-going HIGH-to-LOW clock transition. Transitions of the JK inputs before or after the active clock are ignored.

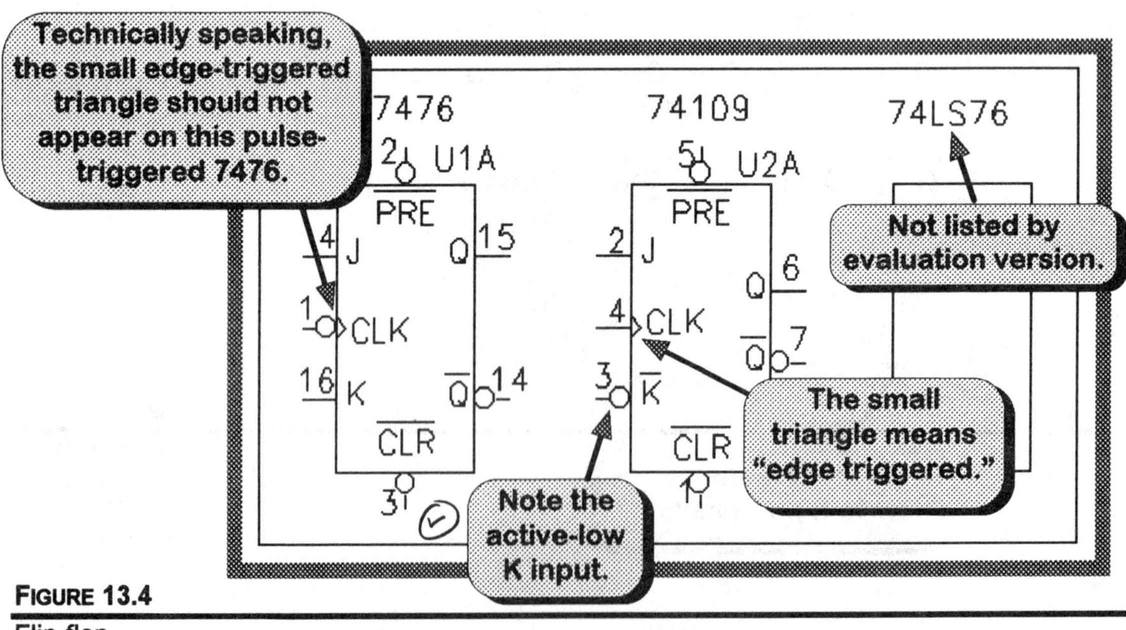

FIGURE 13.4
Flip-flop comparison

The 7476 master/slave JK flip-flop

As shown by Figure 13.4, the 7476 offers both asynchronous (PRE and CLR) and synchronous (JK) inputs. In Table 13.2, synchronous input (JK) data is loaded into the master upon the positive-going clock transition and transferred to the slave on the negative-going clock transition.

Clock pulse.

CLK	J	K	Action
⎍	0	0	none
⎍	1	0	set (Q high)
⎍	0	1	reset (Q low)
⎍	1	1	toggle (Q changes)

TABLE 13.2
The 7476 function table

SIMULATION PRACTICE

NOTE: If you experience difficulty with any digital IC (such as receiving only unknowns on outputs), set the initial state to zero (**Analysis**, **Setup**, **Digital Setup**, **All 0**, **OK**, **Close**).

The D-latch

The 7475 Level-Triggered D-Latch

1. Draw the 7475 D-latch circuit of Figure 13.5. Set the attributes for digital *clock* (CLK) device DSTM1 as shown, and set the values for digital *stimulus* (S) device DSTM2 as follows:

Frequency = 1MEG	*Duty cycle = .5*
Initial value = 0	*Time delay = .25µs*

2. When programmed as specified in step 1, Figure 13.6 shows the input signals to the D-latch. Below the signals, predict (draw) the Q output waveform in the space provided. (Remember, the 7475 is level-triggered.)

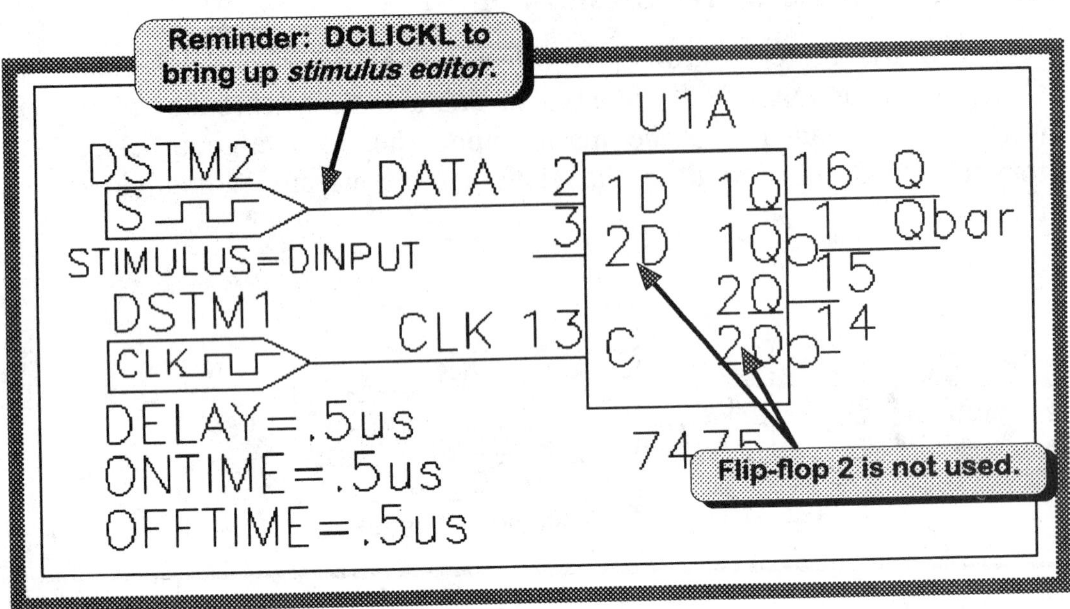

FIGURE 13.5
7475 D-latch circuit

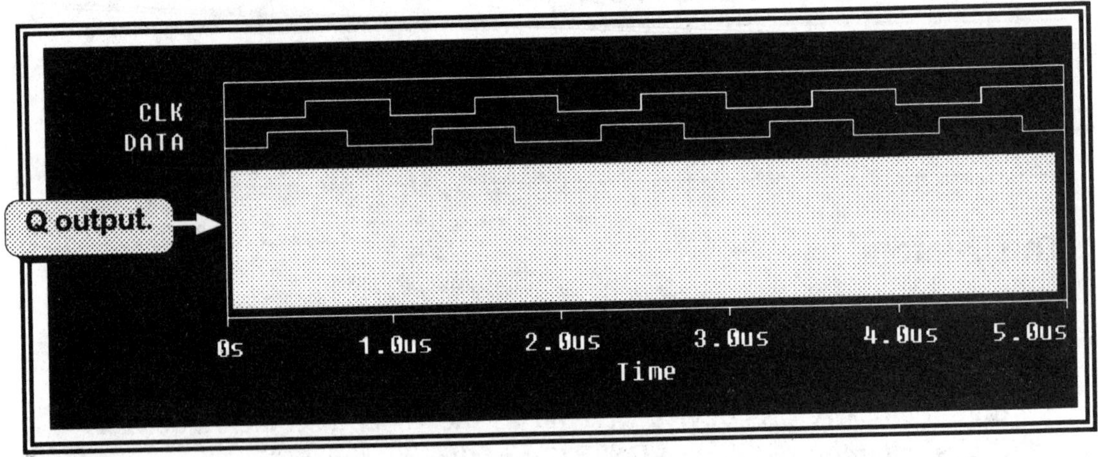

FIGURE 13.6
7475 D-latch waveforms

3. Generate the Q digital output waveform using PSpice. Were your predictions correct? (Make any necessary corrections to your predicted waveform.)

Yes **No**

The 7474 Edge-Triggered D-Latch

4. Substitute the 7474 D-latch for the 7475 and produce the circuit of Figure 13.7. (DSTM1 and DSTM2 remain unchanged.)

5. As with the 7475, sketch your predicted Q output waveform on the graph of Figure 13.8.

6. Generate the Q output waveform using PSpice. Were your predictions correct? (Make any necessary corrections to your predicted waveform.)

   Yes **No**

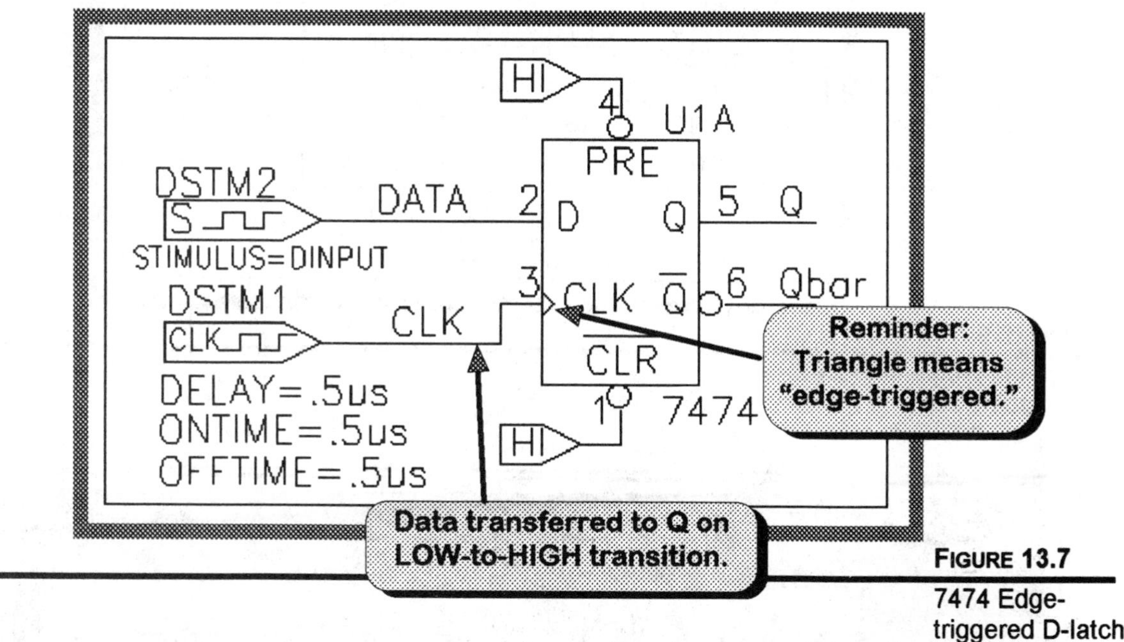

FIGURE 13.7
7474 Edge-triggered D-latch

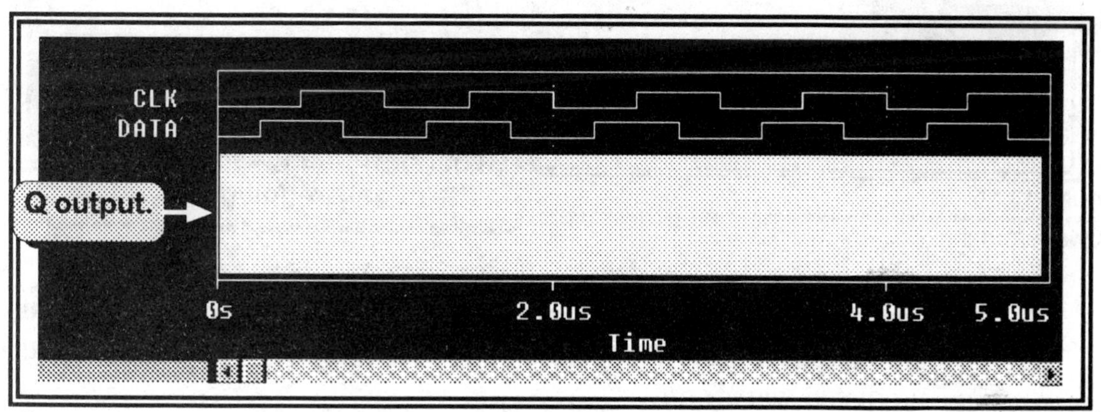

FIGURE 13.8
7474 D-latch waveforms

The JK flip-flop

7. Draw the test circuit of Figure 13.9 and set the attributes as shown.

8. As before, sketch your predicted Q output waveform on the graph of Figure 13.10.

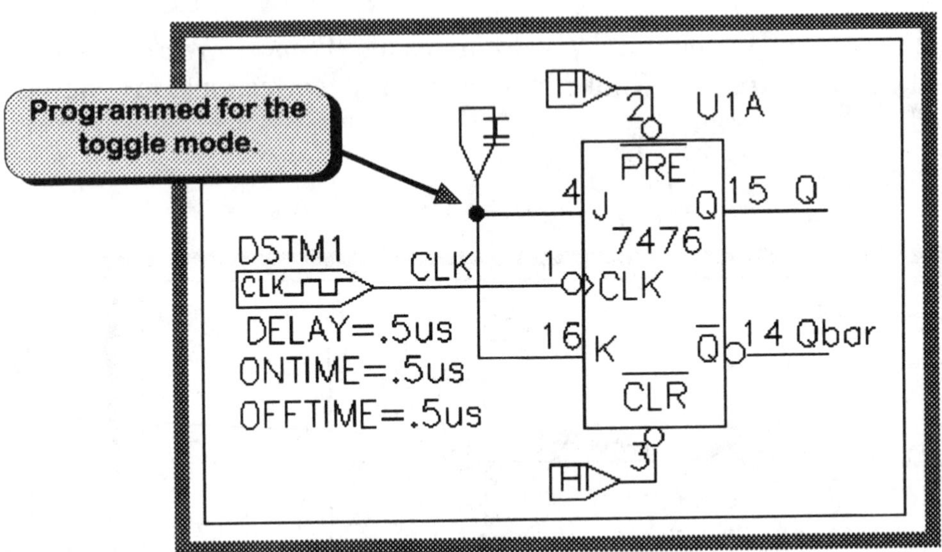

FIGURE 13.9
JK flip-flop programmed for toggle mode

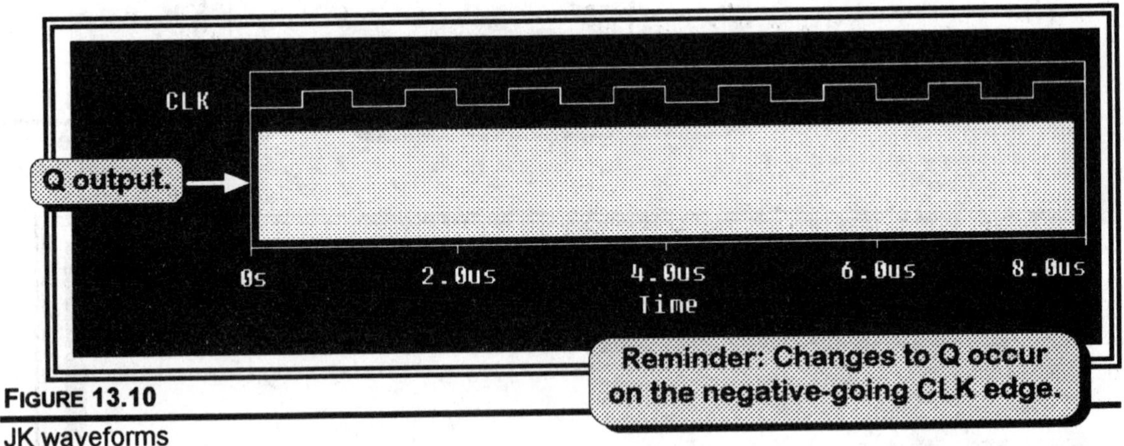

FIGURE 13.10
JK waveforms

9. Generate the Q output waveform using PSpice. Were your predictions correct? (Make any necessary corrections to your predicted waveform.)

Yes **No**

10. Reviewing your results, what is the frequency relationship between input and output?

    f(in) = __________ × f(out)

11. Modify your circuit as shown in Figure 13.11, and sketch your predicted Q and Qbar output waveforms on the graph of Figure 13.12.

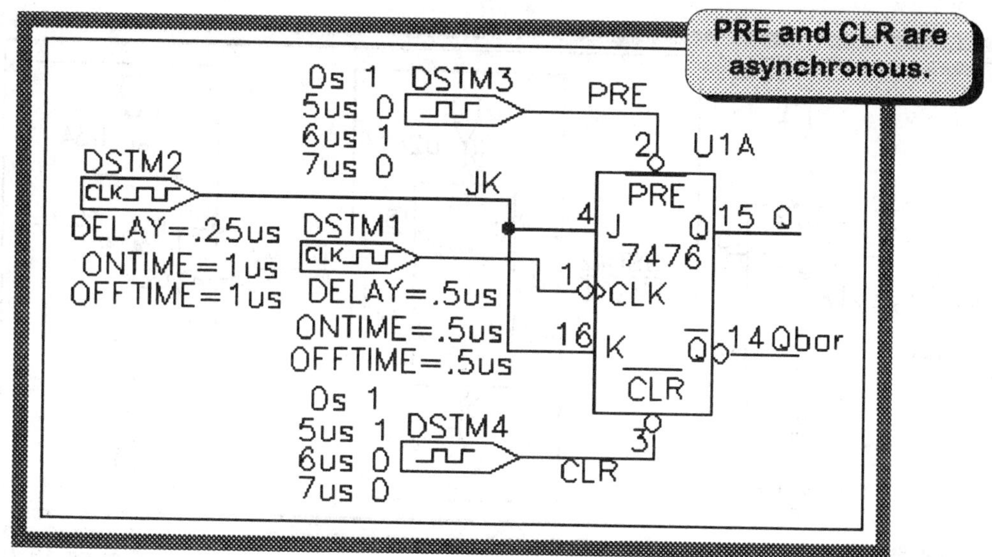

FIGURE 13.11
JK flip-flop test circuit

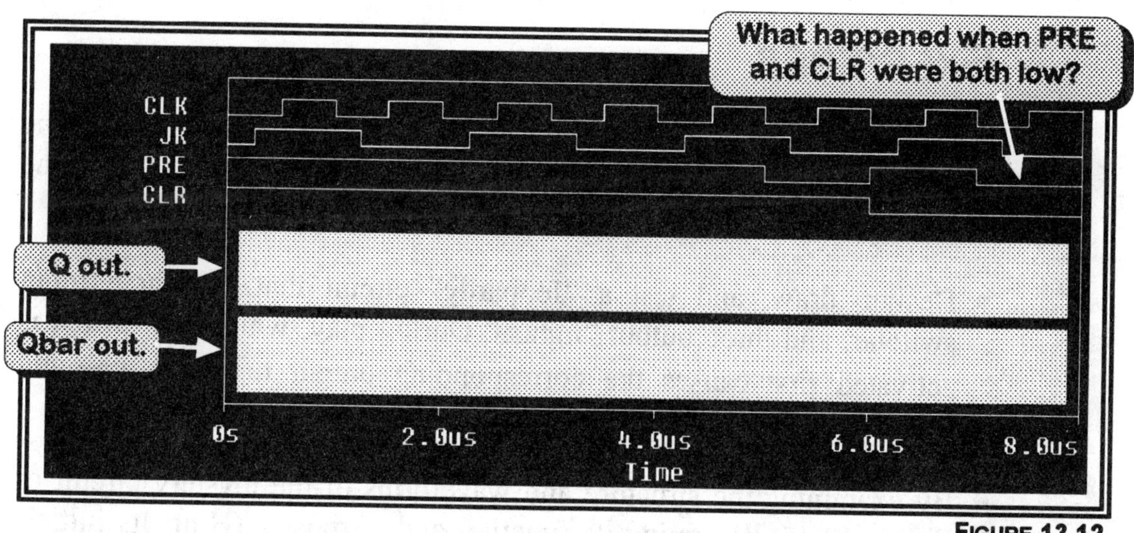

FIGURE 13.12
JK test waveforms

Advanced activities

12. *Using only basic gates*, design an edge-triggered D flip-flop. (Hint: Make use of the clock-edge detection circuit of Figure 11.16.)

13. Sketch the Q output for any or all of the test circuits of Figure 13.13. Using PSpice, test your predictions.

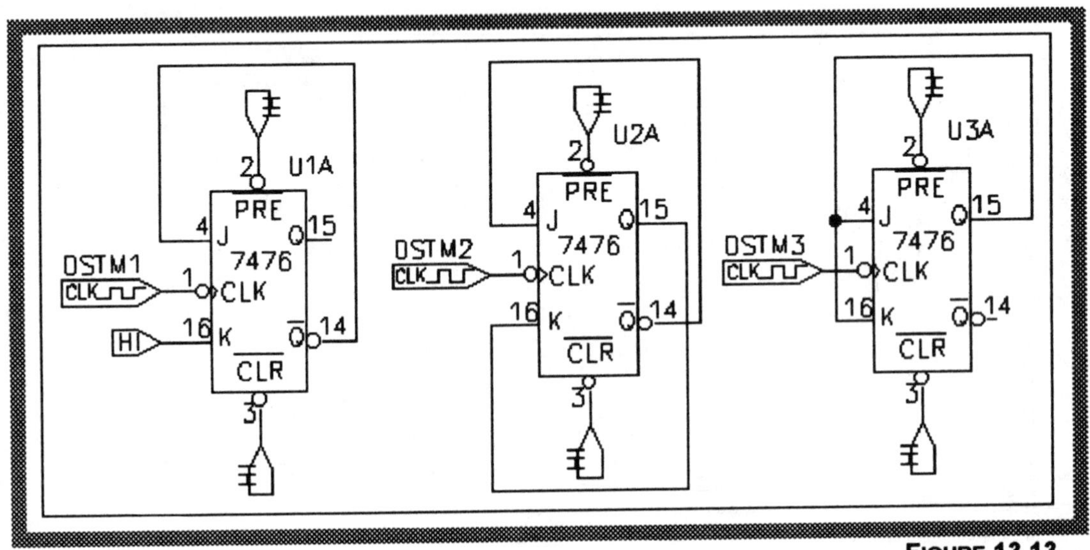

FIGURE 13.13
JK test circuits

EXERCISES

- Draw and test the parity circuit of Figure 13.14. Does it determine even or odd parity? [As serial data enters the Data input (D1) at the clock rate (D0), the circuit keeps a running total of parity.]

- Design, draw, and test a "Jeopardy" circuit that has three input switches and three output "lights." During play, whichever switch is activated first causes the corresponding output to go active—and deactivates the other two switches.

- By examining the structure and waveforms of the mystery circuit of Figure 13.15, determine its function and purpose. (Hint: Its initials are "SR.")

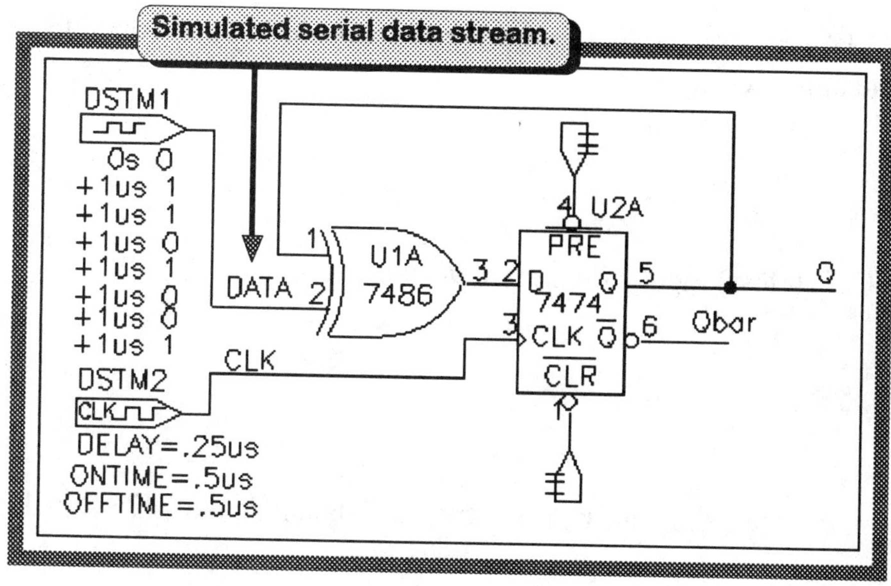

FIGURE 13.14
Parity test circuit

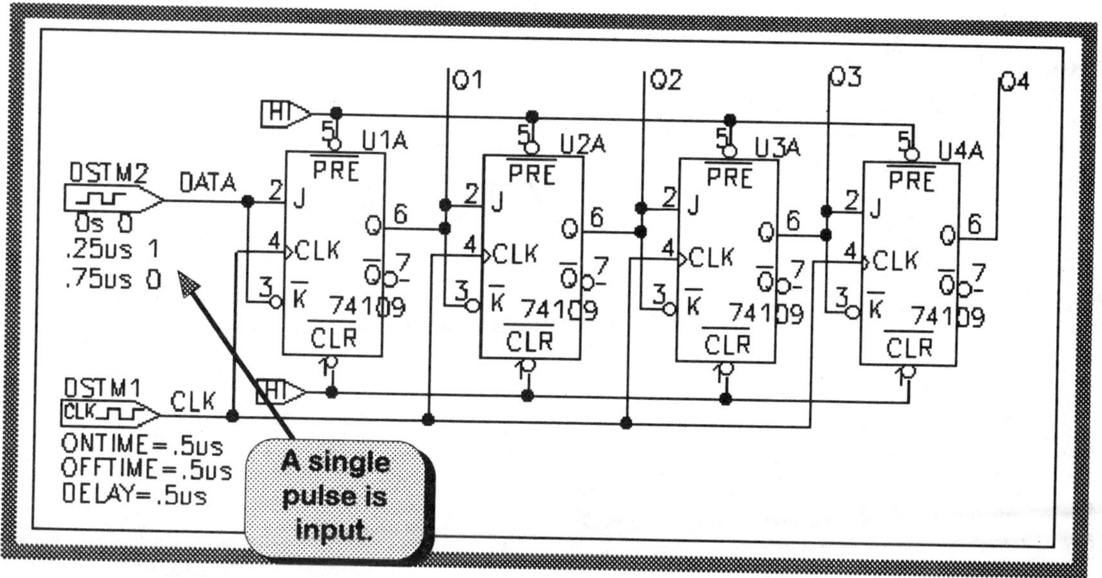

FIGURE 13.15
Flip-flop mystery circuit

QUESTIONS AND PROBLEMS

1. What is the difference between *level-triggered* and *edge-triggered*?

2. With a 7476 JK master/slave flip-flop, what happens during the rising clock pulse and the falling clock pulse?

3. Why is the CLR (clear) input called *asynchronous*?

4. By adding an inverter, show how to construct a D-latch from a 7476 flip-flop.

CHAPTER 14

Digital Worst Case Analysis

Ambiguity and Timing Violations

OBJECTIVES

- To apply worst case analysis techniques to digital circuits.
- To identify and correct a variety of timing violations.
- To analyze the effects of ambiguity.

DISCUSSION

As a production engineer, you note a puzzling situation: How can it be that all the circuits came off the same assembly line, but 20% of them do not work? As with their analog counterparts, we suspect that the problem is caused by component *tolerances*. From one circuit to the next, there is a varying combination of individual component tolerances, and in 20% of the cases it is causing a problem.

In this chapter we will investigate the following tolerance parameters:

- *Propagation delay* — the time span between an input signal transition and the resulting output response.
- *Setup* — the time span during which a signal must be held steady *before* an action is taken.
- *Width* — the width of a clock pulse.
- *Hold* — the time span during which a signal must be held steady *after* an action is taken.

Ambiguity

As shown in Figure 14.1, PSpice places all digital signals into one of six states:

- 0 for *low*
- 1 for *high*
- R for *rising*
- F for *falling*
- X for *unknown*
- Z for *high impedance*

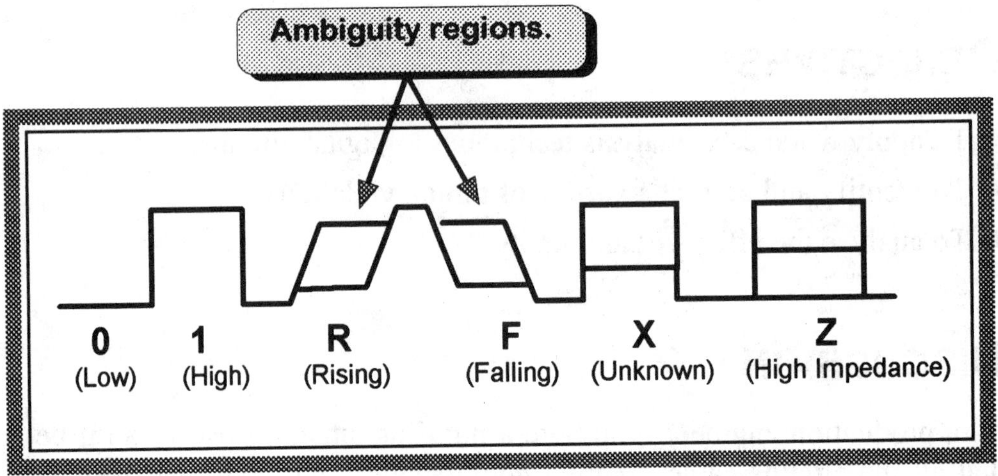

FIGURE 14.1
The six digital states

All R and F regions are known as *ambiguity regions* because the exact time of transition is not precisely known—all we know for sure is that it falls somewhere in the region. The major source of ambiguity is the variation in a component's *propagation delay*.

For example, Table 14.1 shows propagation delay data for the 7408 AND gate. As shown, the propagation delay is not a constant; it is a range of values between the *minimum* and *maximum* extremes. Most propagation values will of course be clustered about the *typical* value. By definition, the ambiguity (uncertainty) times fall between the minimum and maximum values.

	RISING	FALLING
MINIMUM	7ns	5ns
TYPICAL	17ns	12ns
MAXIMUM	27ns	19ns
AMBIGUITY	20ns (27ns – 7ns)	14ns (19ns – 5ns)

Ambiguity is the difference between maximum and minimum propagation delays.

TABLE 14.1
7408 propagation delay summary

Worst case analysis

As we did with analog circuits in volume I, we will handle the problem of tolerance variations in digital circuits by way of *worst case analysis.* However, instead of displaying voltage differences, we will display the ambiguity regions.

For example, based on the data of Table 14.1, Figure 14.2 shows the output ambiguity region for just the risetime portion of the input pulse. When the input signal rises at the 40ns point, the output goes high anywhere from 7ns to 27ns later. The 20ns ambiguity region represents the interval between the earliest and the latest time that the low-to-high transition can occur.

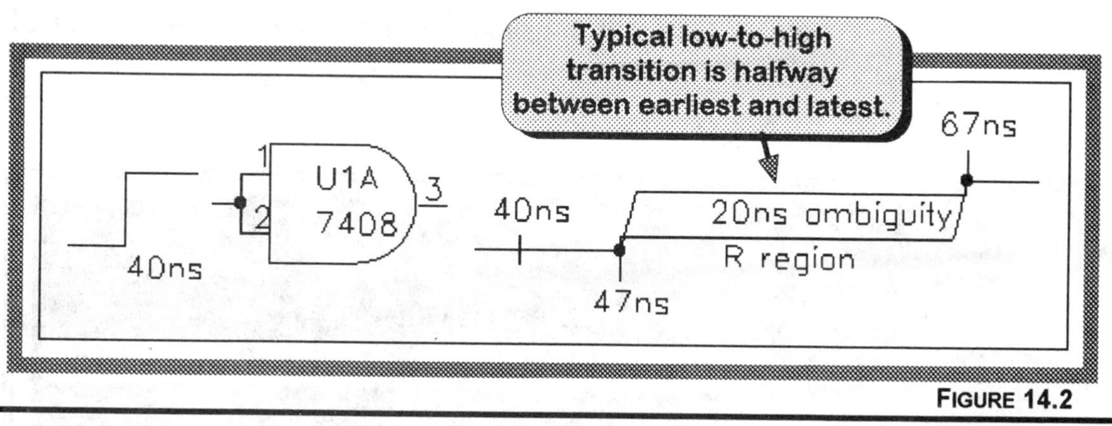

FIGURE 14.2
Risetime ambiguity region

When using worst case analysis to uncover race (timing) problems, all possible combinations of propagation delays are generated and worst case *ambiguity regions* are automatically displayed.

SIMULATION PRACTICE

Propagation delay

1. Draw the test circuit of Figure 14.3 and set the clock stimulus for 125MEGHz (1/80ns) as shown.

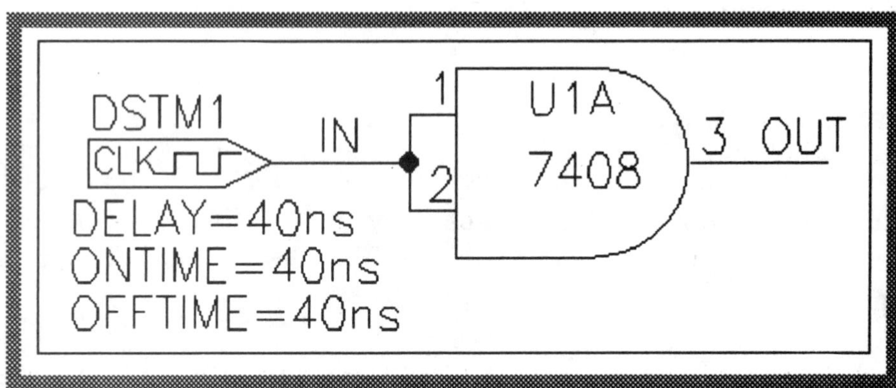

FIGURE 14.3
Propagation delay test circuit

2. Set up the timing analysis for *minimum* (*Sets up the simulation analysis for action* toolbar button, **Digital Setup**, **Minimum**, **OK**, **Close**), enable the transient mode, and generate the waveform of Figure 14.4.

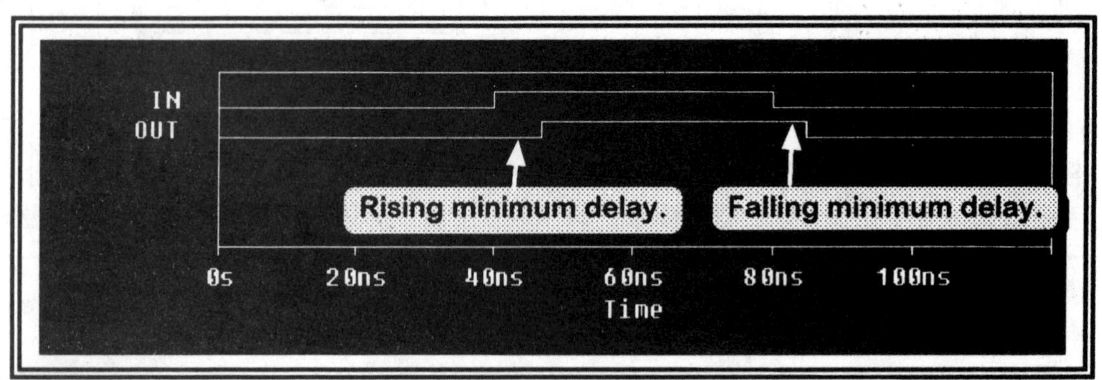

FIGURE 14.4
7408 minimum propagation delays

3. Are the *minimum* rising and falling propagation delays the same as shown in Table 14.1? (Suggestion: Use the cursors.)

   Yes **No**

4. Repeat for the *typical* and *maximum* cases (as outlined in step 2). Do they agree with Table 14.1?

   Yes **No**

Ambiguity

5. This time, repeat the same process, but use *Worst-Case[Min/Max]* analysis to generate the curves of Figure 14.5.

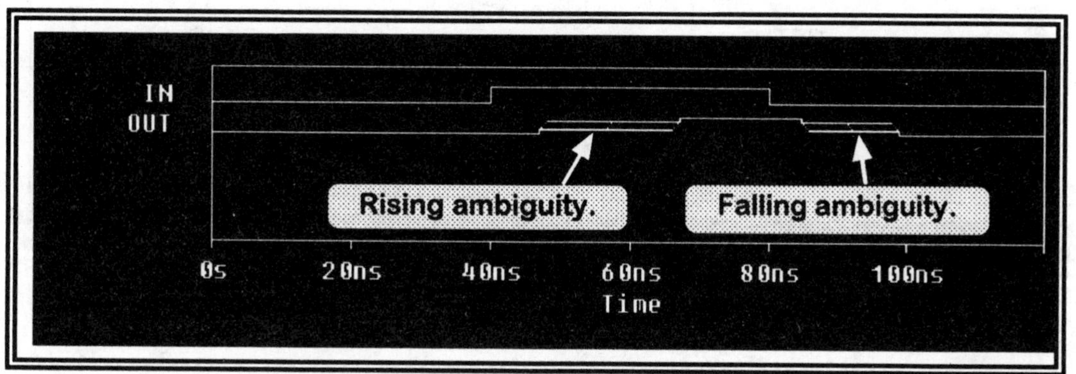

FIGURE 14.5
Worst case run, showing ambiguity regions

6. Examine the results and answer the following:

   (a) Does the risetime ambiguity region match the predictions of Table 14.1?

   Yes **No**

   (b) Does the falltime ambiguity region match the predictions of Table 14.1?

   Yes **No**

Convergence hazard

The first hazard we will study is called a *convergence hazard.* It occurs when two or more signals with overlapping ambiguity regions converge at a common point (such as a gate). The overlap period results in an ambiguous output period.

7. Draw the test circuit of Figure 14.6 and set the commands as shown. (The timing analysis is still *Worst Case*.)

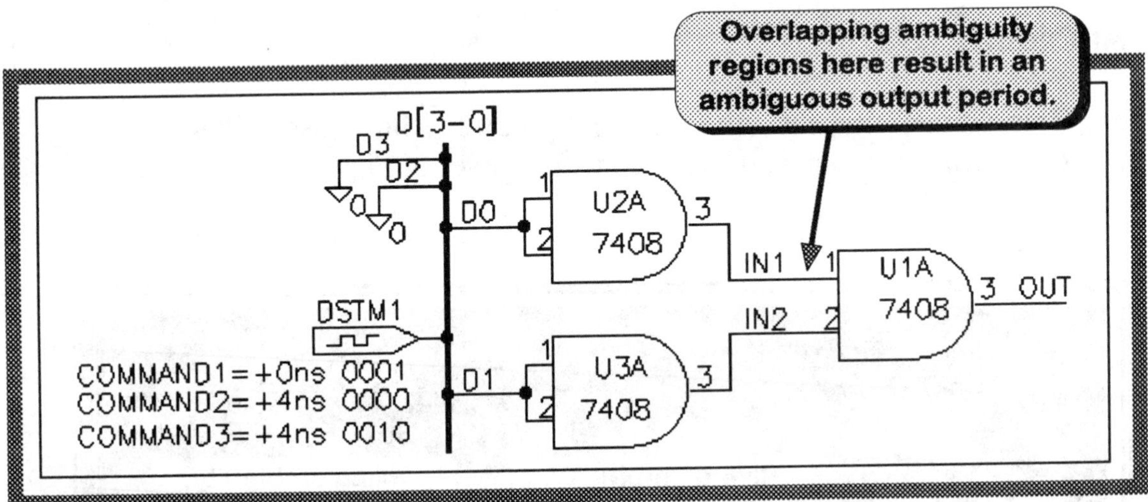

FIGURE 14.6
Test circuit for convergence hazard

8. Figure 14.7 shows the input waveforms to the AND gate array. Assuming ideal AND gates (no propagation delay), draw the expected output waveform in the space provided.

   Is the expected output waveform zero at all times?

   Yes **No**

9. Using worst case analysis, run the simulation. Because a violation was found by PSpice, **OK** in the *Simulation Messages* dialog box to bring up the *Simulation Message Summary* dialog box.

   What type of hazard occurred? ____________________

   When did the hazard occur? ____________________

   What device sensed the hazard? ____________________

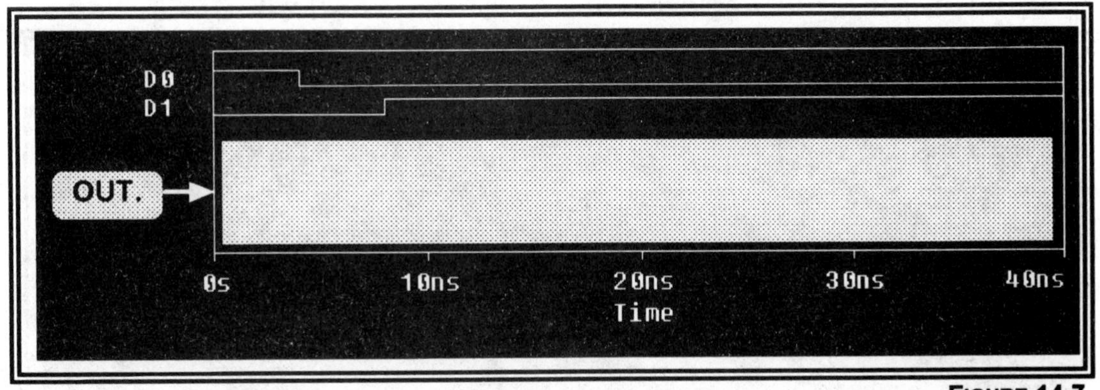

FIGURE 14.7
Convergence hazard input waveforms

10. Display the various options available under *Minimum Severity Level.* In which of the categories listed below did the convergence error fall? (Circle your answer.)

    WARNING **SERIOUS** **FATAL**

11. Activate the various options available under *Sort By* (except *Section*) and note the changes. (Return to *time*.)

12. Whenever a hazard occurs, we are given the option of viewing the waveforms associated with the hazard.

    To exercise this option: **Plot**, **Close** to generate the special hazard Probe curves of Figure 14.8.

13. The convergence hazard occurred at 15ns because that is when the input ambiguity regions overlapped. During this overlap period, the output waveform is uncertain.

    (a) What is the state of the output ambiguity region? Circle your answer.

    R **F** **X**

    (b) From Table 14.1, the minimum rising propagation delay is 7ns. Does the output signal ambiguity occur 7ns after the ambiguity convergence hazard (where the ambiguity regions of the two input signals first overlap)?

    Yes **No**

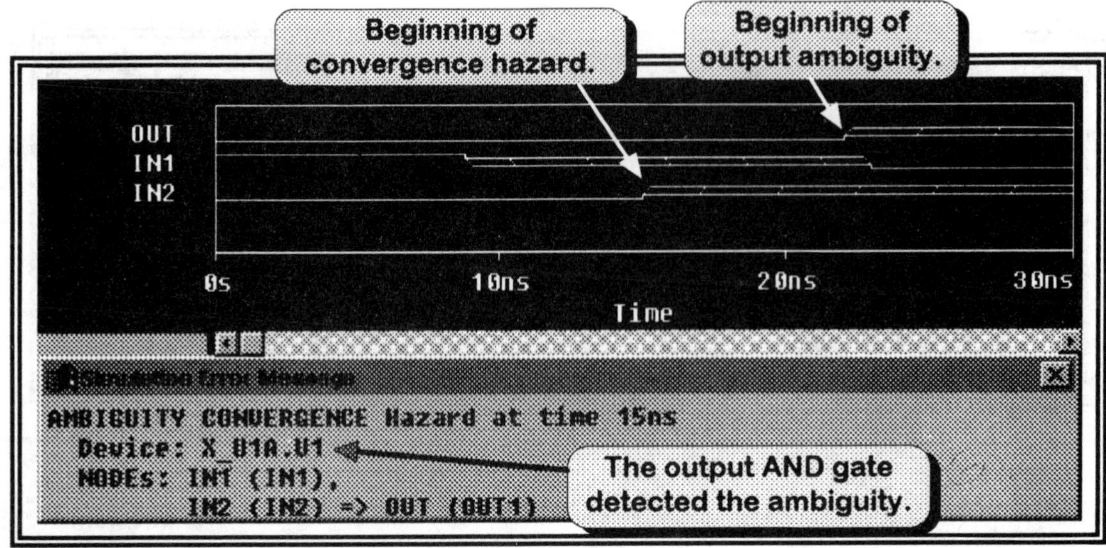

FIGURE 14.8
Convergence hazard waveform set

14. Review the *Simulation Error Message* (bottom of Figure 14.8):

    At what input nodes did the signals converge? ______________

    At what output node did the hazard occur? ______________

15. Referring back to Figure 14.6, double the input signal spacing from +4ns to +8ns and rerun the worst case analysis.

    (a) Was the ambiguity convergence hazard eliminated? (Was the input ambiguity region overlap reduced to below the minimum rising delay of 7ns?)

    Yes **No**

    (b) Is the output signal zero at all times (as predicted for ideal gates in step 8)?

    Yes **No**

Cumulative ambiguity hazard

A *cumulative ambiguity hazard* occurs when signals propagate through layers of gates. As the signal passes through each gate, ambiguity accumulates (the ambiguity regions widen). When the rising edge ambiguity overlaps the falling edge ambiguity, an unknown ("X") output region is created and a hazard is predicted.

16. Draw the test circuit of Figure 14.9 and set the commands as shown. (The timing analysis is still *Worst Case.*)

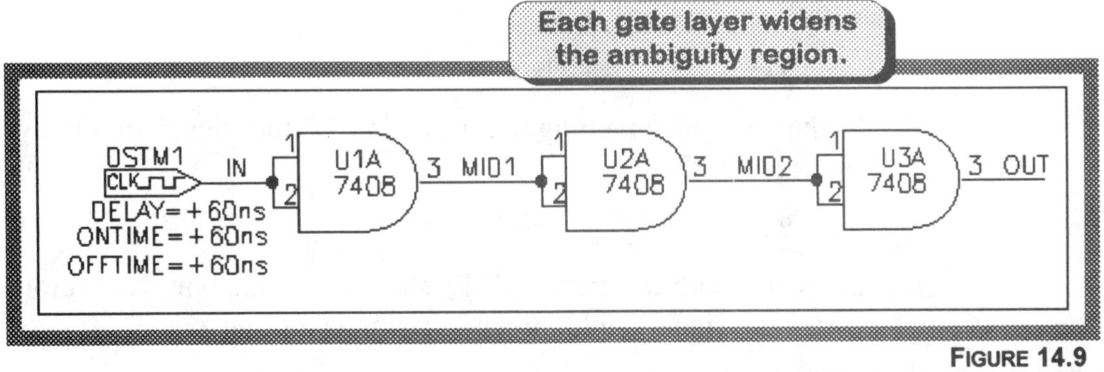

FIGURE 14.9
Cumulative ambiguity test circuit

17. Perform a worst case analysis and generate the hazard curves of Figure 14.10. (Hint: Add curves as necessary and expand the X-axis.)

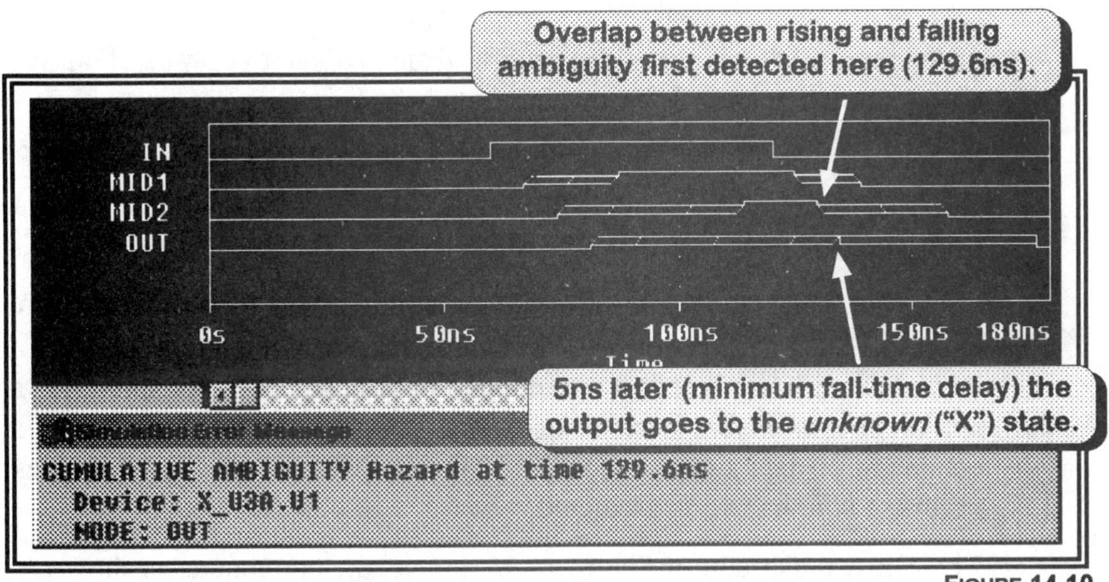

FIGURE 14.10
Cumulative ambiguity waveform set

18. Analyze the curves and answer the following:

    (a) As the input waveform propagates through the gates, does the ambiguity accumulate? (Do the ambiguity regions grow wider?)

    Yes **No**

    (b) Is the regular (unambiguous) portion of the signal gradually squeezed out?

    Yes **No**

    (c) Does the output signal (OUT) show an unknown (X) region where the signal could either be rising or falling?

    Yes **No**

19. The cumulative ambiguity hazard occurred at 129.6ns because that is when PSpice first determined that the rising and falling ambiguity regions would overlap (converge).

    Does the unknown (X) region occur in the output signal 5ns (the minimum falling propagation delay) after the hazard is detected?

    Yes **No**

20. Increase the input pulse width from +60ns to +80ns. Did the cumulative hazard disappear?

    Yes **No**

Setup, hold, and width hazards

The *setup*, *hold*, and *width* hazards commonly occur in circuits that are clocked—such as flip-flops. If the width of a clock signal is too low, a *width* hazard is predicted; if a command or data signal is not stable for a sufficient time *before* a clock occurs, a *setup* hazard is predicted; if a command or data signal is not stable for a sufficient time *after* a clock occurs, a *hold* hazard is predicted.

21. Draw the circuit of Figure 14.11 and set the commands as shown.

22. Run PSpice (*Worst Case*) and bring up the *Simulation Errors* dialog box, and note the number of errors. **OK** to bring up the *Simulation Message Summary* dialog box and note that the first three hazards are *Setup* (20ns), *Hold* (24ns), and *Width* (40ns).

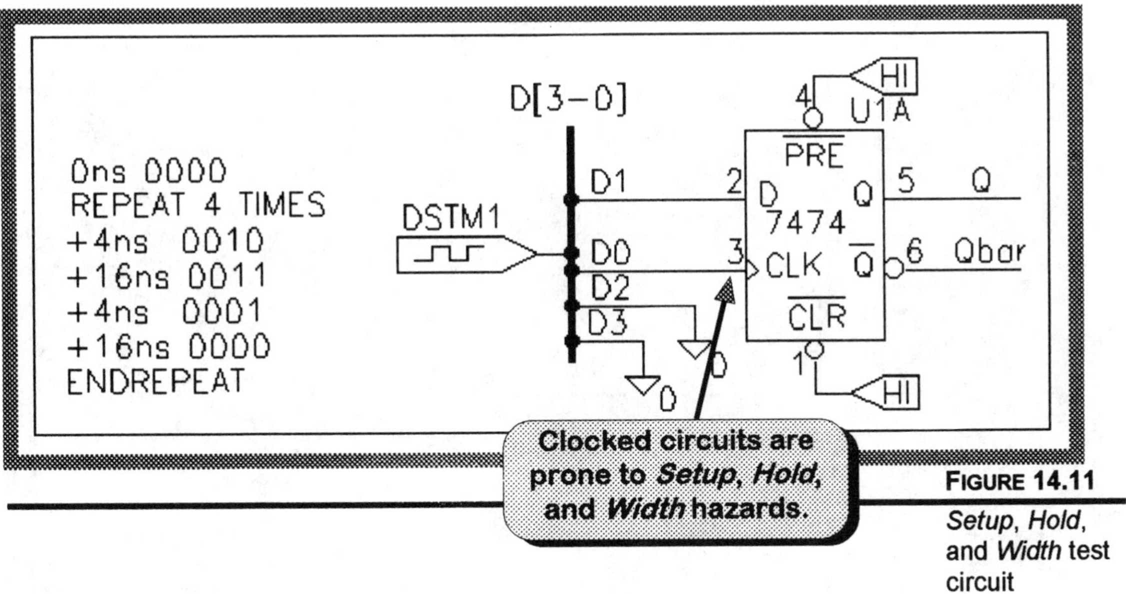

FIGURE 14.11
Setup, *Hold*, and *Width* test circuit

23. To create three windows corresponding to the first three hazards, **Plot**, **Plot**, **Plot**, **Close**. (We have created three data files of the same name, but identified by three consecutive letters.)

Setup

24. First, select the *Setup* window of Figure 14.12. (**Window**, select the first window listed with the present file name and the lowest letter designation).

25. The *Setup* hazard occurred at 20ns because that is when the system first determined that the time between a data change (D1 at 4ns) and a clock transition (D0 at 20ns) was below the minimum allowed.

    With the help of the *Simulation Error Message* box, determine each of the following:

    Measured D/CLOCK *Setup* = __________

    Minimum required *Setup* (TSUDCLK) = __________

    Additional *Setup* time required (minimum – measured) = __________

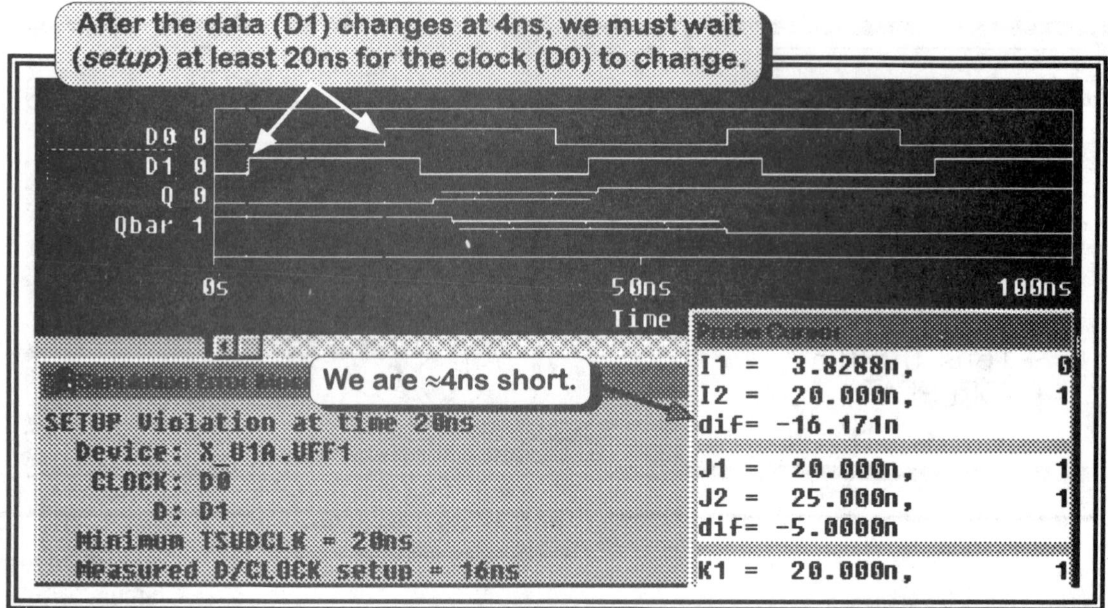

FIGURE 14.12
Setup violation waveform set

Hold

26. Next, select the *Hold* window (**Window**, select the next lettered file). The *Hold* hazard occurred at 24ns because that is when the system first determined that the data input (D1 at 24ns) changed too quickly *after* the clock (D0 at 20ns) went active high. Record each of the following:

Measured D/CLOCK *Hold* = __________

Minimum required *Hold* (THDCLK) = ____________

Additional *Hold* time required (minimum – measured) = ________

Width

27. Finally, select the *Width* window. The *Width* hazard occurred at 40ns because that is when the system first determined that the width of the clock signal (D0) was too small. Record each of the following:

Measured CLOCK pulse *Width* = ___________

Minimum required *Width* (TWCLKH) = ____________

Additional *Width* time required (minimum – measured) = _______

28. Double the command times in Figure 14.11 from 4ns/16ns to 8ns/32ns. Were all the violations eliminated?

    Yes **No**

Advanced activities

29. Perform a worst case analysis on the oscillator of Figure 14.13 (from Chapter 11) and generate the output graph of Figure 14.14.

> Because of the continuous accumulation of ambiguity in the feedback loop, the output signal quickly reaches the X (unknown) state and ceases to operate. Yet we know this does not happen in real life. *We conclude that worst case analysis cannot be used in circuits with asynchronous feedback.*
>
> To overcome the problem, set all active devices in the feedback loop (the 7414s) to typical (TYP) by changing model parameter MNTYMXDLY from 4 (*Worst Case*) to 2 (*Typical*). (**DCLICKL** to bring up each *Part Name* dialog box). All other devices, if any, will continue to operate in the worst case mode. Does the circuit now operate properly? (Return to MNTYMXDLY = 4 when done.)

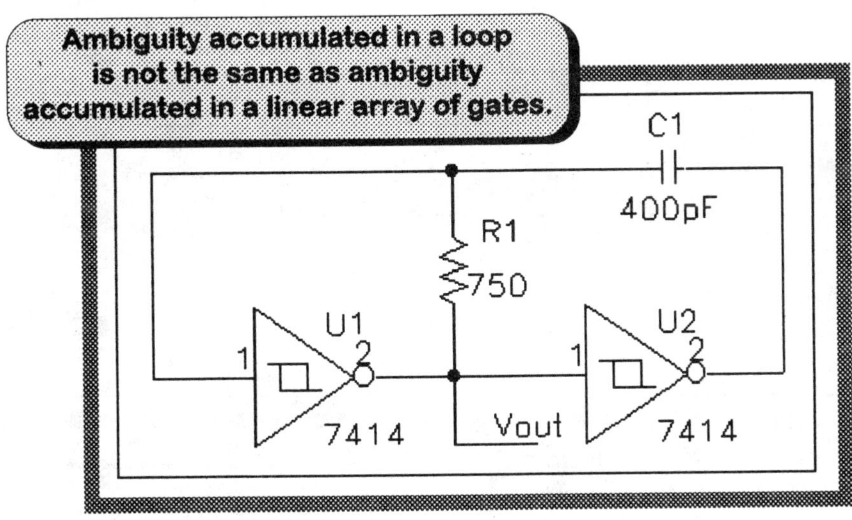

FIGURE 14.13
Digital oscillator

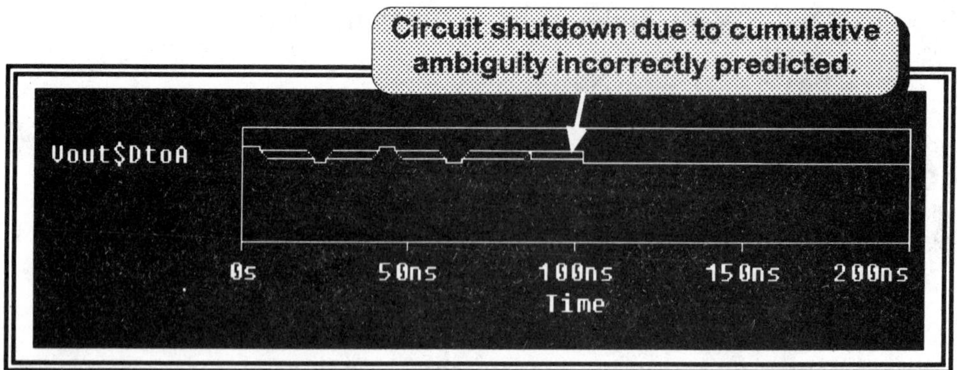

FIGURE 14.14
Ambiguity build-up in oscillator

Exercise

- Referring to Figure 14.15, predict the highest safe frequency of operation (no unknown states in the output signal). Verify your predictions using PSpice (*worst case analysis*).

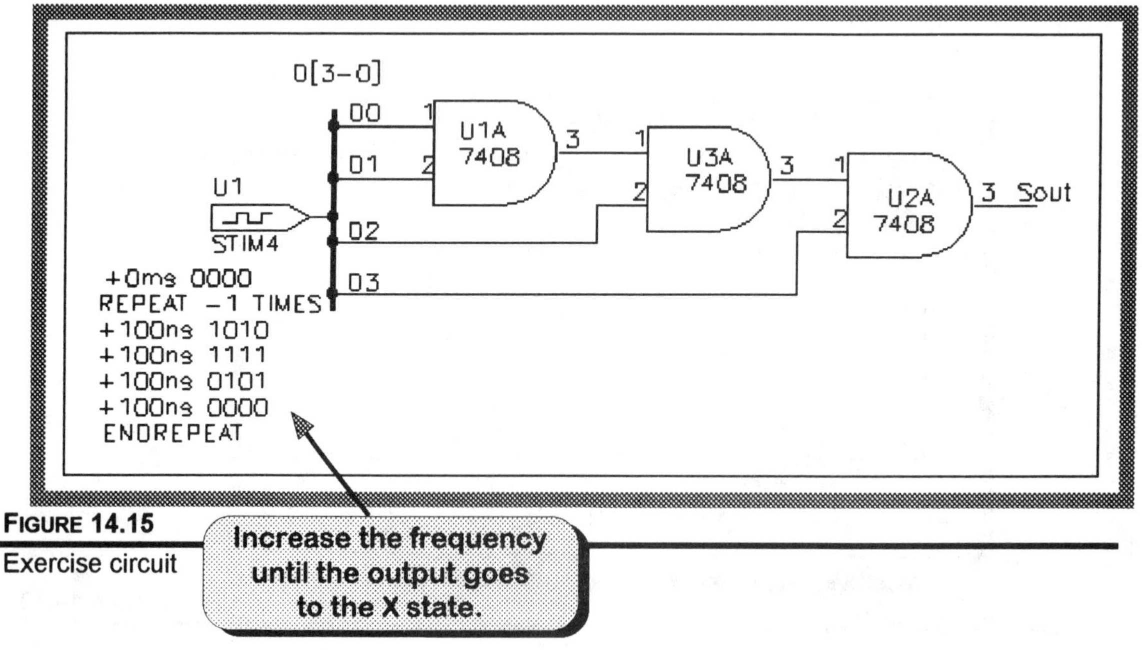

FIGURE 14.15
Exercise circuit

QUESTIONS AND PROBLEMS

1. What is the difference between *Setup* and *Hold*?

2. On a digital waveform, what does the X state indicate?

3. Why is it unrealistic to perform a worst case analysis on a feedback loop?

4. The input waveform shown in Figure 14.16 has input rising and falling ambiguities of 1ns. After passing through the 7404, which adds additional ambiguities, draw the output waveform (including ambiguity regions).

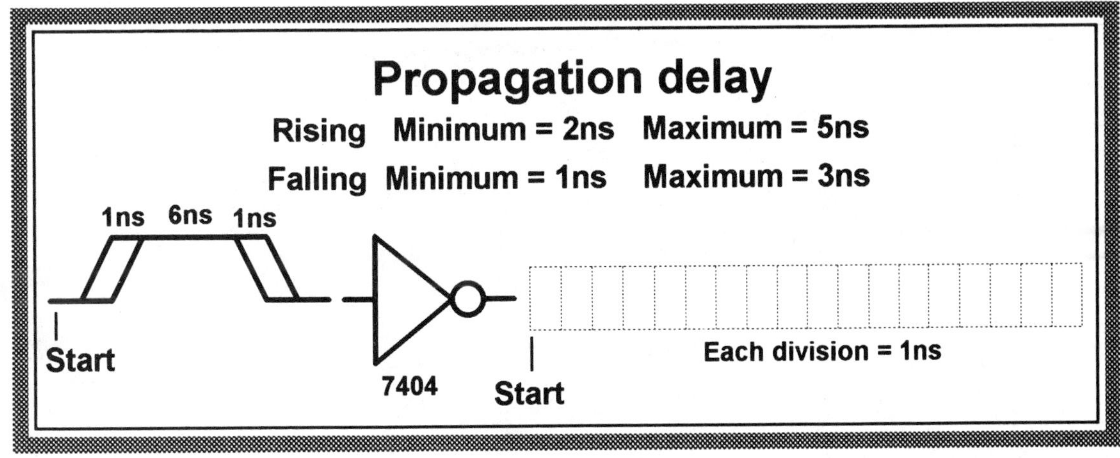

FIGURE 14.16

Ambiguity study

CHAPTER 15

Critical Hazards

Persistence

OBJECTIVES

- To apply worst case analysis techniques to digital circuits.
- To identify and correct a variety of timing violations.
- To analyze the effects of ambiguity.

DISCUSSION

All the violations and hazards of the previous chapter were warnings. As such, they may or may not cause a serious problem. They are simply events that the designer must examine in order to determine if a change in circuit design is warranted. As an example, consider the circuit of Figure 15.1.

When the circuit is processed under PSpice, the waveform set of Figure 15.2 shows that a setup warning was properly identified at 20ns. This is because the data line transition (D1) occurred only 10ns before the active high clock pulse (D0)—and a minimum of 20ns is required.

However, this probably will not cause a problem in this case because a second clock pulse comes along at 80ns and properly latches the data. (Note that the ambiguity in the Q signal occurs only after the first clock pulse.) Therefore, the design engineer may very well decide to ignore this particular setup warning.

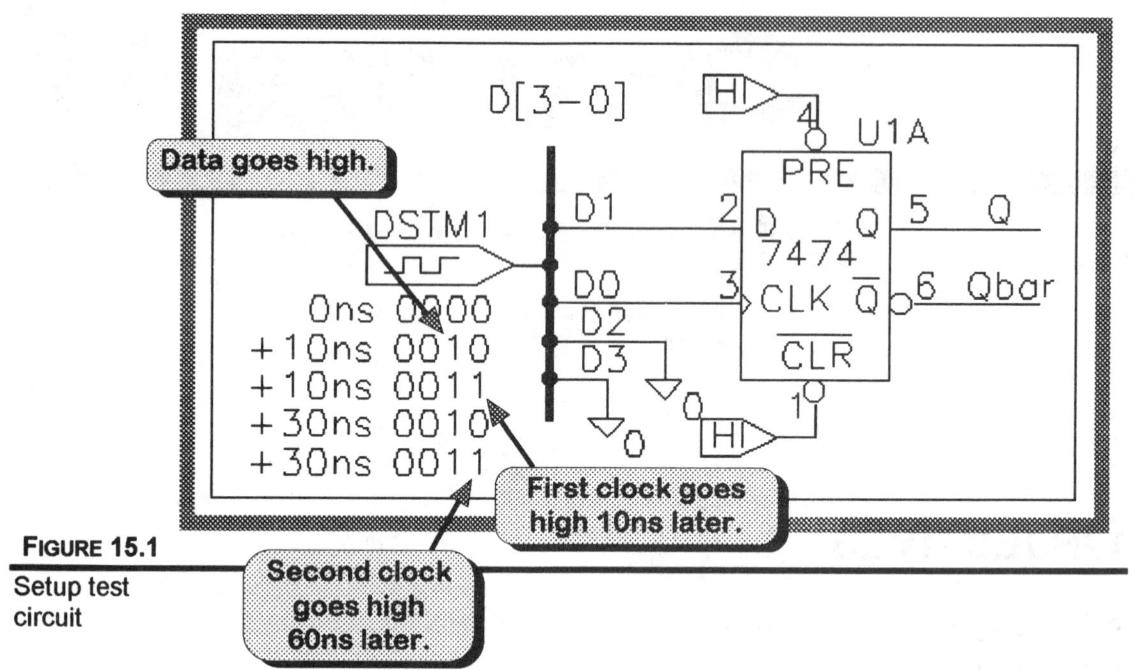

FIGURE 15.1
Setup test circuit

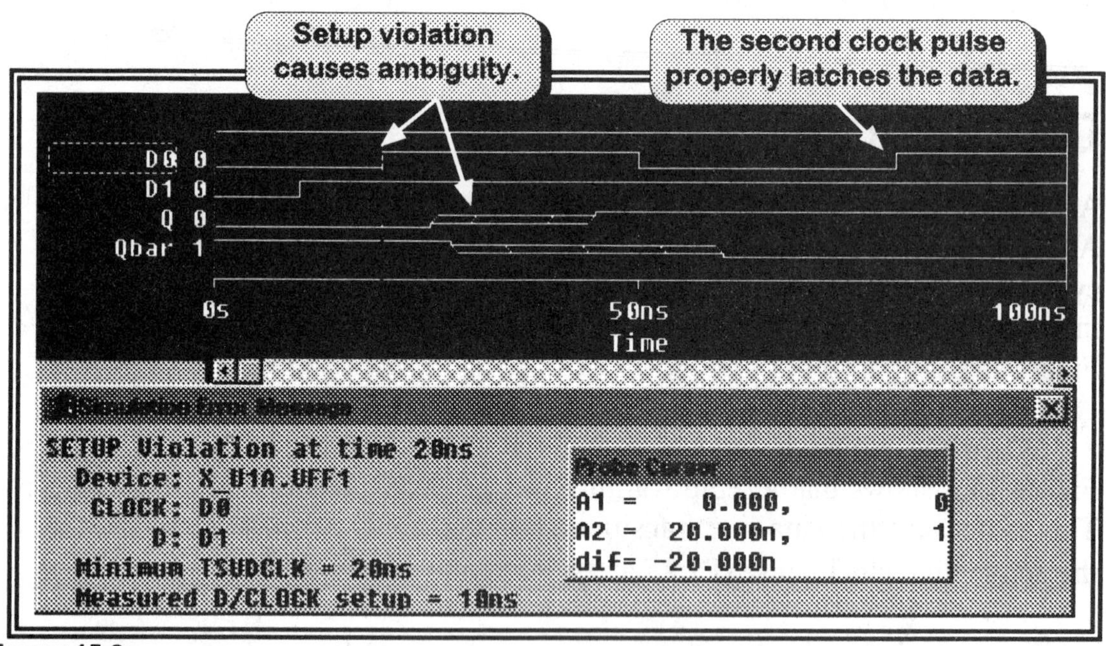

FIGURE 15.2
Setup warning waveform set

Persistent hazard

A *persistent hazard*, on the other hand, is a timing violation or hazard that can cause an incorrect state to be latched into an internal circuit (such as a flip-flop), or one that is passed on to a primary circuit output.

These more serious persistent hazards, which usually cannot be ignored, are a major subject of this chapter.

Design methodology

Because of violations and hazards, the development of a digital circuit is a two-phase process: *design* and *verification.*

- During *design*, all tolerances are set to typical and we concentrate on the *state response* of the circuit.
- During *verification*, all tolerances are set to worst case and we concentrate on violations and hazards. Starting at a hazard point, we work backward until the cause of the hazard is uncovered and corrected.

SIMULATION PRACTICE

External ports

1. Draw the CONVERGENCE hazard test circuit of Figure 15.3. Be sure to attach an EXTERNAL_OUT *port* symbol (from library *port.slb*) to the circuit output. (The EXTERNAL_OUT ports represent data outputs to other circuitry.)

2. Run a *worst-case* simulation and **Plot**, **Close** (from the *Simulation Message Summary* dialog box) to generate the hazard waveform set of Figure 15.4. As shown, two plots are generated:

   - The upper plot displays the persistent hazard at 22ns due to the propagation of a timing hazard to an output port.
   - The lower plot displays the *cause* of the persistent hazard—a convergence hazard at 15ns. (Does 22ns – 15ns = 7ns, the *minimum rise time delay* for a 7408?)

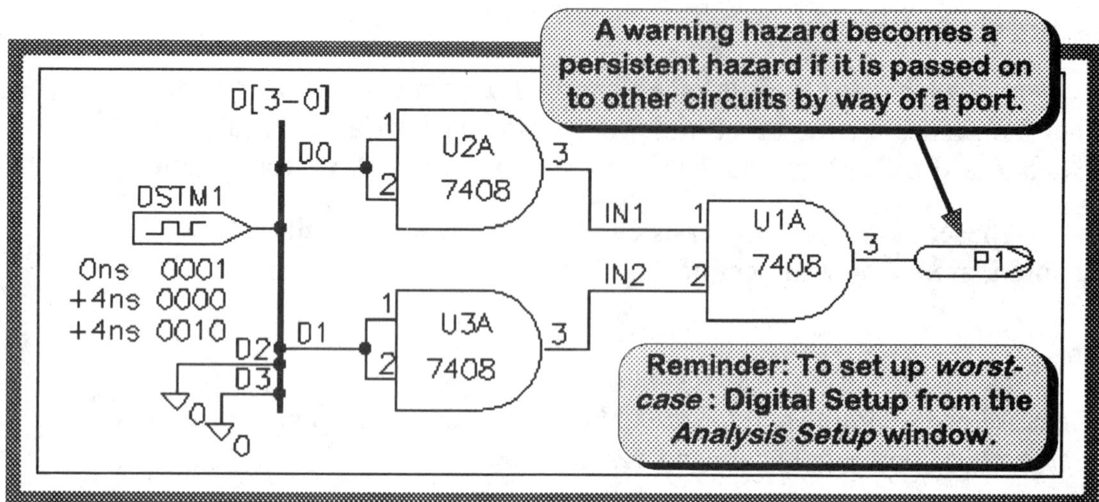

FIGURE 15.3

EXTERNAL_OUT
port added

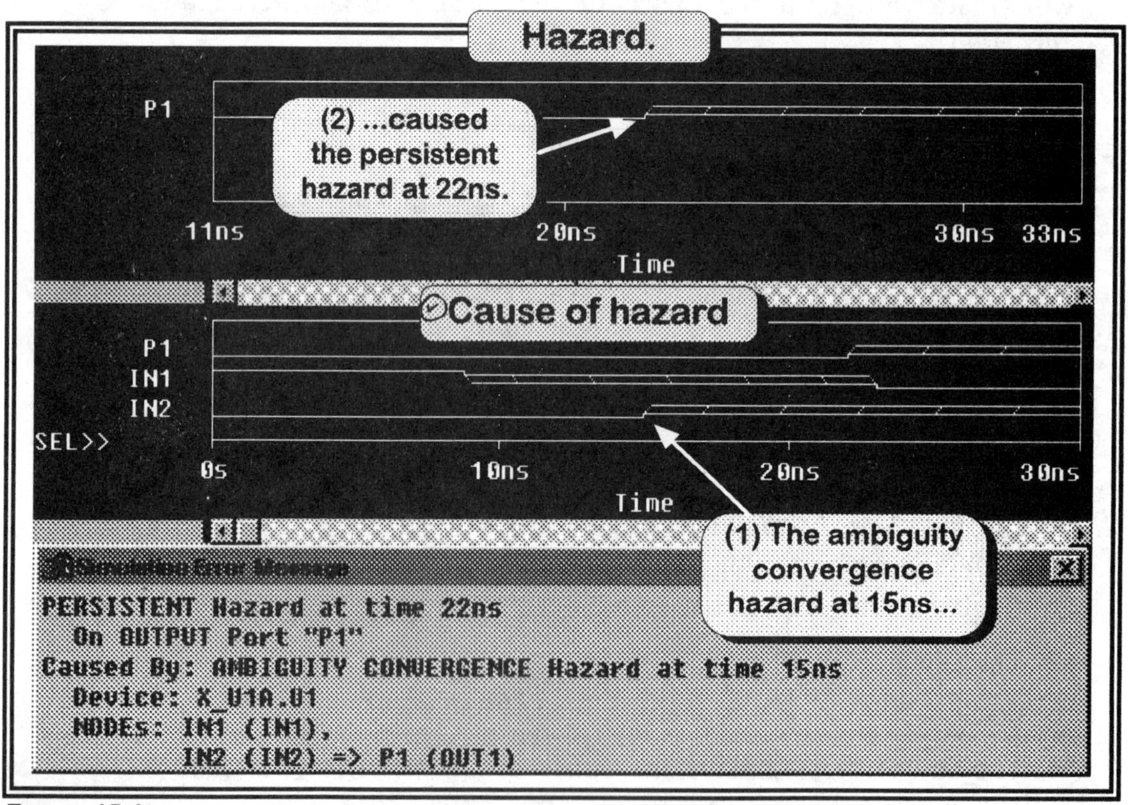

FIGURE 15.4

Persistent hazard
and cause

3. Increase the timing from 4ns to 8ns. Was the hazard eliminated, and did the output of the circuit (P1) remain low?

   Yes **No**

Latched hazard

4. Draw the circuit of Figure 15.5 and set the stimulus commands as shown.

5. Run the *worst-case* simulation and plot the first persistent hazard, generating the waveform set of Figure 15.6.

   - The upper plot displays a persistent hazard at 60ns because an ambiguity state (R) was present at the data input (Data) to the flip-flop when the clock (CLK) went active high.

     As a result, an unknown value was latched into the flip-flop, and its output (Q) went indeterminate (state X) at 67ns. This could cause a serious malfunction in the overall circuit.

   - The lower plot shows us that the cause of the persistent hazard was the ambiguity convergence hazard at the output of the AND gate array at 49.8ns.

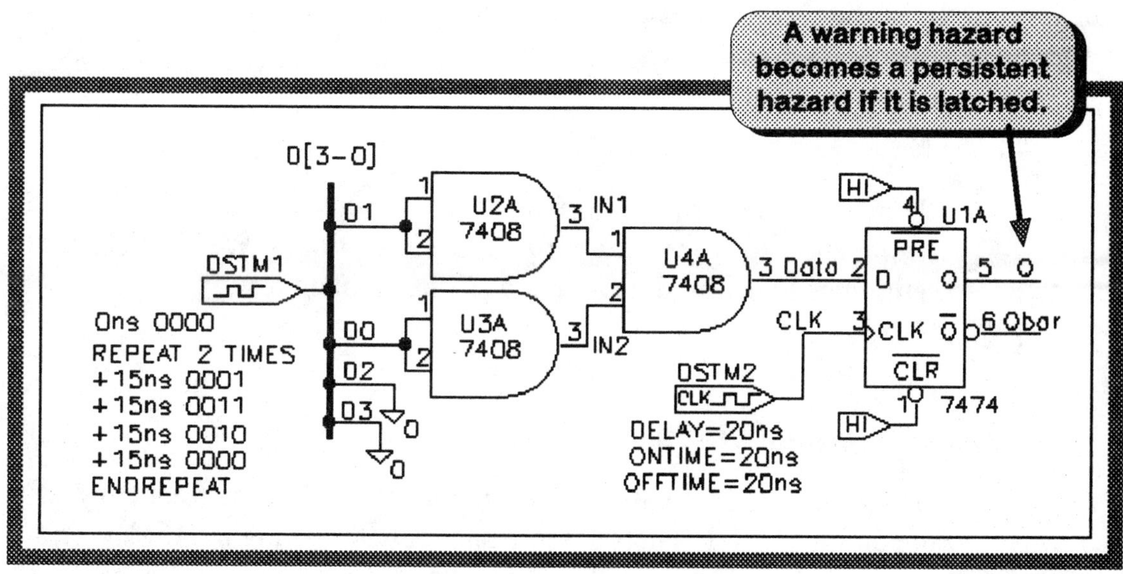

FIGURE 15.5
Internal latch test circuit

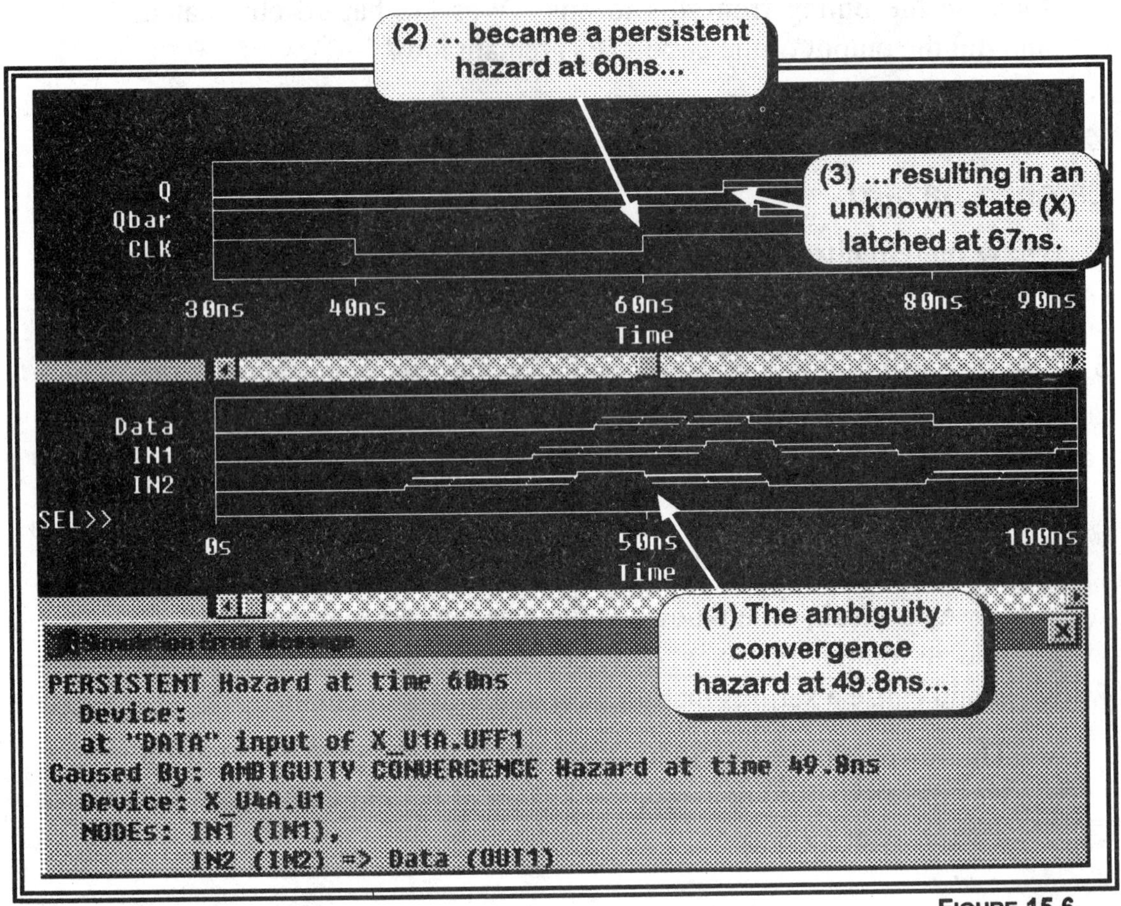

FIGURE 15.6
Persistent hazard and cause

6. Increase all DSTM1 times from 15ns to 60ns and all DSTM2 times from 20ns to 40ns and rerun the simulation. Were all hazards eliminated and did the *Data* and *Q* lines remain low?

   Yes **No**

Advanced activities

7. Using worst case analysis, determine the maximum frequency of operation of the D-latch-based ripple counter of Figure 15.7. (As the frequency is increased, what is the first hazard that appears?)

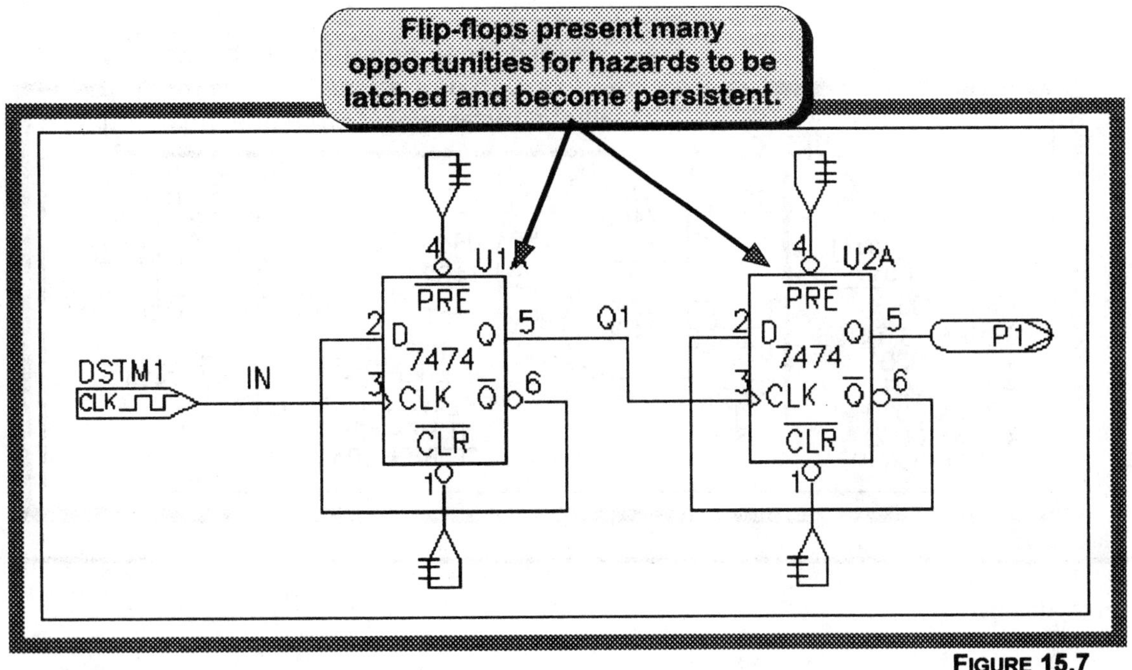

FIGURE 15.7
Worst case test circuit

EXERCISES

Perform the two-step design methodology outlined in the discussion on the circuit of Figure 15.8.

- Set the timing state to *typical*, run the simulation, and generate a waveform set. Do the waveforms indicate that the state response of the circuit is correct?
- Switch to worst case mode and generate a hazard waveform set. Report how you corrected the hazard.

QUESTIONS AND PROBLEMS

1. Reviewing Figure 15.2, what is the setup time of the data (D1) for the *second* clock pulse (D0)?

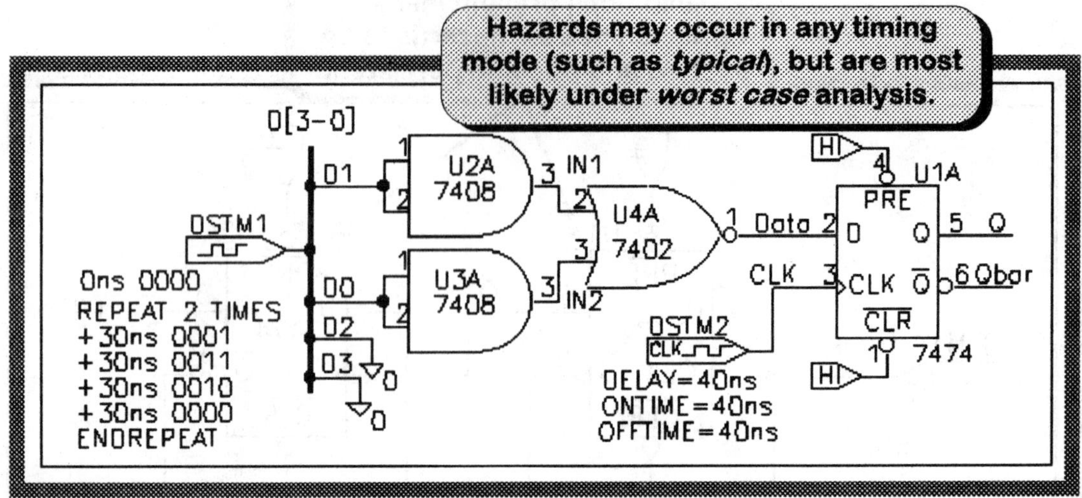

FIGURE 15.8
Design methodology test circuit

2. What two basic steps should every design engineer perform on a digital circuit?

3. What two conditions does PSpice look for when tagging a hazard as *persistent*?

4. Does PSpice display both the *cause* and *effect* of a persistent hazard?

CHAPTER 16

Counters and Shift Registers
Synchronous Versus Asynchronous

OBJECTIVES

- To compare *synchronous* versus *asynchronous* digital circuitry.
- To design and test a variety of counters and shift registers.

DISCUSSION

A digital counter is a device that keeps track of clock cycles (counts). Counters are fashioned from a series of flip-flops strung together in sequence, with one flip-flop feeding the next.

The number of different binary states defines the *modulus* (MOD) of the counter. For example, if a counter counts from 0 to 3, it is a MOD-4 counter. A MOD-4 counter must be constructed from 2 flip-flops.

Synchronous versus asynchronous

Digital counters are classified as either *synchronous* or *asynchronous*.

- If the output of one flip-flop is used to clock the next flip-flop, as shown in the MOD-4 counter of Figure 16.1(a), the counter is asynchronous. Asynchronous counters are also called *ripple counters* because the clock signal ripples from one flip-flop to the next.
- If all the flip-flops are clocked at the same time, as shown in the MOD-4 counter of Figure 16.1(b), the counter is synchronous. Synchronous counters are much faster and more precise than asynchronous counters because they don't have to wait for the signal to propagate through the flip-flops.

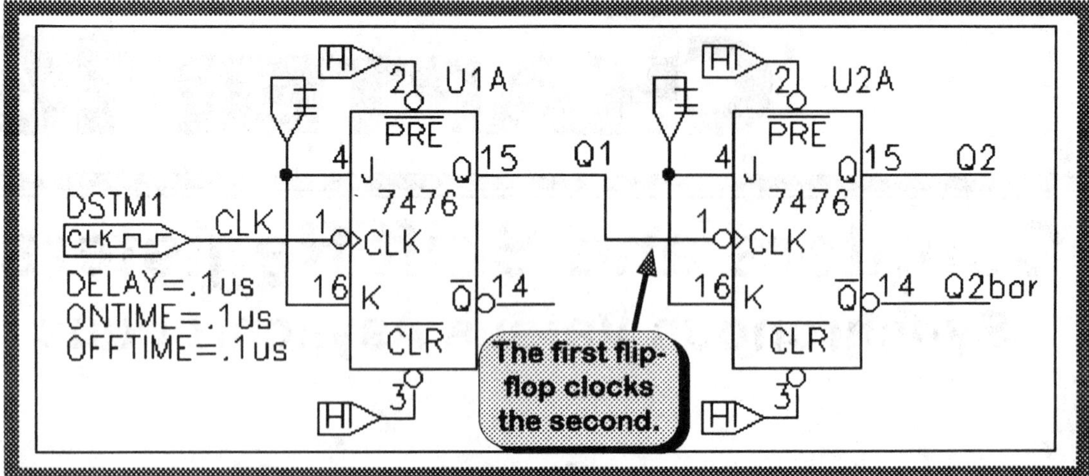

(a) MOD-4 binary ripple counter

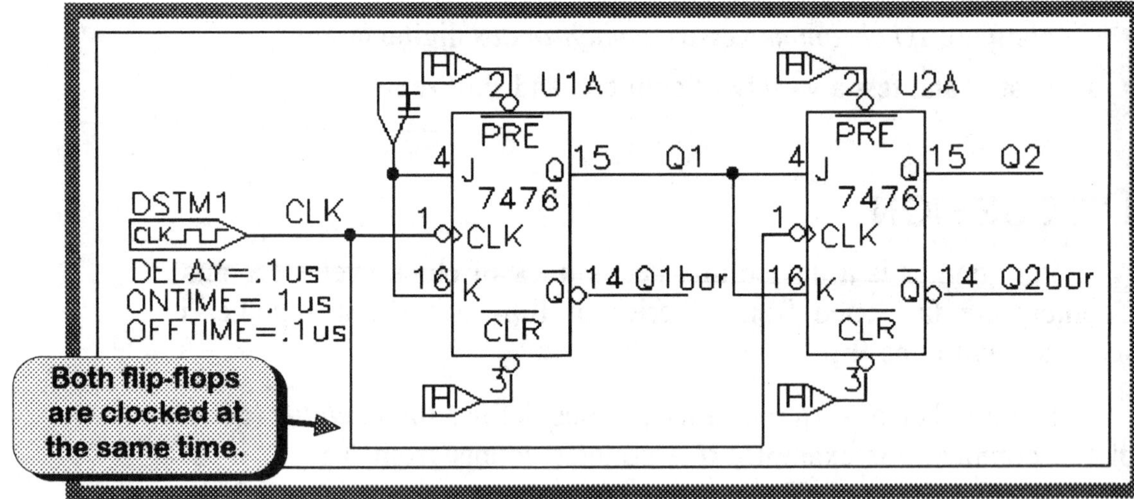

(b) MOD-4 synchronous counter

FIGURE 16.1

Two-stage binary counters
(a) MOD-4 binary ripple counter
(b) MOD-4 synchronous counter

Odd modulus counters

The natural modulus of a counter is 2^N, where N is the number of flip-flops. For example, we see from Figure 16.1 that the natural count of a two-flip-flop counter is 4 (0 to 3).

Suppose, however, we desired an unnatural modulus—such as 3 (0 to 2)? As shown by Figure 16.2, we use feedback and feedforward to properly program the flip-flops.

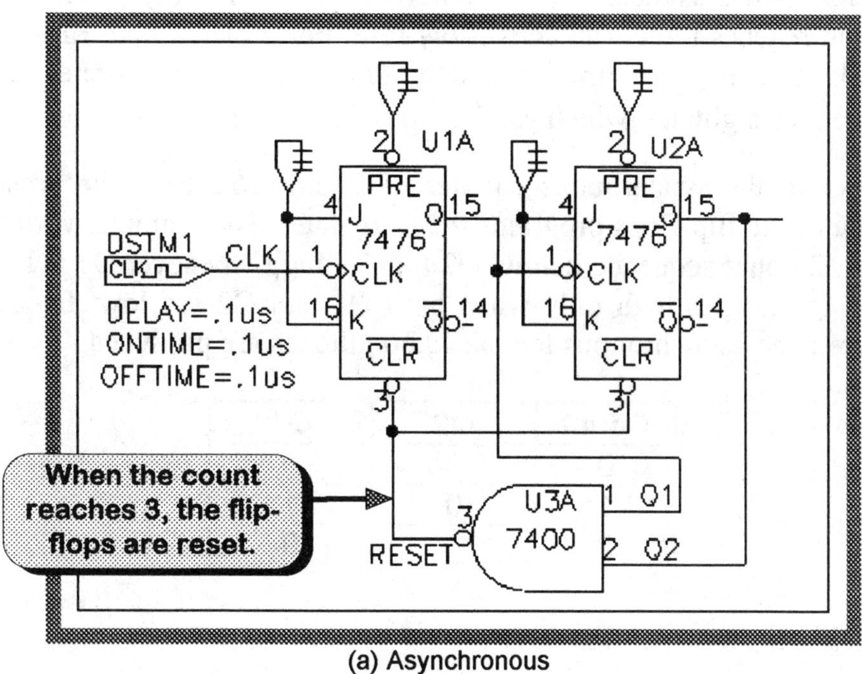

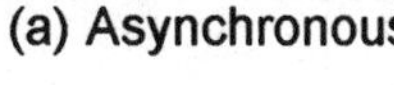
(a) Asynchronous

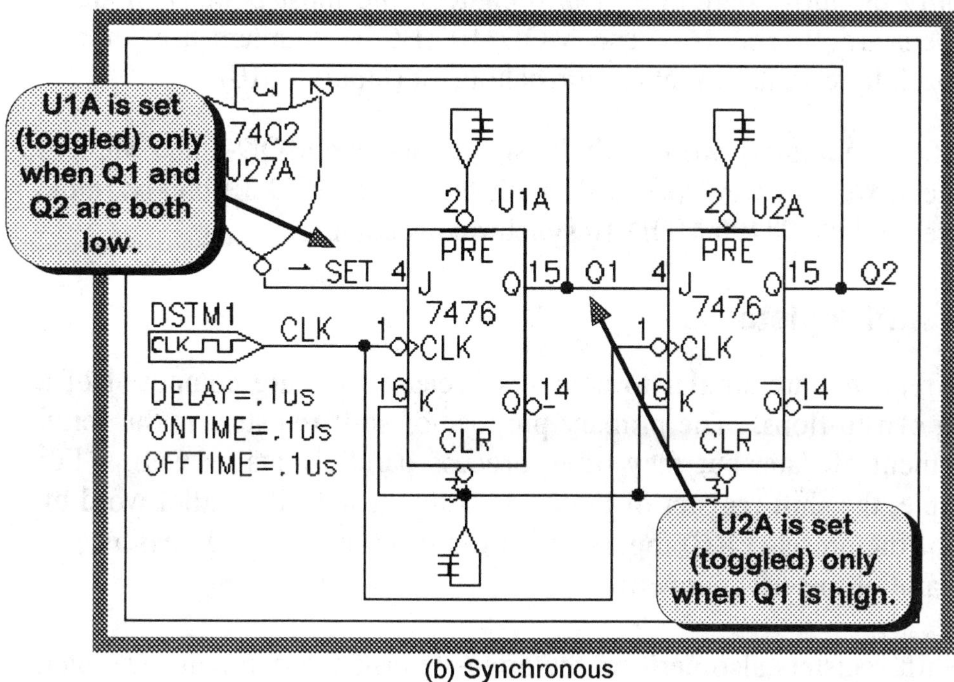

(b) Synchronous

FIGURE 16.2
MOD 3 counters
(a) Asynchronous
(b) Synchronous

- Refer to the asynchronous counter of Figure 16.2(a). When the count reaches three, an active-low reset pulse clears both flip-flops back to zero. The brief period of time that the three-state exists generates a glitch—which can be a problem at high frequencies.
- Refer to the synchronous counter of Figure 16.2(b). The present state of the flip-flops programs the next state. For example, with the 0, 1, 2 count sequence below, Q2 toggles high whenever Q1 is high, and Q1 toggles high only when both Q1 and Q2 are low. Because there is no asynchronous feedback, no glitches are produced.

Count	Q2	Q1
0	0	0
1	0	1
2	1	0

Integrated circuit counters

Fortunately, a wide variety of integrated circuit counters are available to simplify circuit design. For 4-bit counters, the most popular modulus numbers are 10 and 16. The MOD 10 (BCD) counters use internal feedback to reset the count upon reaching 10 (binary 1010).

After preliminary work with the simple discrete MOD-3 and MOD-4 counters, we will work primarily with the 7490 MOD-10 binary ripple counter and the 74190 MOD 10 synchronous counter.

The shift register

Shift registers are similar to counters because they are composed of a string of flip-flops. The primary purpose of shift registers is the serial movement of data, but they often process parallel data as well. For example, the shift register of Figure 16.3 inputs a 4-bit parallel word by way of the preset (PRE) inputs and shifts it out in serial. In essence, it is a parallel-to-serial converter.

Shift registers also perform the opposite task and come in integrated circuit form. For example, during simulation practice, we will use the 74165 and 74194 shift register ICs to design a complete serial communication system.

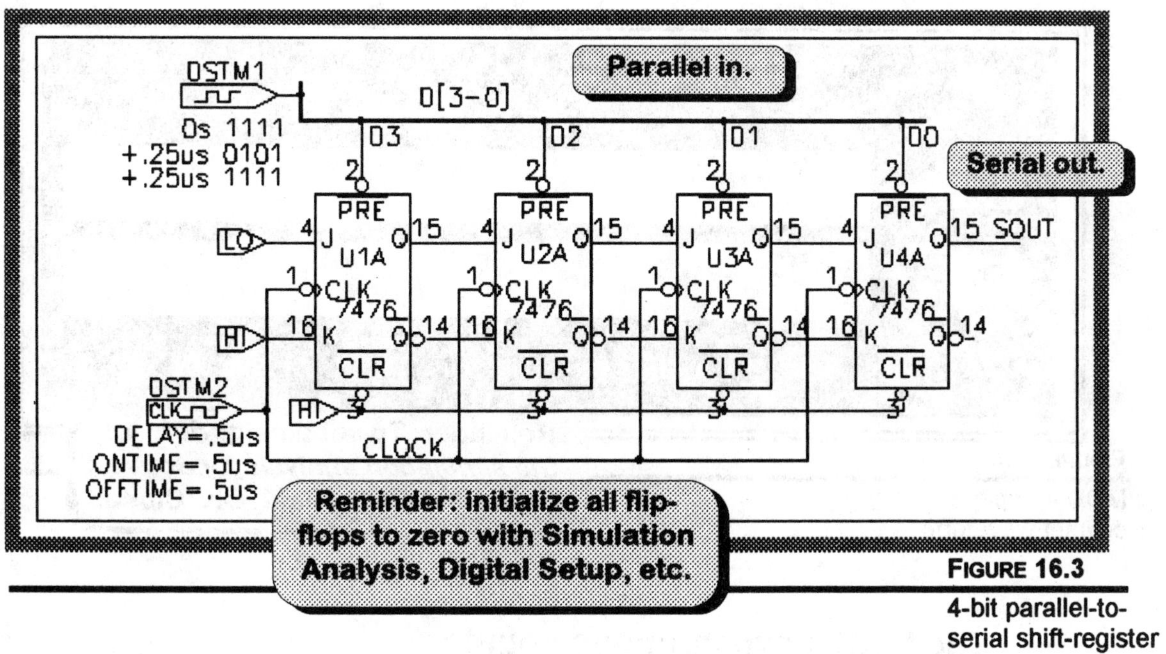

FIGURE 16.3
4-bit parallel-to-serial shift-register

SIMULATION PRACTICE

The MOD-4 binary ripple counter

1. Draw the MOD-4 (2-bit) binary ripple counter circuit of Figure 16.1(a) and set the attributes as shown.

2. On the graph of Figure 16.4, predict (draw) the Q1 and Q2 waveforms in the spaces provided.

3. Predict the frequency relationship between CLK and Q2.

   CLK frequency = ________ × **Q2** frequency

4. Using *typical* timing (**Simulation Analysis**, **Digital Setup**), run PSpice and generate the CLOCK, Q1, and Q2 waveforms. Did the count go from 0 to 3, and was CLK four times the frequency of Q2?

   Yes **No**

Analysis Setup

5. Switch to *worst-case* timing and repeat. Did the ambiguity regions widen from Q1 to Q2?

   Yes **No**

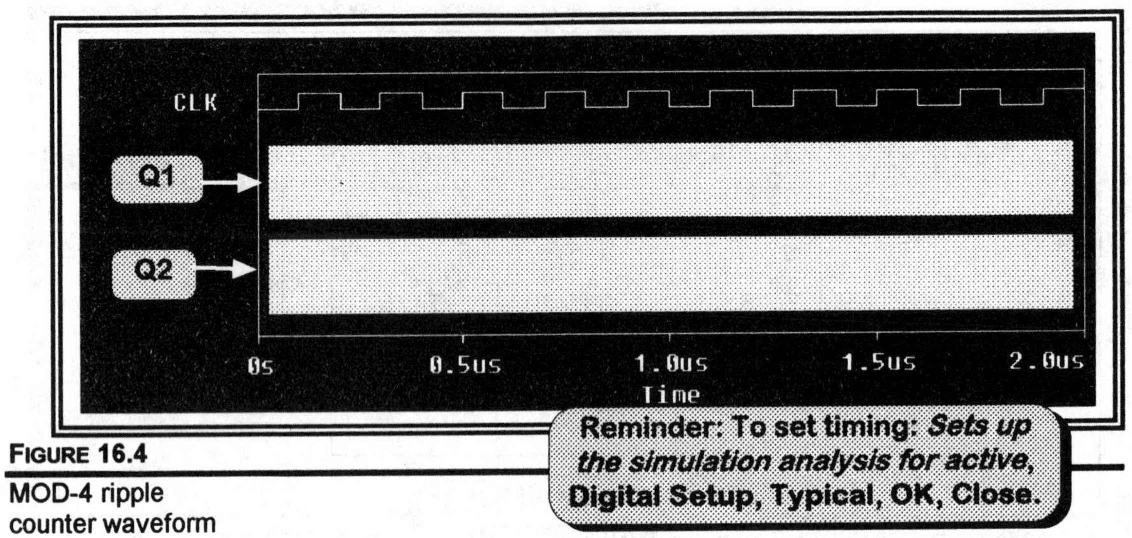

FIGURE 16.4
MOD-4 ripple counter waveform

The MOD-4 synchronous counter

6. Repeat steps 1 through 5 for the synchronous counter of Figure 16.1(b). Were the results the same, except that the ambiguity regions of Q1 and Q2 remained constant (did not widen)?

   Yes **No**

The MOD-3 (divide-by-three) counters

7. Using *typical* timing, generate the Q1, Q2, and RESET waveforms for the MOD-3 ripple counter of Figure 16.2(a).

   (a) Did the count go from 0 to 2?

   Yes **No**

   (b) Was a short RESET glitch observed when the count went briefly to 3?

   Yes **No**

8. Using *typical* timing, generate the CLK, Q1, Q2, and SET waveforms for the MOD-3 synchronous counter of Figure 16.2(b).

   Were the results the same as for the MOD-3 ripple counter of step 7, except no glitches were observed?

   Yes **No**

The 7490 MOD-10 (divide-by-ten) ripple counter

The 7490A cascades a MOD-2 (1 flip-flop) and a MOD-5 (3-flip-flop) ripple counter to create a MOD-10 counter. The MOD-5 counter uses feedback to reduce its natural modulus from 8 to 5. For user flexibility, the connection between the MOD-2 and MOD-5 counters is done externally.

Because of internal ripple delays, decoded outputs are subject to spikes.

9. Draw the asynchronous counter/decoder combination circuit of Figure 16.5 and set all attributes as shown. (Is the clock speed set higher for this integrated-circuit counter?)

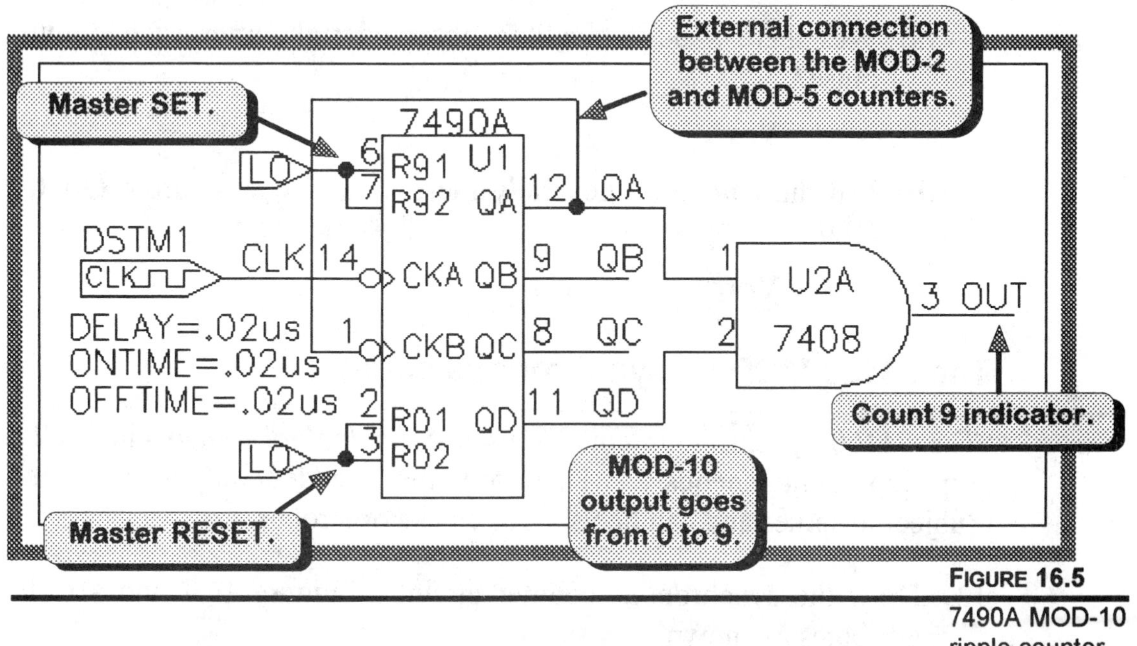

FIGURE 16.5
7490A MOD-10 ripple counter

10. Using *typical* timing, generate the graph of Figure 16.6.

    (a) Did the count go from 0 to 9, and did the output decoder detect the 9 count properly?

    Yes **No**

    (b) Was there any evidence of glitches?

    Yes **No**

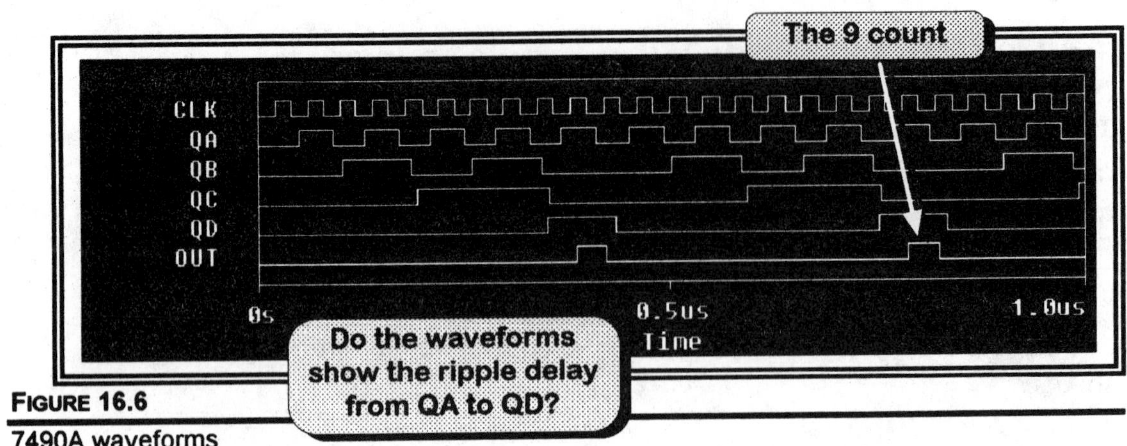

FIGURE 16.6
7490A waveforms

11. Switch to *worst-case* analysis and repeat. (**Cancel**, when the *Simulation Messages* window appears.)

    (a) Was there a definite glitch observed right after each output pulse?

    Yes **No**

    (b) Did the ambiguity generally increase as we went from QA to QD?

    Yes **No**

The 74160 MOD-10 synchronous counter

The 74160 synchronous counter uses *carry-look-ahead* to clock all flip-flops simultaneously. Therefore, the decoded outputs are not subject to spikes and can be used for clocks or strobes.

12. Draw the synchronous counter circuit of Figure 16.7 and set all attributes as shown.

13. Using *typical* timing, generate two cycles (0 to 2μs) of output waveforms (QA, QB, QC, QD, and RCO).

    (a) Does the counter cycle properly from 0 to 9, and does RCO give the terminal count?

    Yes **No**

    (b) Is there any evidence of delay from QA to QD?

    Yes **No**

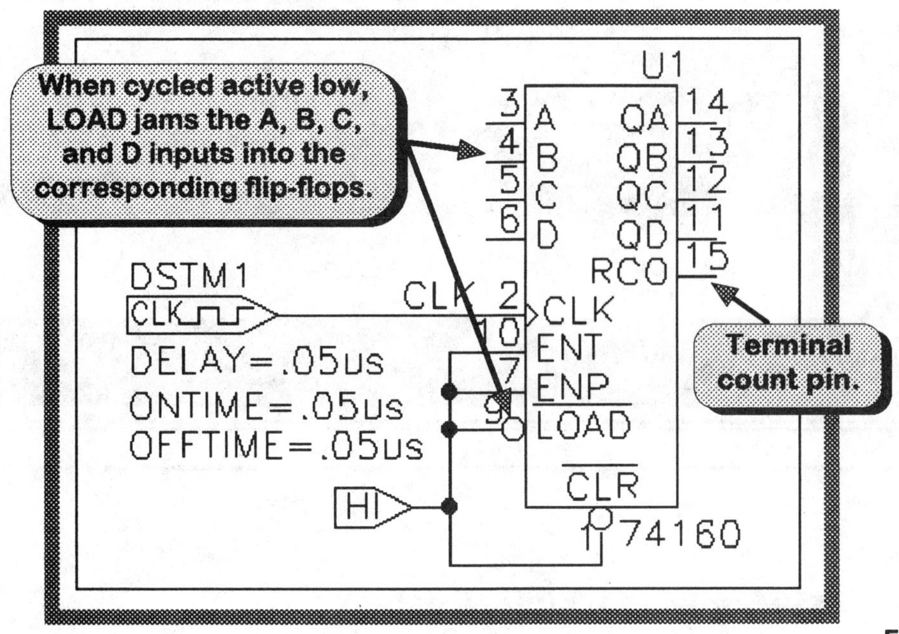

FIGURE 16.7
74160 synchronous counter

14. Switch to *worst-case* analysis and repeat.

    (a) Were glitches observed anywhere in the RCO (terminal count) signal?

    Yes **No**

    (b) Did the ambiguity generally stay constant as we went from QA to QD?

    Yes **No**

The 4-bit shift register

15. Draw the 4-bit parallel-to-serial shift register of Figure 16.3 and generate the waveforms of Figure 16.8.

16. In the space provided, predict the SOUT waveform. (Note that zeros are walked in from the left as the data is transmitted from the right.)

17. Generate the waveforms using PSpice and compare to your predictions. Did you analyze the circuit correctly?

    Yes **No**

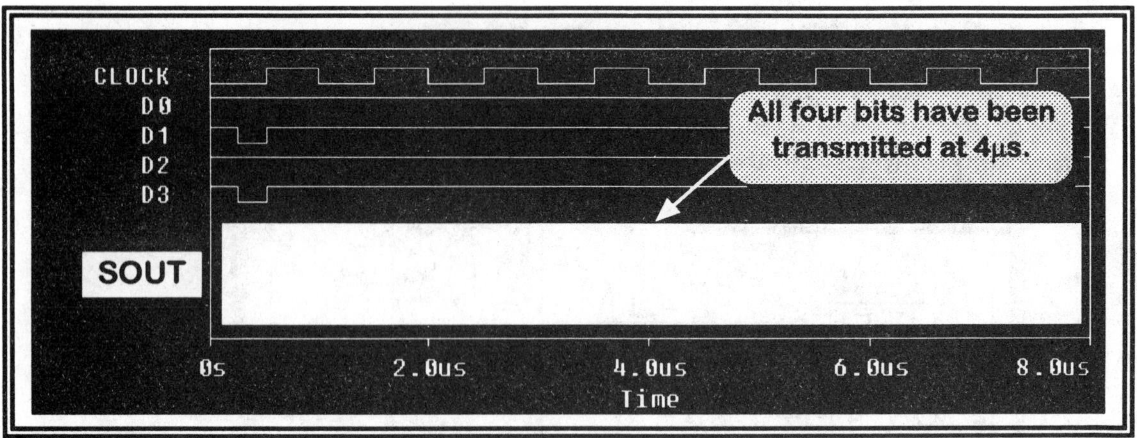

FIGURE 16.8
Shift-register waveforms

The 74194 universal shift register

The integrated circuit shift register we have chosen for study (the 74194) is called "universal" because it can be programmed for input or output in serial or parallel. The 74194 is programmed by way of the S0 and S1 pins according to Table 16.1.

In this section, we use a pair of 74194s to design a complete serial communication system. A transmitting 74194 will load parallel data and transmit in serial; a receiving 74194 will input in serial and convert to parallel for output.

Operating mode	S1	S0
Hold	0	0
Shift left	1	0
Shift right	0	1
Parallel load	1	1

TABLE 16.1
Programming the 74194 universal shift register

18. Draw the communication circuit of Figure 16.9. Be sure to set the voltage markers in the following order: CLOCK, IN[3-0], LOAD, SERIAL, and OUT[3-0].

19. Run the simulation (typical timing) and generate the waveforms of Figure 16.10. (Note that placing markers on bus strips causes waveforms to be displayed in hexadecimal format.)

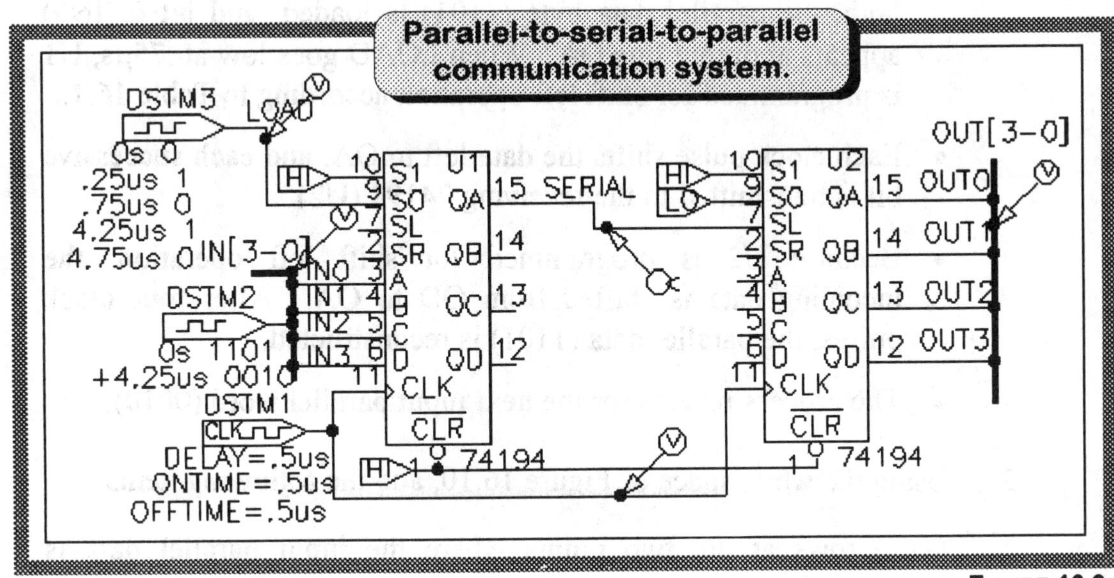

FIGURE 16.9

74194 serial communication test system

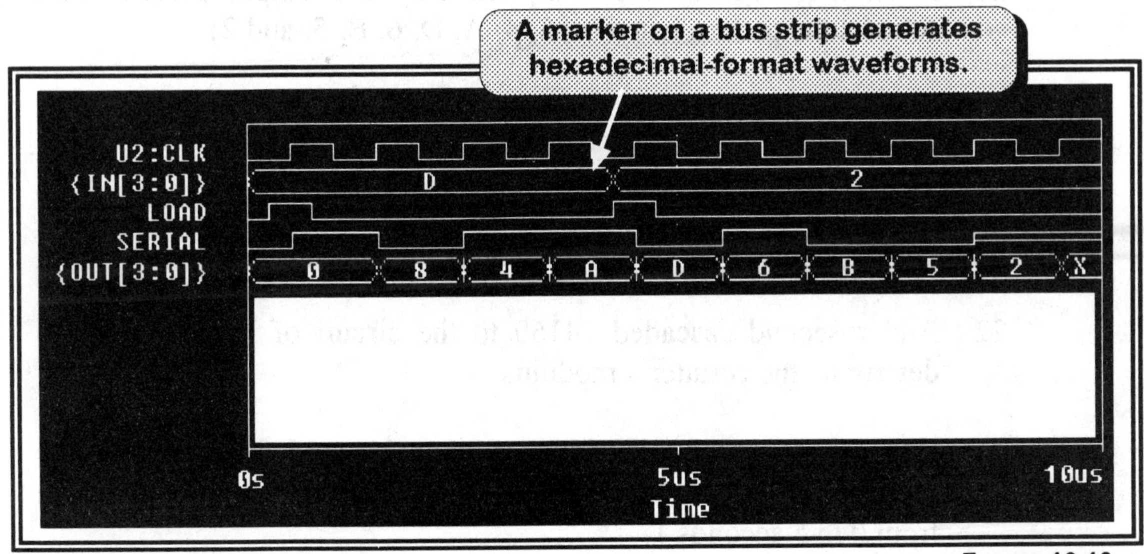

FIGURE 16.10

74194 system waveforms

From input to output (left to right), here is how the system works:

- Parallel data byte 1101 (D) is placed on the parallel input lines (A to D) of the transmitting 74194 (U1) at 0μs. At .25μs, S0 goes high and programs U1 for loading. At .5μs the clock goes high, the parallel data byte (1101) is loaded, and bit 0 (IN0) appears at output line QA. When LOAD goes low at .75μs, U1 is programmed for shift left operation according to Table 16.1.
- Each clock pulse shifts the data left to QA, and each successive bit is transmitted to the receiving 74194 (U2).
- Because U2 is programmed for shift left operation, the incoming data is shifted from QD to QA. After four clock pulses, the parallel data (1101) is reconstructed.
- The process repeats for the next input parallel word (0010).

20. Using the white space of Figure 16.10, add the following items:

    (a) Arrows at the two points where the input parallel data is loaded.

    (b) Brackets below each serial word (D and 2).

21. Reviewing Figure 16.10, explain why the output parallel data follows the pattern listed (0, 8, 4, A, D, 6, B, 5, and 2).

Advanced activities

22. Add a second cascaded 74160 to the circuit of Figure 16.7 and determine the counter's modulus.

23. What might the circuit of Figure 16.11 be used for? What is the purpose of each component? (Hint: Display the output waveform from 0 to 5 seconds.)

24. What changes would you make in the communication circuit of Figure 16.9 to reverse the flow of data?

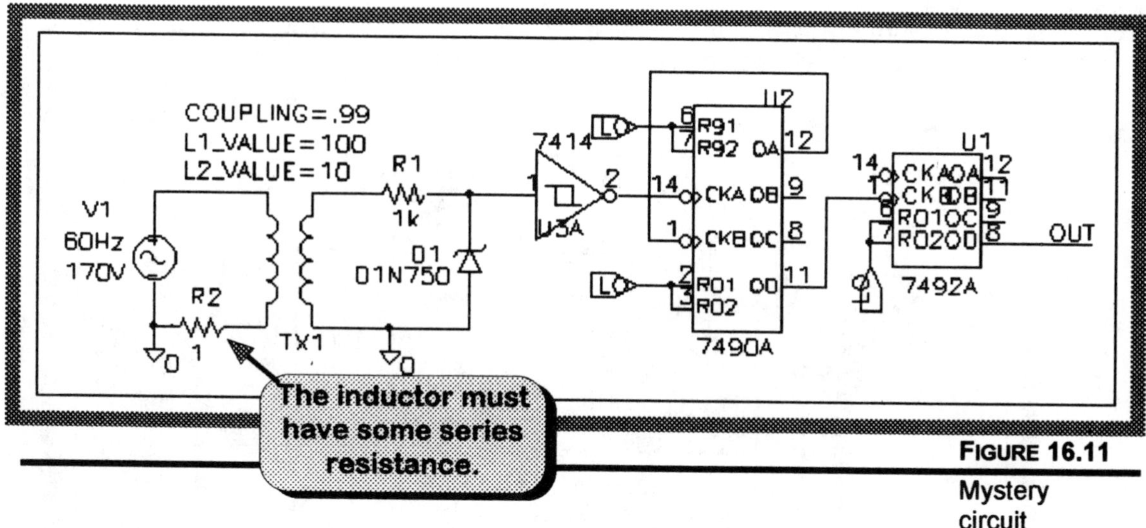

FIGURE 16.11
Mystery circuit

EXERCISES

- Determine the count sequence of the *Johnson counter* of Figure 16.12.
- To rotate a stepper motor, we require the 4-bit sequence of Table 16.2. Show how to use a 74194 universal shift register to generate both sequences.

Clockwise rotation	Counterclockwise rotation
1000	0001
0100	0010
0010	0100
0001	1000

TABLE 16.2
Stepper motor logic sequences

QUESTIONS AND PROBLEMS

1. How many flip-flops would be required to design a counter that counts from 0 to 63?

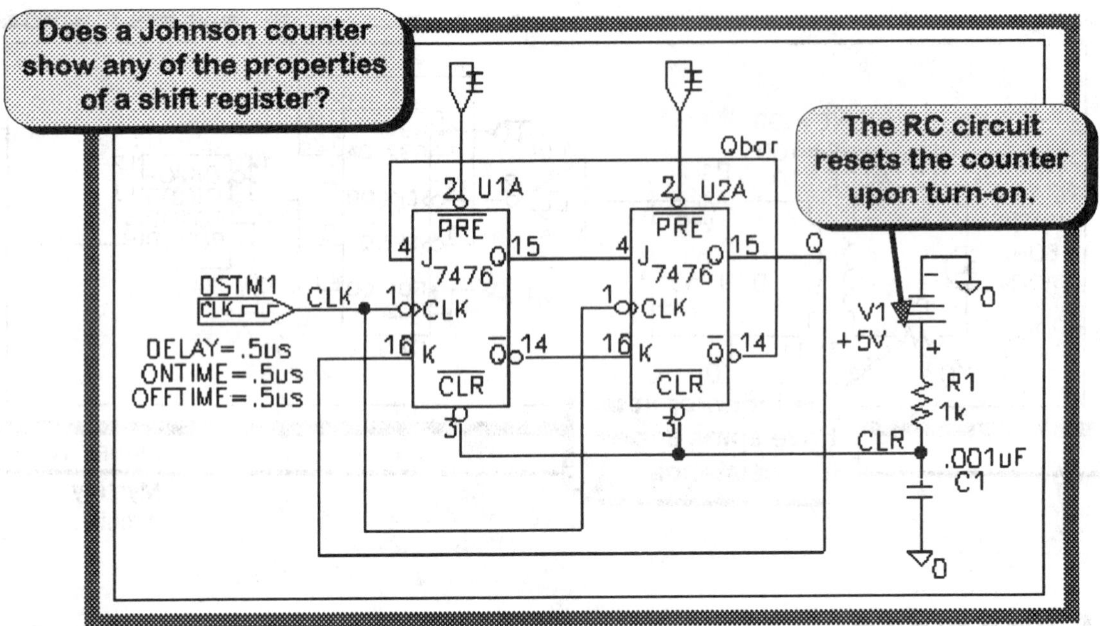

FIGURE 16.12
Johnson counter

2. What is the difference between *synchronous* and *asynchronous* counters?

3. Because of the minimum width requirements of the clock, the maximum frequency for both the 7490A asynchronous and 74060 synchronous counters were the same. What then is the advantage of the synchronous counter?

4. Using the 74160 synchronous counter of Figure 16.7, design a MOD 7 (2 to 8) counter.

CHAPTER 17

Digital Coding

Comparators, Coders, and Multiplexers

OBJECTIVE

- To design and test circuits that use comparators, encoders, decoders, and multiplexers.

DISCUSSION

In this chapter we investigate three popular types of digital integrated circuits:

- *Comparators*, which input two digital numbers and report if the first is greater than, equal to, or less than the second.
- *Decoders*, which convert a multi-bit input code into a single output line, and *encoders*, which reverse the process.
- *Multiplexers*, which select one signal from a number of choices, and *demultiplexers*, which reverse the process.

SIMULATION PRACTICE

Comparators

1. Draw the simple 2-bit comparator test circuit of Figure 17.1 and set the attributes as shown. (The 7493A counter is used for testing and is not part of the comparator.)

2. On the plot of Figure 17.2, draw the predicted output signal. (Hint: The inputs match when $A_0 = B_0$ and $A_1 = B_1$.)

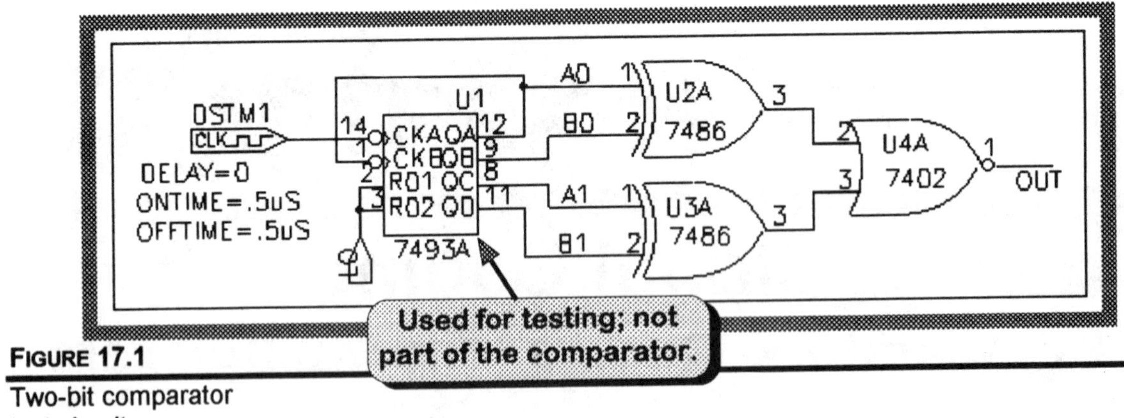

FIGURE 17.1
Two-bit comparator test circuit

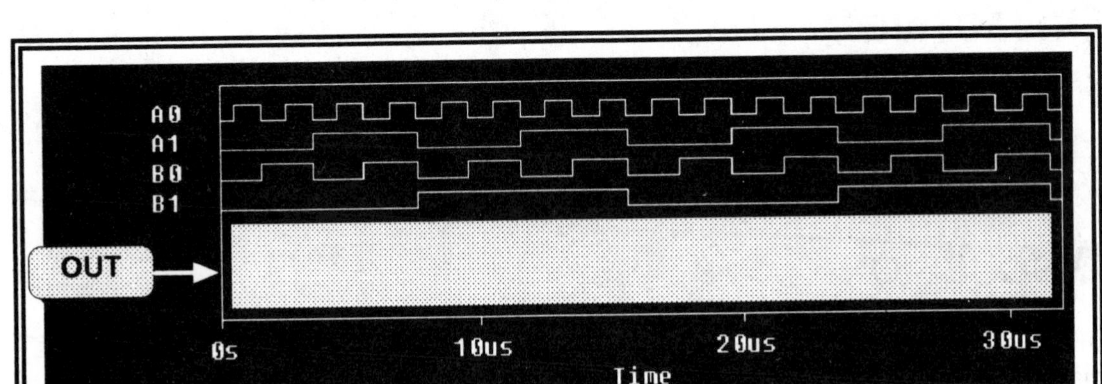

FIGURE 17.2
Two-bit comparator waveforms

3. Generate the waveforms using PSpice and compare them to your predictions. Were they the same? (Make any necessary corrections.)

   Yes **No**

4. The 7485 *4-bit magnitude comparator* of Figure 17.3 is a popular integrated circuit version of a comparator. Draw the circuit and set the attributes as shown.

> The 7485 compares the 4-bit values of the A and B inputs and activates the appropriate output line (A<B, A=B, A>B). By using the *A<B_IN*, *A>B_IN*, and *A=B_IN* lines, any number of 7485s can be cascaded.

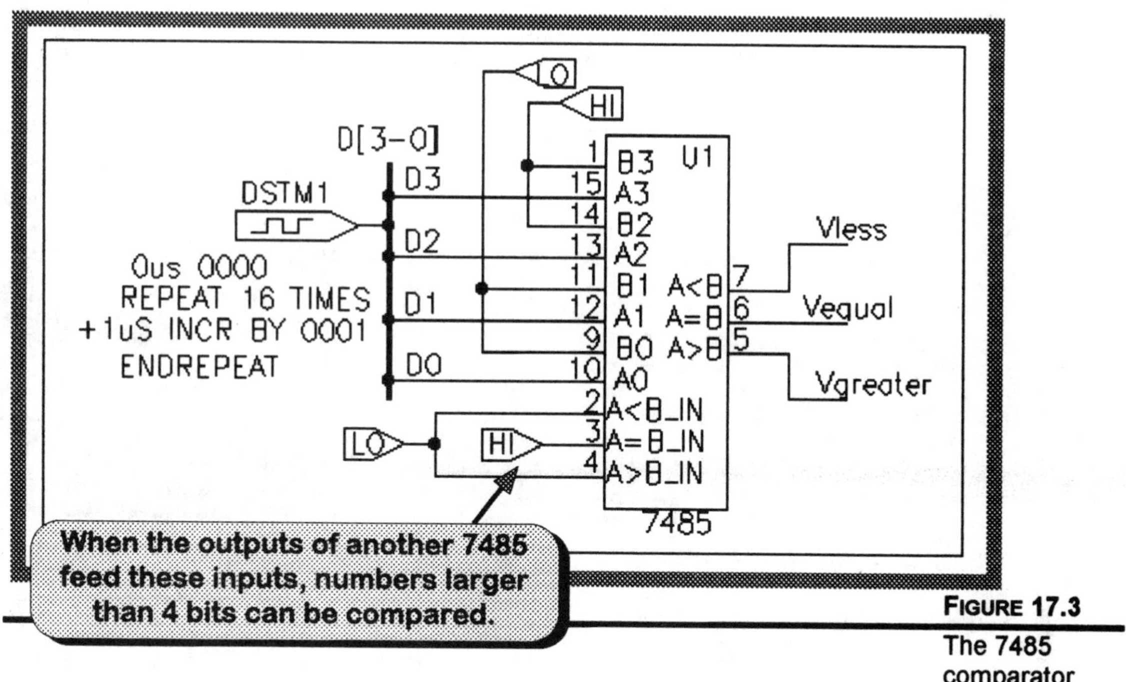

FIGURE 17.3
The 7485 comparator

5. Given the input waveforms of Figure 17.4 (and the B inputs set as shown), predict (draw) the output waveforms (*Vless*, *Vequal*, and *Vgreater*) in the spaces provided.

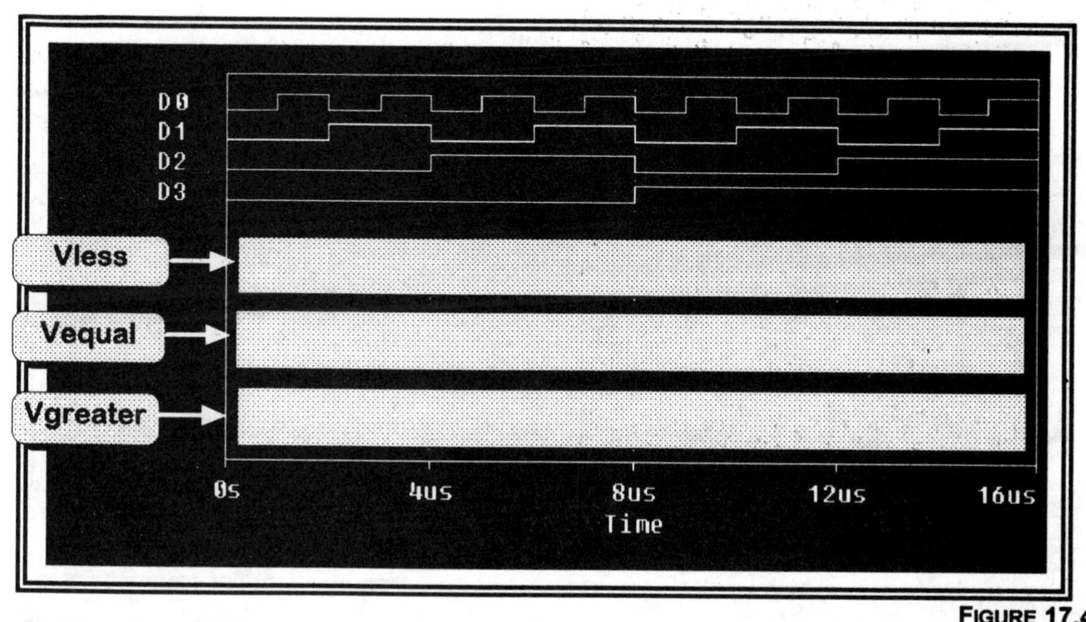

FIGURE 17.4
7485 comparator waveforms

6. Generate the waveforms using PSpice and compare them to your predictions. Are they the same? (Make any necessary corrections.)

   Yes **No**

Decoders

7. Following the previous pattern, first draw the simple 2-bit decoder of Figure 17.5. On the graph of Figure 17.6, predict the circuit output. After testing under PSpice, were your predictions correct?

   Yes **No**

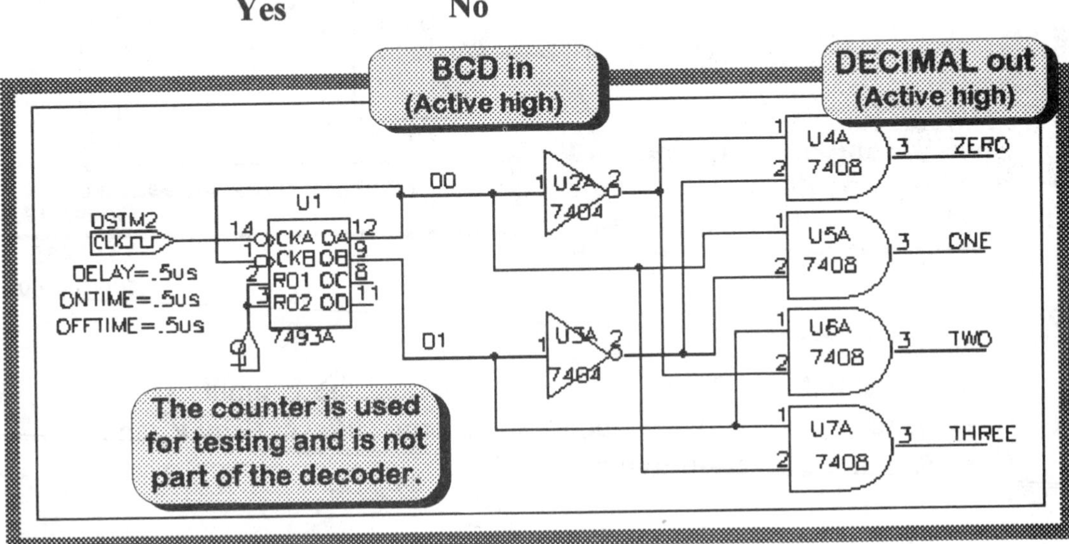

FIGURE 17.5
Two-bit decoder test circuit

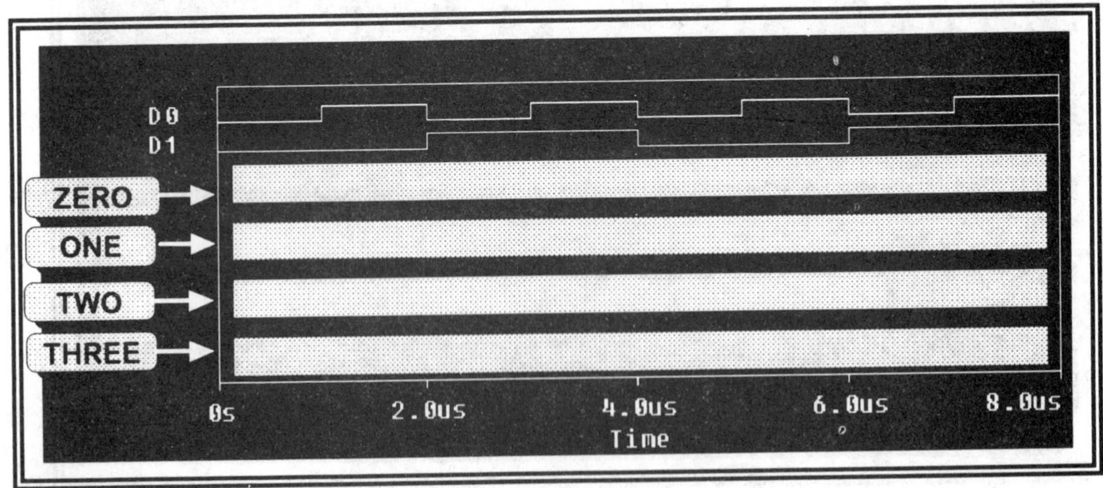

FIGURE 17.6
Two-bit decoder waveforms

8. Repeat using the 7442 IC decoder of Figure 17.7 and waveform set of Figure 17.8. Were your predicted waveforms correct?

Yes **No**

The 7442 is a BCD-to-decimal decoder that converts active-high BCD inputs from 0000 to 1001 to active-low output lines.

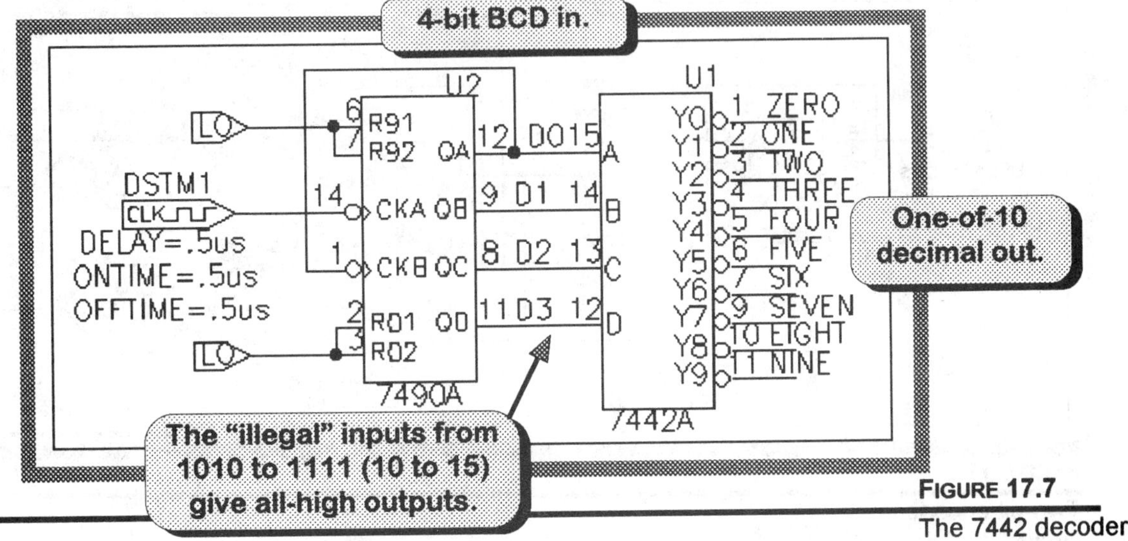

FIGURE 17.7
The 7442 decoder

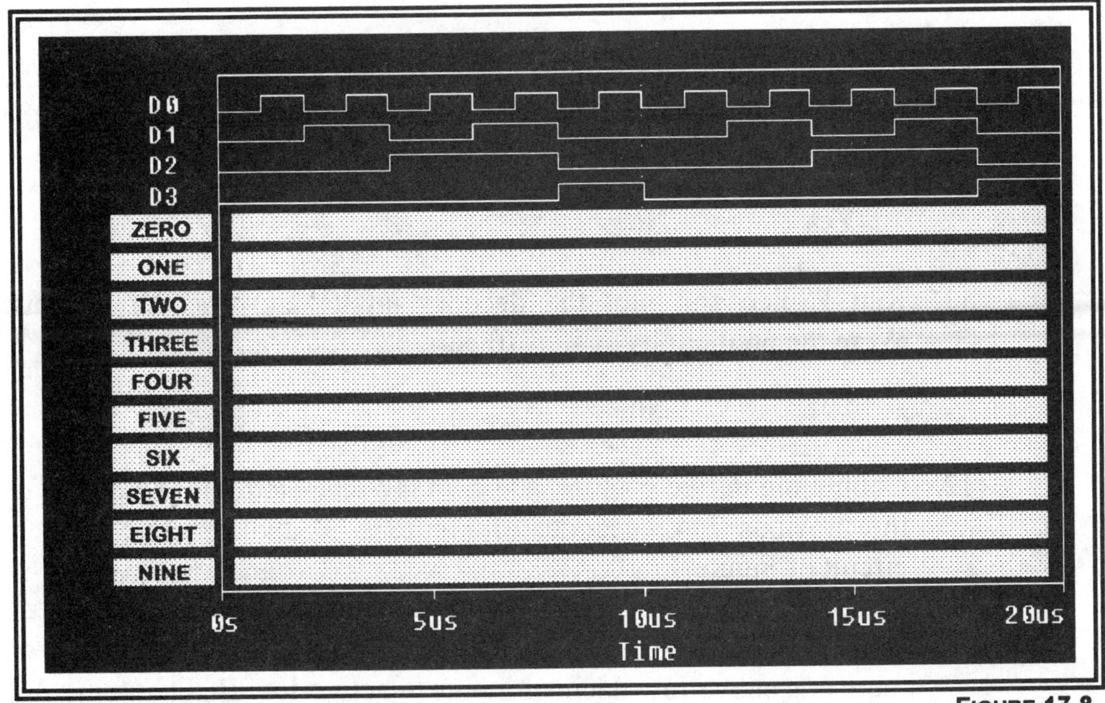

FIGURE 17.8
7442 waveforms

Encoders

9. This time, when we draw and test encoder circuits, we're going to "pull a fast one." As shown by Figure 17.9, we add our 2-bit encoder to the output of the previous 2-bit decoder of Figure 17.5.

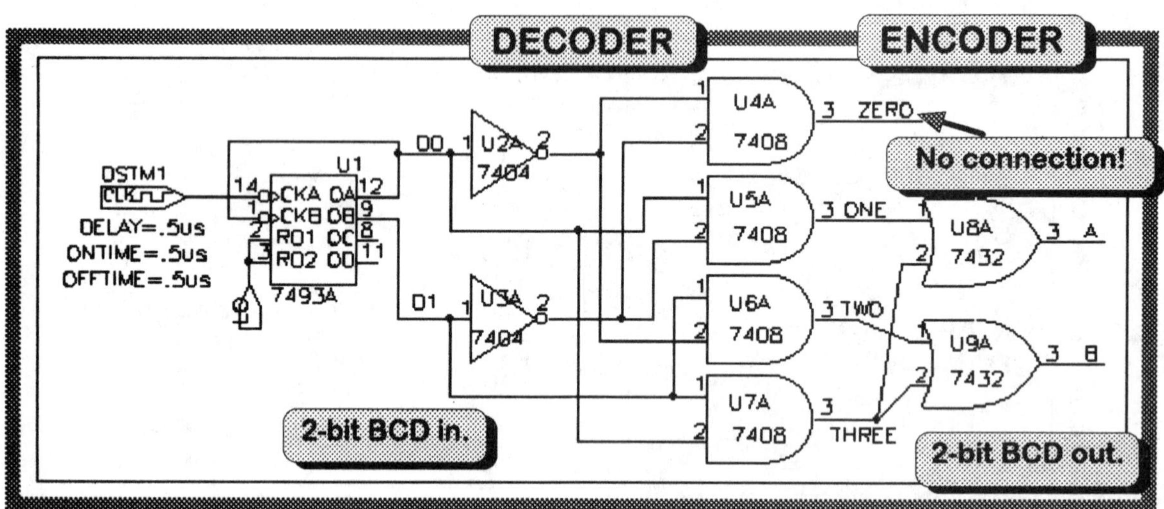

FIGURE 17.9

Two-bit encoder circuit

10. Since all inputs and outputs are active-high, would you expect the output of the encoder to be a perfect copy of the input to the decoder?

    Yes **No**

11. Run PSpice and plot the decoder inputs and encoder outputs. Were the predictions of step 10 correct?

    Yes **No**

12. Continuing in like manner, add a 74147 IC encoder to the output of the previous 7442 decoder of Figure 17.7 and generate the test circuit of Figure 17.10.

> The 74147 is a decimal-to-BCD (10-line-to-4-line) encoder that converts active-low decimal inputs to active-low BCD outputs.

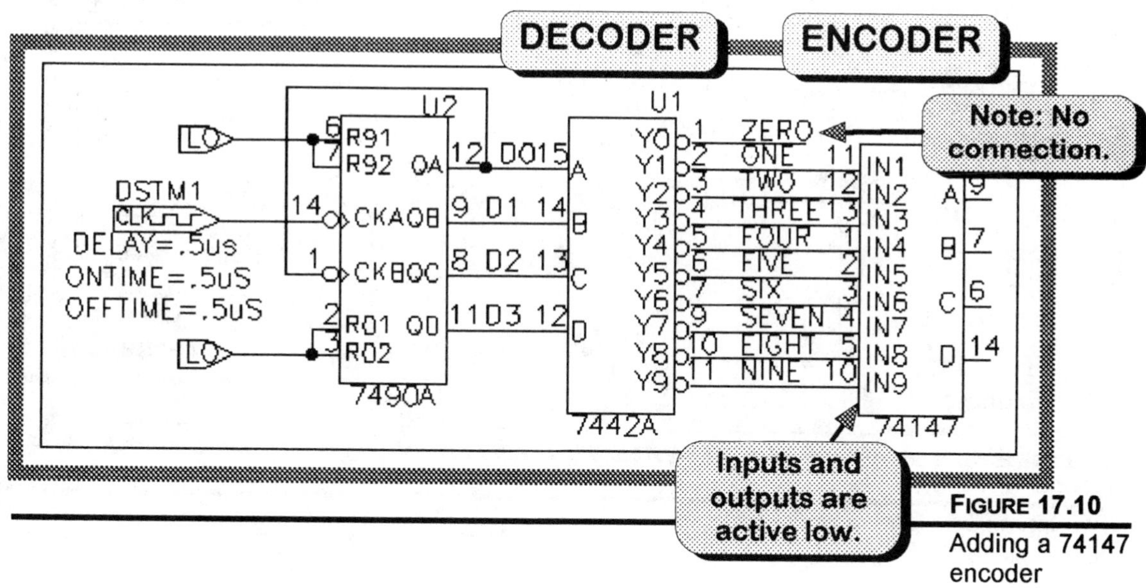

FIGURE 17.10
Adding a 74147 encoder

13. Would you expect the active-low outputs of the 74147 encoder to be a perfect inverted copy of the active-high inputs to the 7442 decoder?

    Yes **No**

14. Run PSpice and plot the decoder inputs and encoder outputs. Were the predictions of step 13 correct?

    Yes **No**

Multiplexer/demultiplexer

15. This time, we go one step further and combine both a 2-bit multiplexer (MUX) with a 2-bit demultiplexer (DEMUX) to begin. Draw the mux/demux circuit of Figure 17.11.

16. Given the circuit inputs of Figure 17.12, predict (draw) the expected outputs at the places indicated.

17. Based on the PSpice-generated waveforms, were your predictions correct?

    Yes **No**

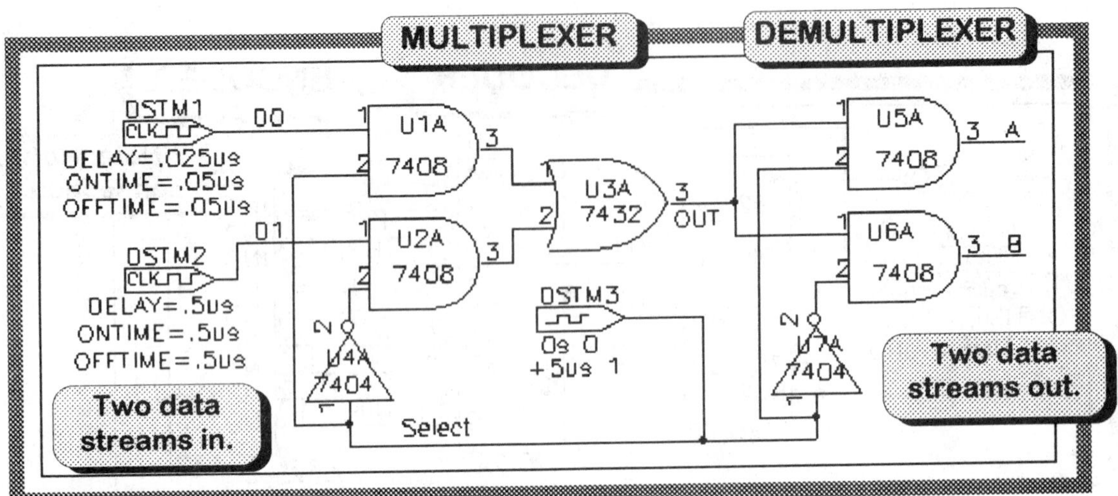

FIGURE 17.11

Two-bit multiplexer/ demultiplexer

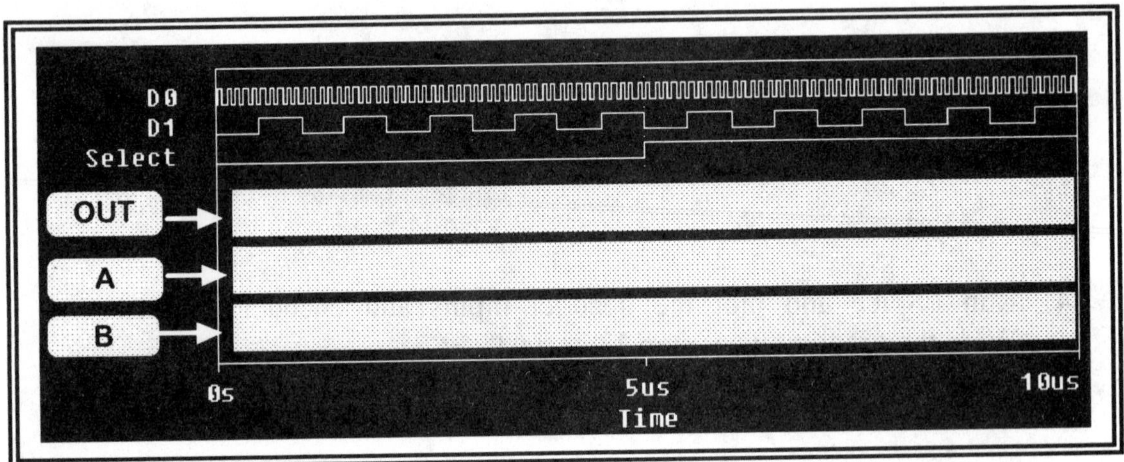

FIGURE 17.12

Two-bit mux/demux waveforms

18. Moving to the IC stage of Figure 17.13, we adopt a 74153 multiplexer to drive a 74155 decoder/demultiplexer.

> The 74153 is a dual 4-line to 1-line multiplexer. We enable the A inputs with a low on EA, and we scan the four input lines by clocking the S0 and S1 select lines with a counter. The 74155 is a dual 1-of-4 decoder/demultiplexer. We select the 1 outputs with a low on 1G. The input to line 1C is routed to the output selected by the AB lines.

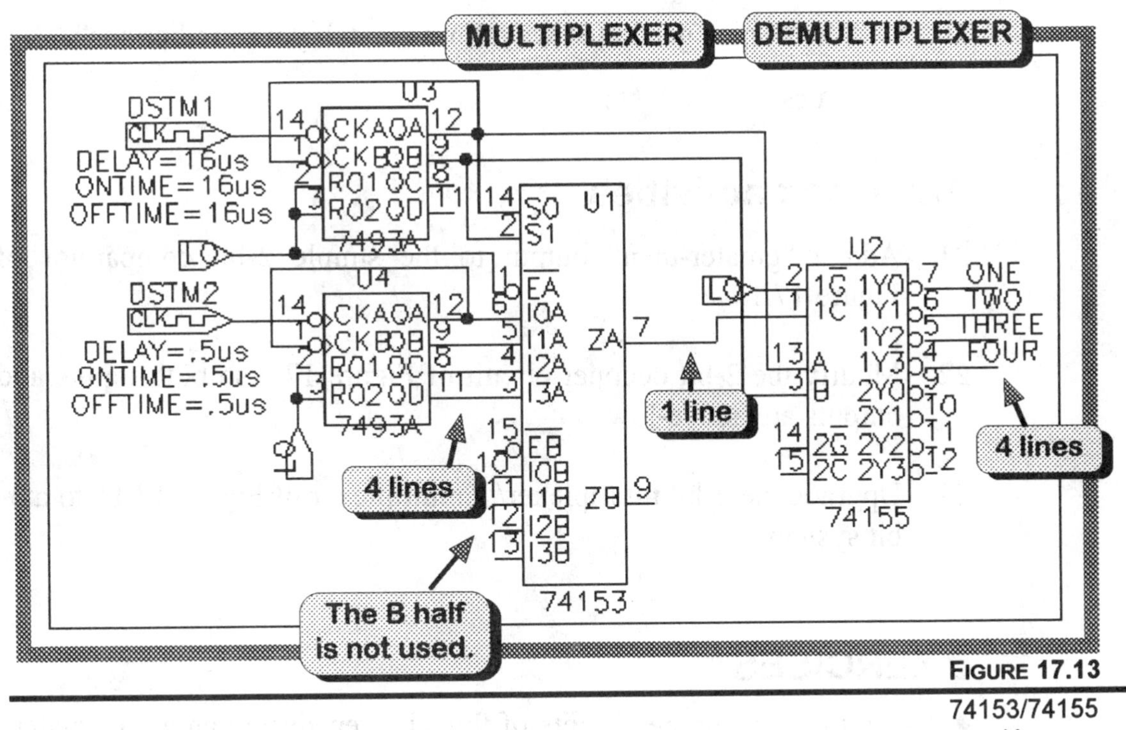

FIGURE 17.13
74153/74155 mux/demux

19. Given the multiplexer inputs shown in Figure 17.14, analyze the circuit and sketch the expected output waveforms in the spaces provided.

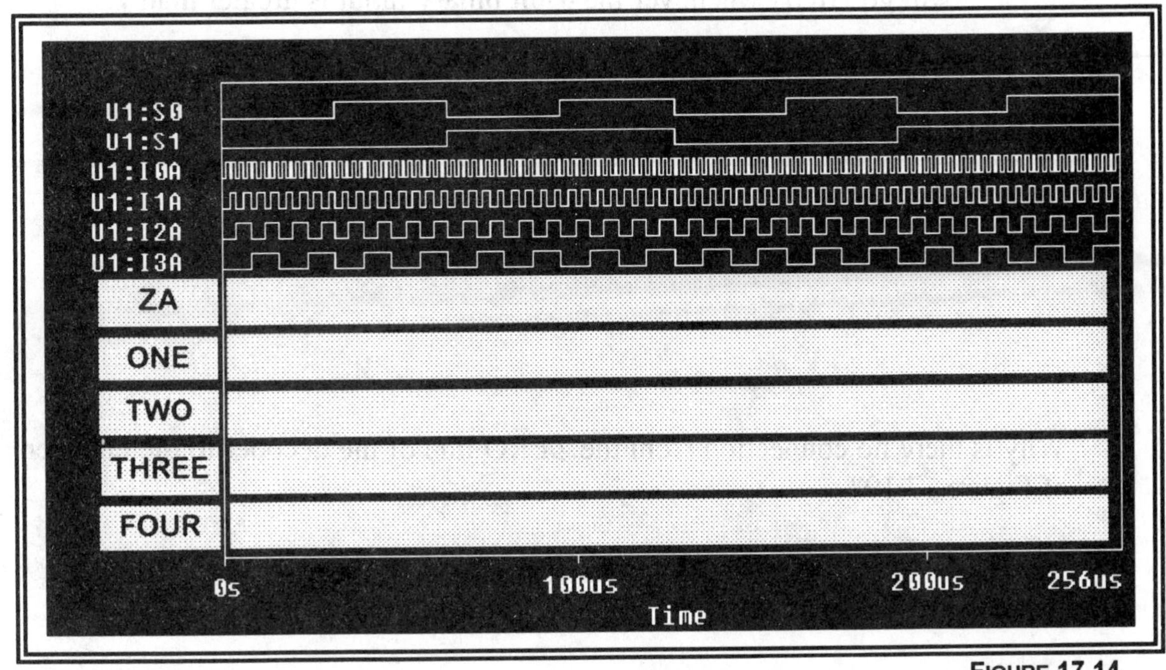

FIGURE 17.14
74153/74155 waveforms

20. Did the PSpice generated waveforms match your predictions?

    Yes **No**

Advanced activities

21. Add a "greater-than" output to the simple 2-bit comparator of Figure 17.1.

22. Modify the 2-bit decoder circuit of Figure 17.5 so both inputs and outputs are active-low.

23. Upgrade the 2-bit multiplexer/demultiplexer of Figure 17.11 to a 3-bit system.

EXERCISES

- Using the appropriate circuits of this chapter, design an alarm system for a room with eight doors and windows. Show how the system might interface to a microprocessor.

- Using the appropriate techniques of this chapter, design a circuit that will go HIGH whenever the 4-bit binary input is greater than 7.

QUESTIONS AND PROBLEMS

1. What is the difference between an *encoder* and a *decoder*?

2. Why is there no connection from the ZERO line of the decoder to the encoder of Figure 17.10?

3. Why is the 74147 called a *priority* encoder? (Hint: Examine the 74147's data sheet in Appendix D.)

4. When using the 74153 multiplexer of Figure 17.13, can two completely independent multiplex operations be going on simultaneously?

CHAPTER 18

The 555 Timer

Multivibrator Operation

OBJECTIVES

- To configure the 555 timer as an astable and monostable multivibrator.
- To configure the 555 timer as a voltage-controlled oscillator.

DISCUSSION

The 555 timer is another popular analog integrated circuit (like the op amp). This versatile chip is used for a wide variety of oscillation and timing needs. The internal schematic of the 555 is shown on Figure 18.1. Note that the 555 uses a combination of analog and digital parts.

By adding external components, the 555 can be configured as any of the following:

- Astable multivibrator
- Monostable multivibrator
- Voltage-controlled oscillator

SIMULATION PRACTICE

Astable multivibrator

1. Draw the circuit of Figure 18.2 and set the attributes as shown.

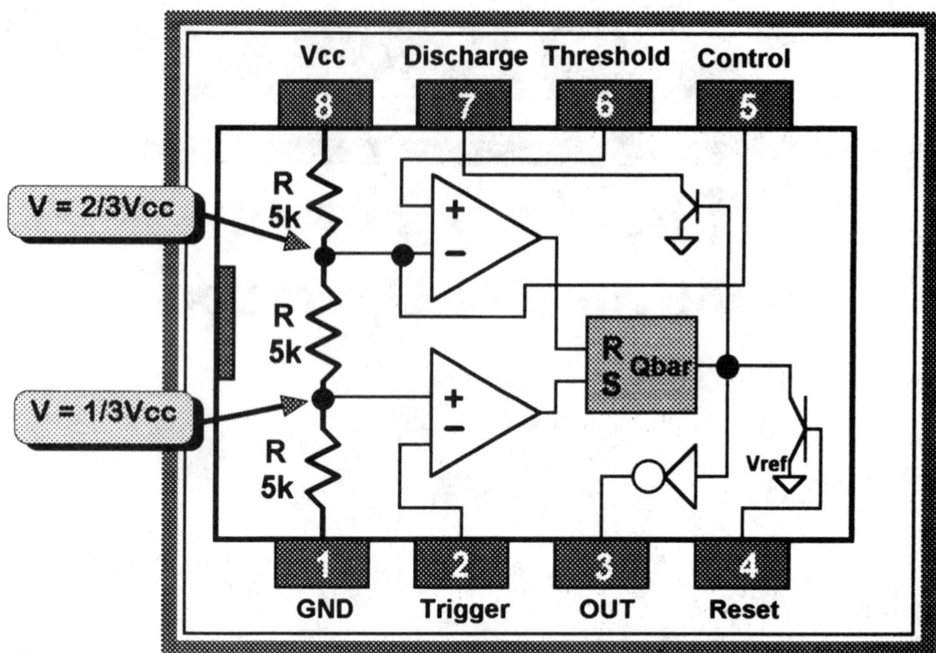

FIGURE 18.1

The 555 timer schematics

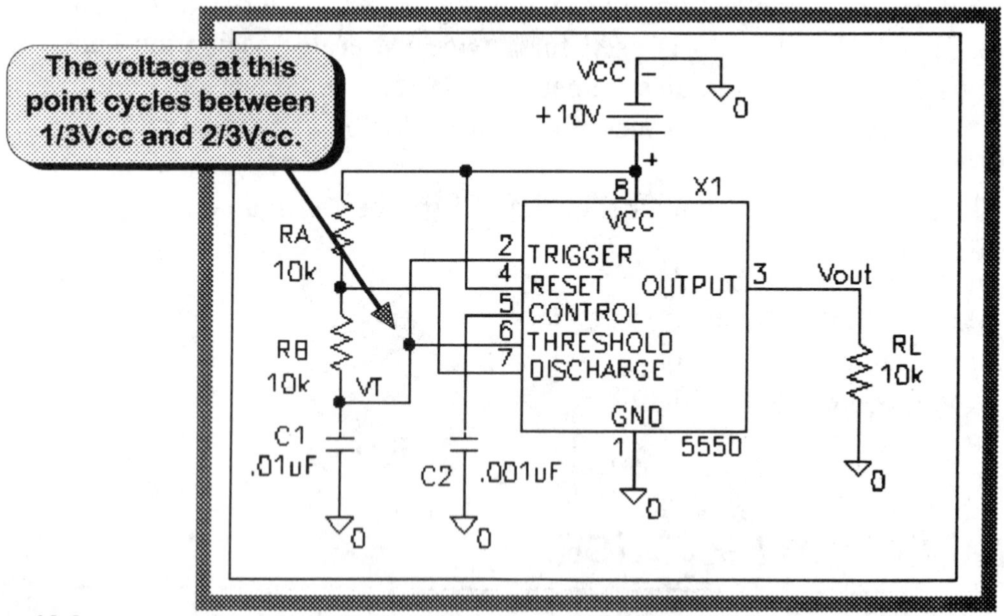

FIGURE 18.2

The astable multivibrator

2. Using the equations below, calculate the expected frequency and duty cycle. Using the results, draw the predicted trigger (VT) and output (V_{OUT}) waveforms on the graph of Figure 18.3. (Be aware that VT starts at 10V, and assumes a regular pattern after the first cycle.)

$$f_O\,(\text{frequency}) = \frac{1.44}{(R_A + 2R_B)C1} = \underline{\qquad\qquad}$$

$$DC\,(\%\ \text{duty cycle}) = \frac{(R_A + R_B)}{(R_A + 2R_B)} \times 100 = \underline{\qquad\qquad}$$

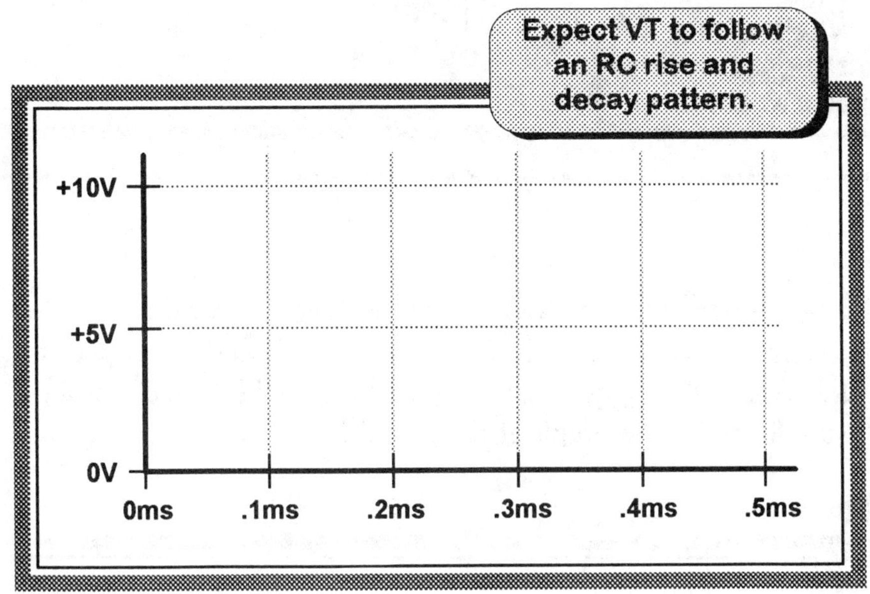

FIGURE 18.3
555 astable waveforms

3. Run a transient analysis (0 to .5ms) and generate the trigger and output waveforms. Do the actual waveforms match the expected values approximately? Make any necessary changes to Figure 18.3.

Yes **No**

Monostable multivibrator

4. By freeing up the trigger (pin 2), we configure the system for monostable operation (Figure 18.4).

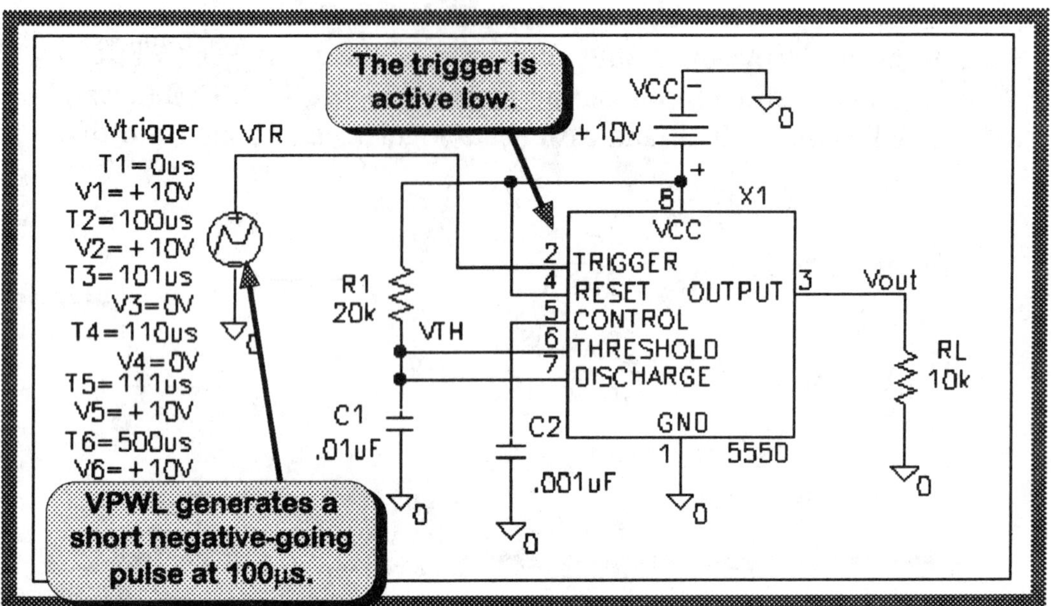

FIGURE 18.4

Monostable configuration

5. When the trigger goes low, the output goes high for a period of time determined by the equation PW = 1.1RC. Using this equation, draw the expected trigger (VTR), threshold (VTH), and output (Vout) waveforms on the graph of Figure 18.5.

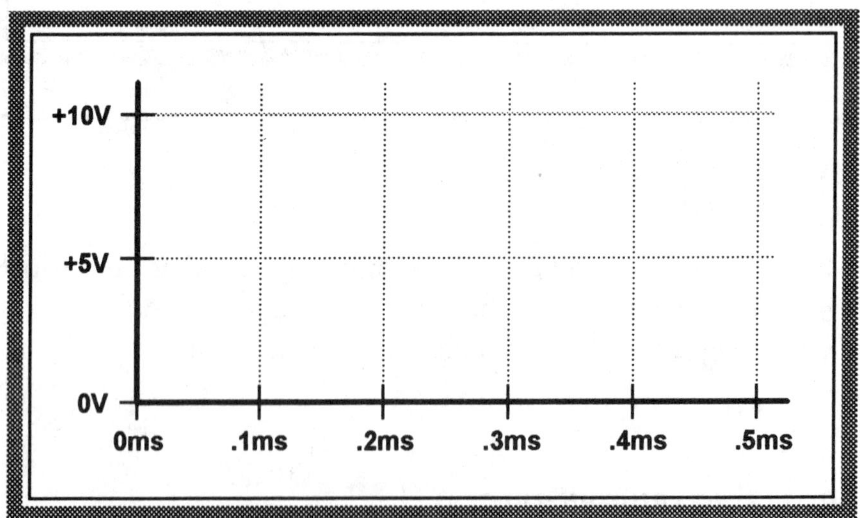

FIGURE 18.5

Monostable waveforms

6. Run a transient analysis (0 to .5ms) and display the *trigger*, *threshold*, and *output* waveforms on separate plots. Do the actual waveforms match the expected values approximately? (Make any necessary corrections.)

   Yes **No**

Advanced activities

7. By varying the voltage into pin 5, we create a voltage-controlled oscillator (VCO). Set up the 555 for VCO operation as shown in Figure 18.6.

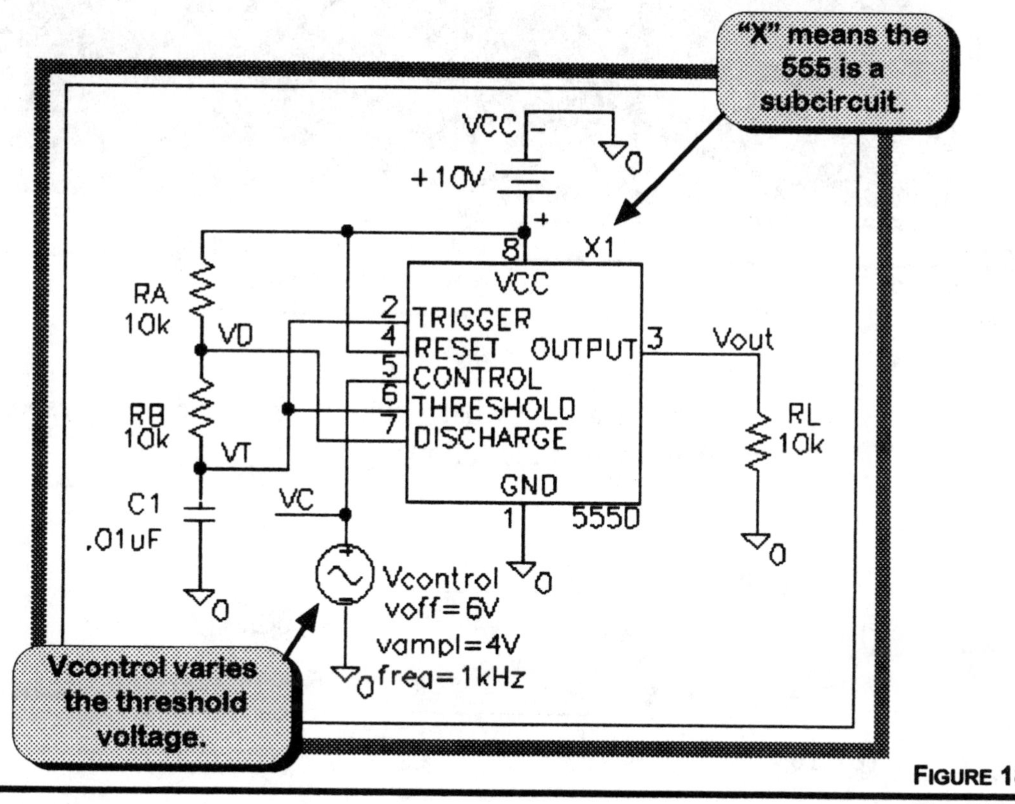

FIGURE 18.6
The 555 as a VCO

8. Run PSpice and generate the control (VC), trigger (VT), and output (Vout) waveforms of Figure 18.7.

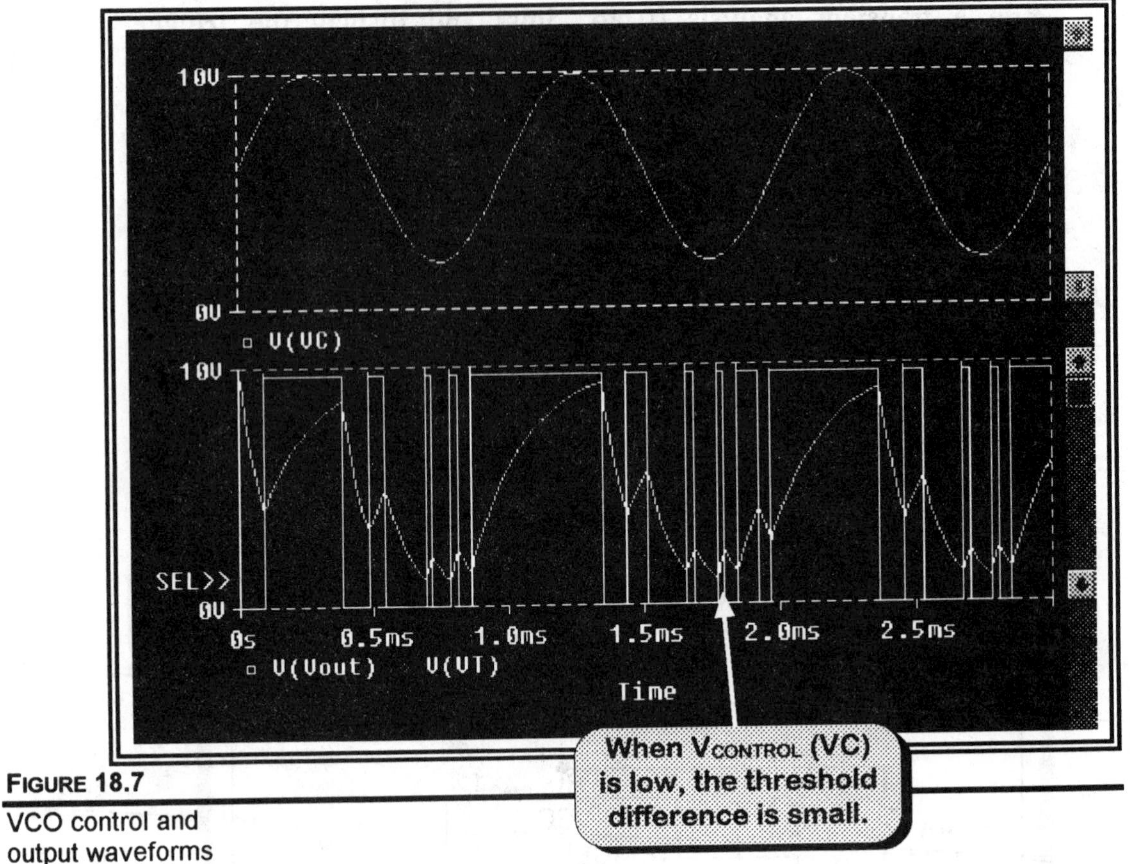

FIGURE 18.7
VCO control and output waveforms

9. Based on the results, what output frequency corresponds *approximately* to each of the following control voltages?

   (a) 2V

   f(out) at 2V ≅ ____________

   (b) 6V

   f(out) at 6V ≅ ____________

   (c) 10V

   f(out) at 10V ≅ ____________

EXERCISE

- Using the 555, design a circuit to delay exactly 1 second from the time of a trigger.

QUESTIONS AND PROBLEMS

1. What does *monostable* mean?

2. Refer to Figure 18.1. What trigger and threshold voltages (as a percentage of VCC) will set and reset the flip-flop?

3. Is the trigger signal of the monostable multivibrator of Figure 18.4 active high or active low?

4. Which of the following 555 signals reacts to a rising high-level voltage?
   (a) Threshold
   (b) Trigger

5. Refer to Figure 18.1. How does the control voltage (pin 5) affect the frequency of operation?

CHAPTER 19

Analog/Digital Conversions

Resolution

OBJECTIVES

- To convert analog signals to digital.
- To convert digital signals to analog.

DISCUSSION

Computers "live" in a digital environment, but people live primarily in an analog world. In order for people and computers to converse, we must have the ability to convert between analog and digital. That is the purpose of this chapter.

The simplest of the two operations is digital-to-analog conversion (DAC). (It's easier for a computer to talk to us than the other way around.) The two DAC methods covered by this chapter are:

- Binary weighted
- Use of integrated circuit

The more difficult analog-to-digital conversion (ADC) is accomplished in a number of ways. The two covered by this chapter are:

- Counter-ramp (staircase)
- Use of integrated circuit

As we will see, the use of an integrated circuit greatly improves convenience and precision.

SIMULATION PRACTICE

DAC

Binary weighted

When using this DAC technique, currents are weighted in the 8, 4, 2, 1 digital pattern, summed together, and converted to voltage.

1. Draw the DAC summing circuit of Figure 19.1.

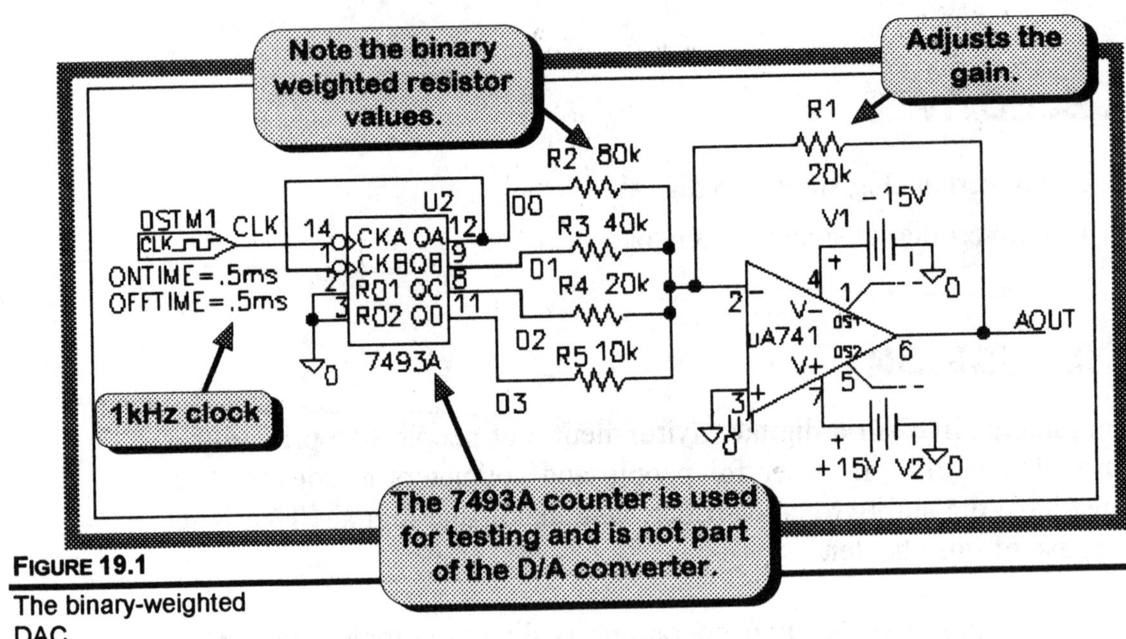

FIGURE 19.1
The binary-weighted DAC

2. Assuming that the Q outputs of U2 (the 7493A) are .1V (logic 0) and 3.5V (logic 1), draw the expected output waveform on the diagram of Figure 19.2. (Remember that the op amp inverts the signal.)

3. Run PSpice and draw the resulting waveform on the same graph (Figure 19.2). Are they approximately the same?

Yes **No**

Reminder: If no signal appears on this or any digital circuit, **CLICKL** on the *Sets up the simulation analysis for active* toolbar button, **Digital Setup**, and set the *Flip-flop initialization* to **All 0**.

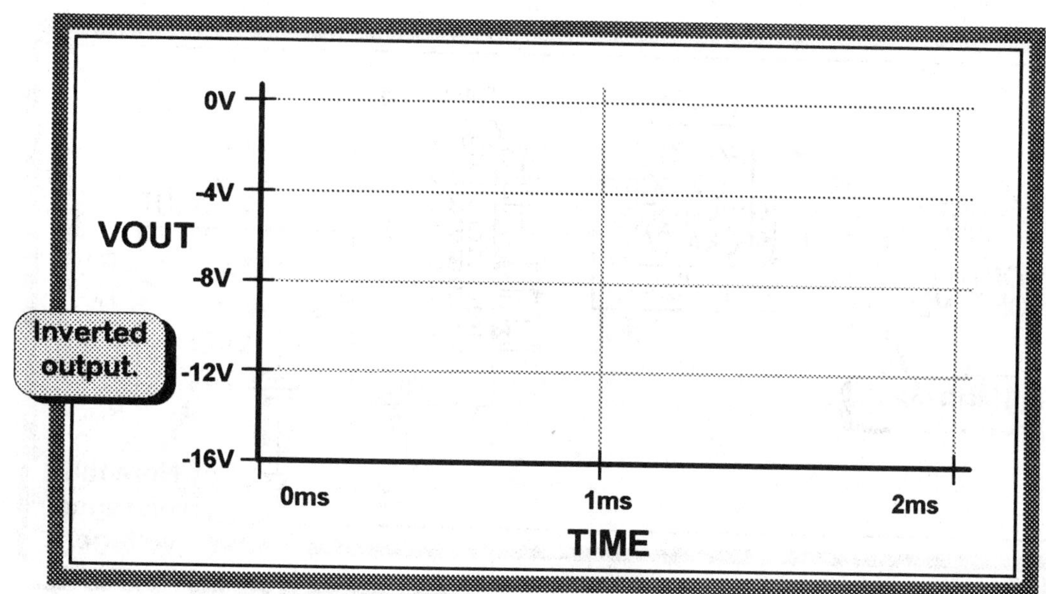

FIGURE 19.2
DAC waveforms

4. What value of R1 would calibrate the DAC so that digital 0 to F (hexadecimal) would correspond to analog 0 to 15V? (Hint: Binary 1,0,0,0 would correspond to –8V.)

   R1 (for calibration) = _______________

5. Substitute the value of R1 determined in step 4 and run a simulation. Is the DAC now calibrated properly? (Be aware that we may hit the 14.7V rail before the final –15V is reached.)

   Yes **No**

Integrated circuit

6. Draw the integrated circuit DAC system of Figure 19.3. (Part **DAC8break** is found in the *breakout* library). Have we set the clock speed of the more precise integrated circuit DAC greater than that of the op amp circuit of Figure 19.1?

   Yes **No**

7. Based on the equation below, predict the output of the DAC and draw your forecast on the graph of Figure 19.4.

$$V_{AOUT} = V(\text{ref},\text{gnd}) \times \frac{(\text{binary value of inputs})}{2^8}$$

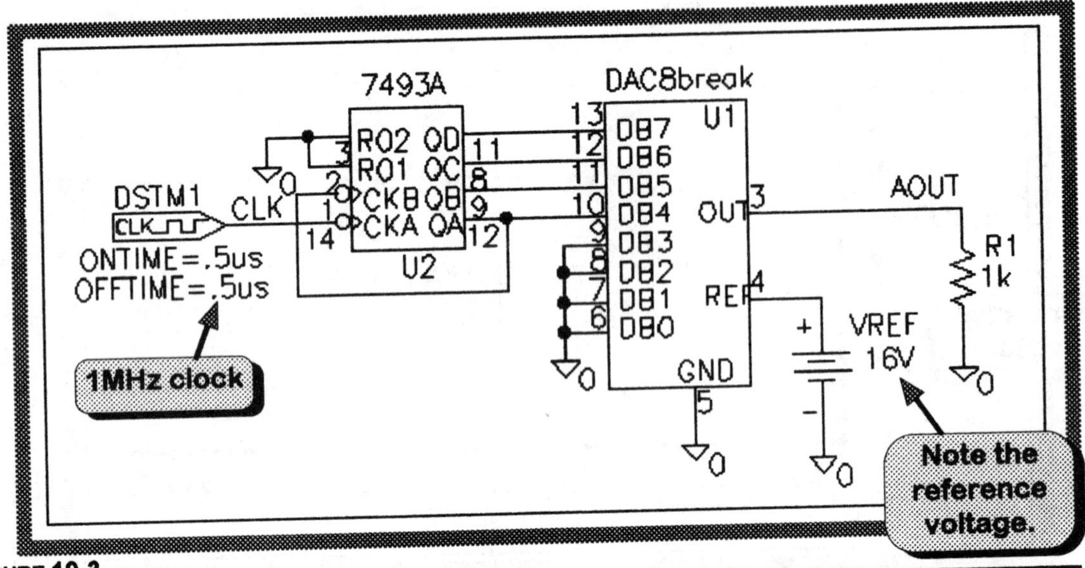

FIGURE 19.3
The integrated circuit DAC

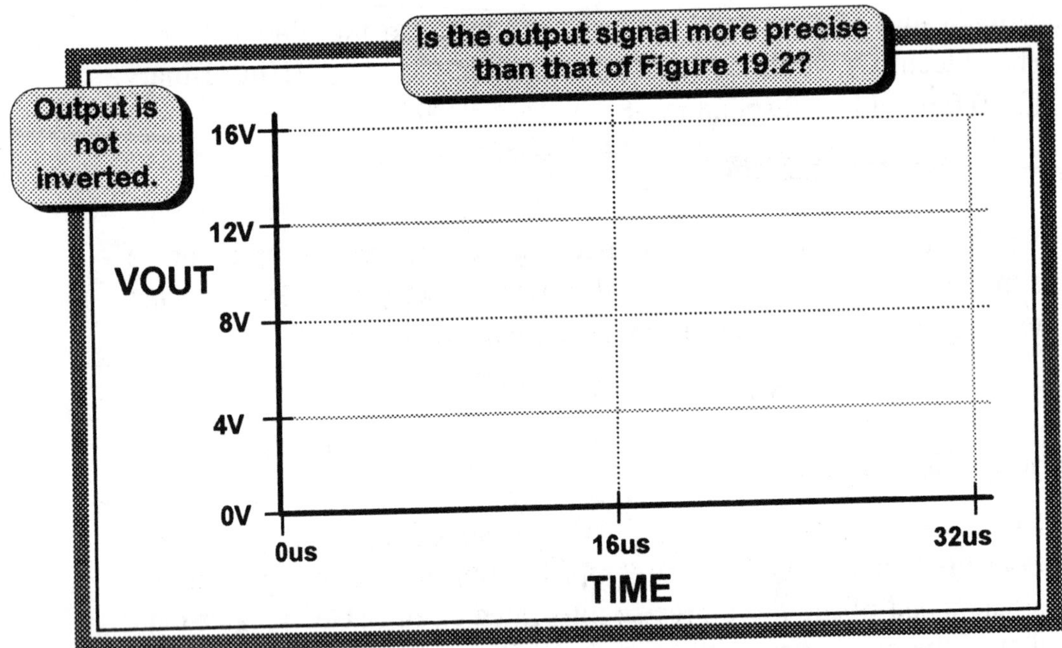

FIGURE 19.4
Integrated DAC waveforms

8. Run PSpice and draw the actual DAC waveform. Did it match your predictions?

   Yes **No**

ADC

The counter-ramp (staircase) method

The staircase method of ADC compares the input analog voltage to a rising staircase voltage. When the two match, the digital equivalent of the staircase voltage is output and latched.

9. Draw the ADC circuit of Figure 19.5.

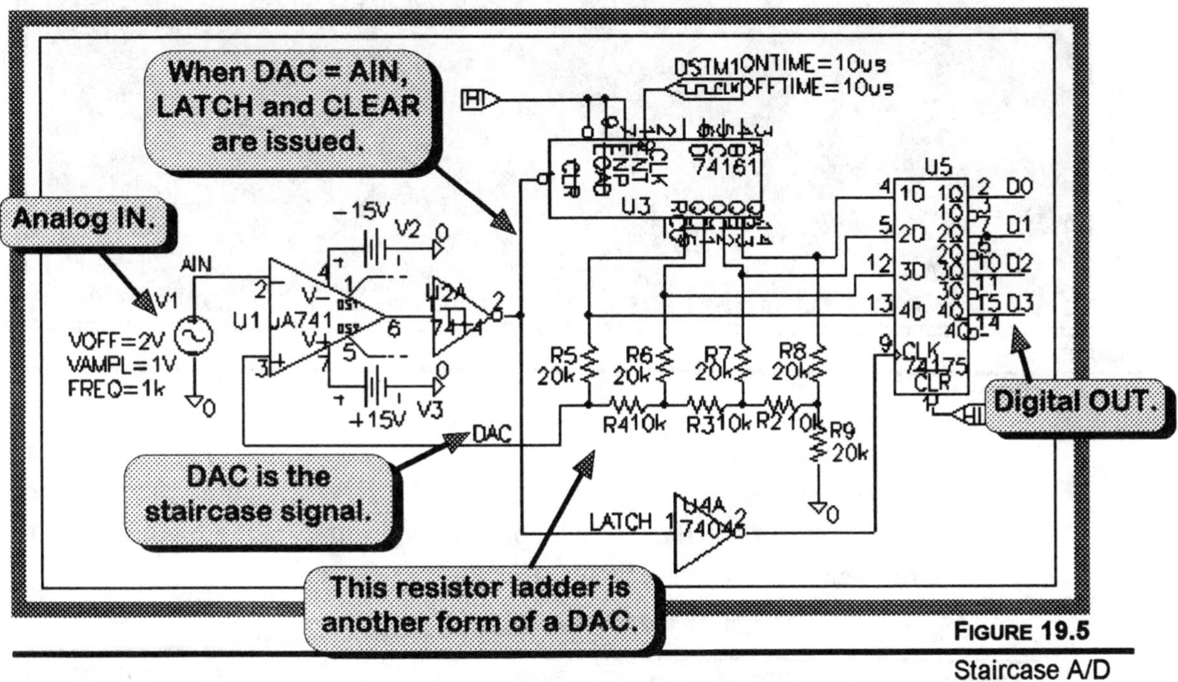

FIGURE 19.5
Staircase A/D

10. Run PSpice and generate the waveforms of Figure 19.6.
    (a) Is a latch signal issued whenever the staircase rises to the analog input level?

        Yes **No**

    (b) Is the conversion time proportional to the size of the analog input signal?

        Yes **No**

    (c) Do the digital output signals match the analog input?

        Yes **No**

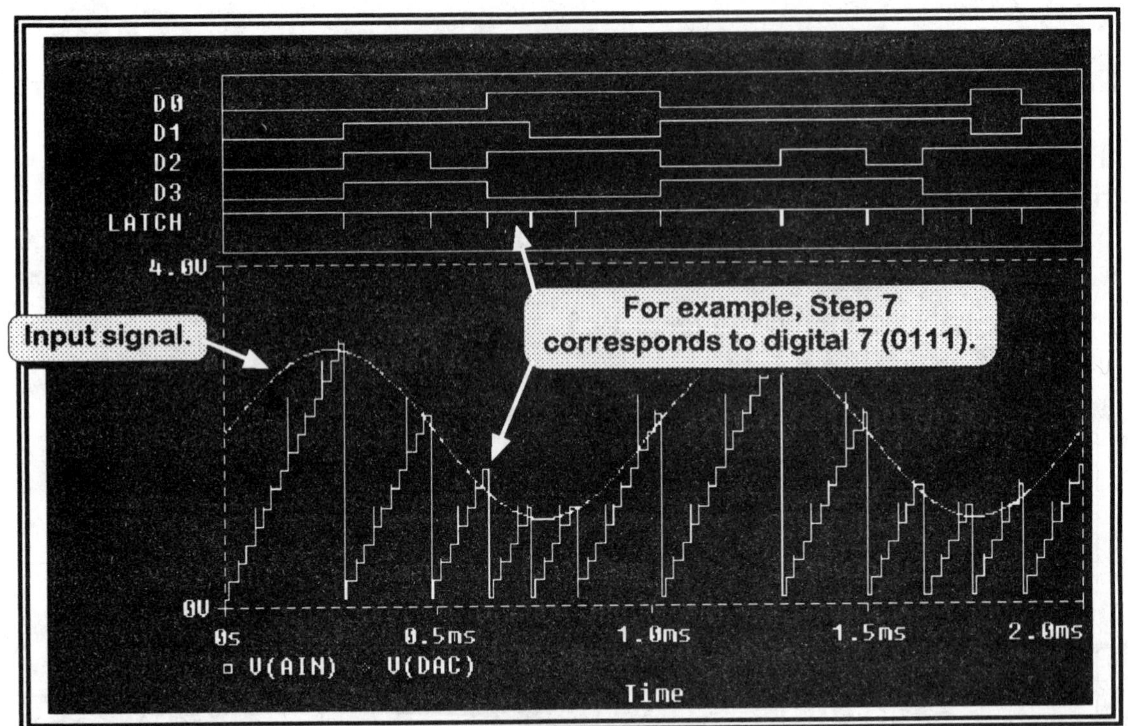

FIGURE 19.6

A/D waveforms for staircase method

11. To test the resolution of the ADC, add a DAC (such as the resistor ladder driven by the 74161 counter of Figure 19.5) to the output and generate the waveforms of Figure 19.7.

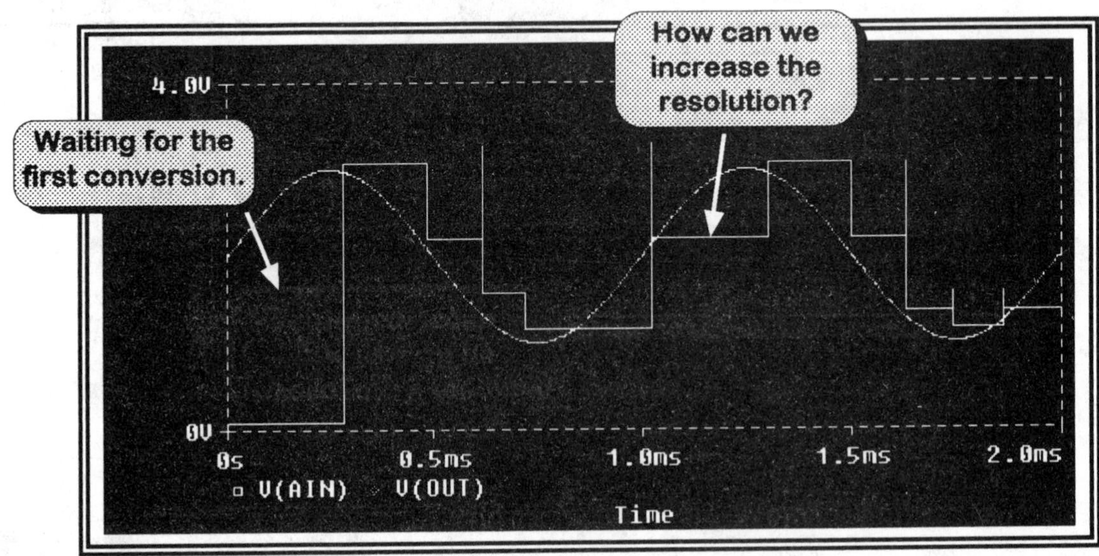

FIGURE 19.7

A/D resolution

12. Is the resolution poor?

    Yes **No**

13. To improve the resolution, double the clock speed of DSTM1. Was the resolution improved?

    Yes **No**

Integrated circuit

14. Draw the integrated circuit ADC system of Figure 19.8. (Part **ADC8break** is from the *breakout* library). As with the DAC8 precision DAC device, have we set all signals at generally higher frequencies than with the ramp method?

    Yes **No**

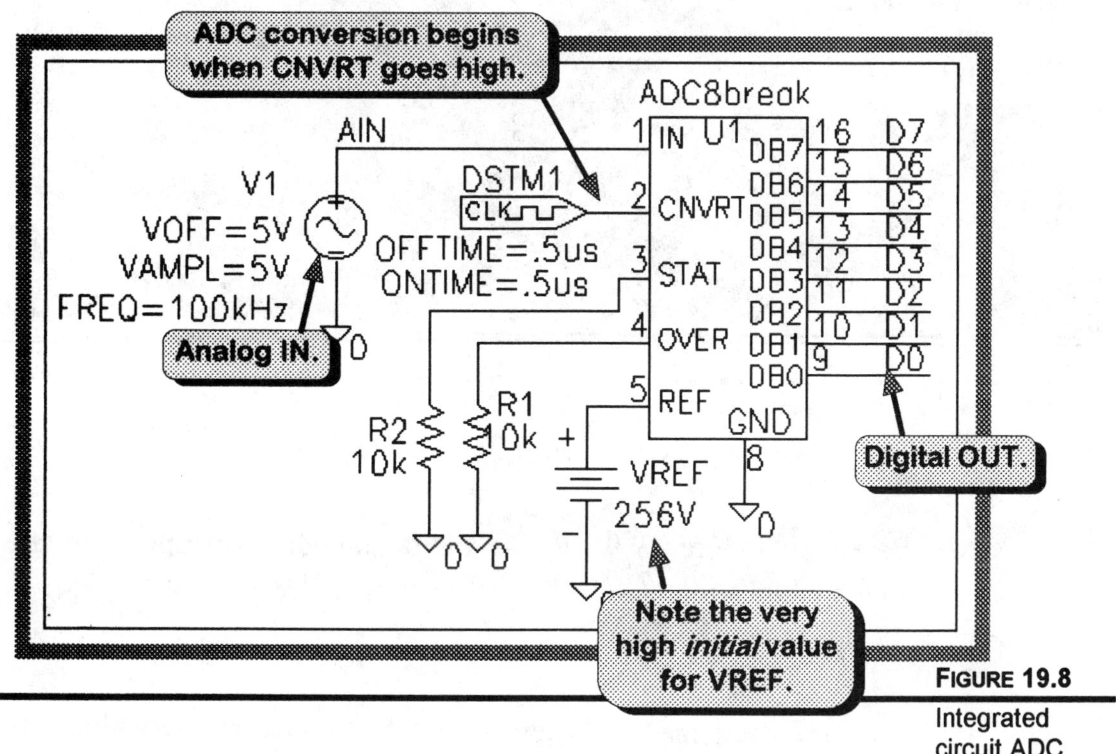

FIGURE 19.8 Integrated circuit ADC

15. Based on the equation below, predict the digital output of the ADC when *AIN* is 10V.

$$\text{Digital output} = \underline{\text{binary}} \text{ value of } \frac{V(\text{in,gnd})}{V(\text{ref,gnd})} \times 2^8 = \underline{\hspace{3cm}}$$

D7 to D0

16. Run PSpice and generate the curves of Figure 19.9.

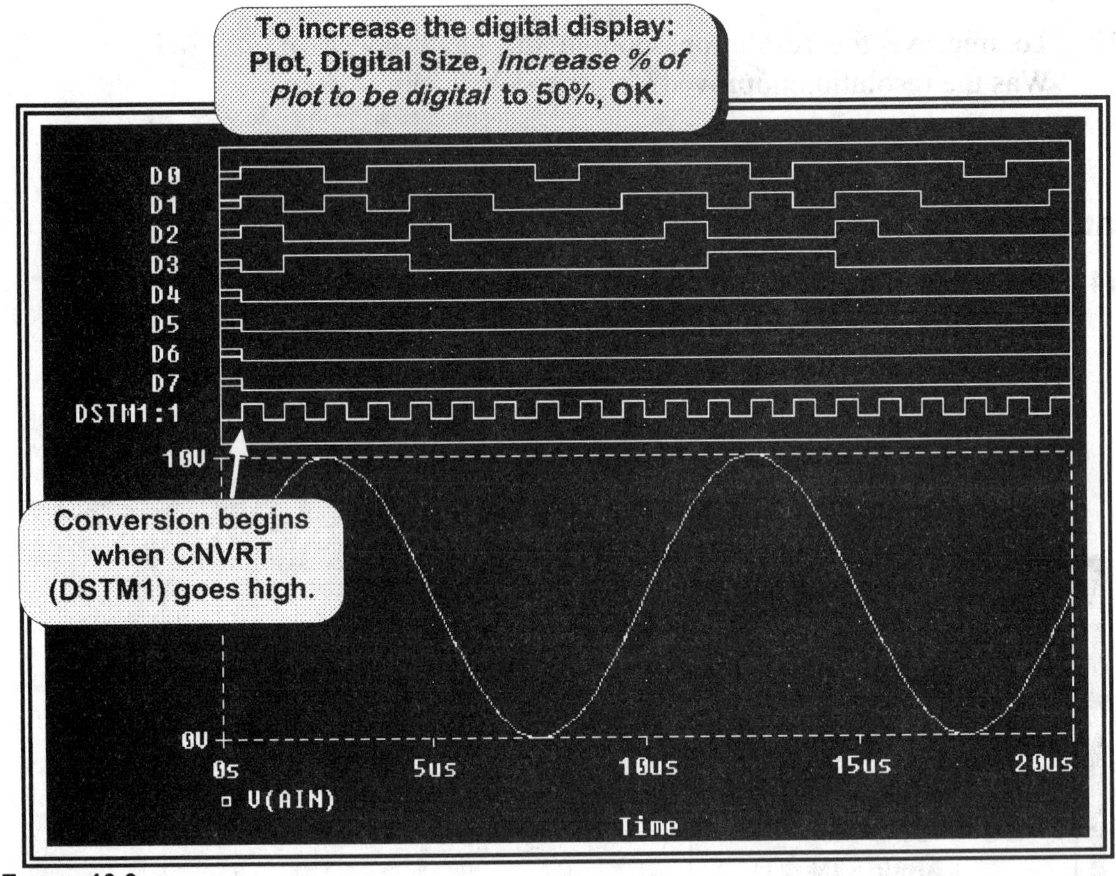

FIGURE 19.9

Integrated circuit A/D waveforms

17. Based on Figure 19.9, do the digital outputs correspond to the analog input values? (Does 10V in equal 00001010 out?)

Yes **No**

18. As before, test the resolution of our A/D converter by adding the DAC8break D/A chip to the output (Figure 19.10).

19. Run PSpice and generate the curves of Figure 19.11. Does the output equal the input, but is the resolution still quite low?

Yes **No**

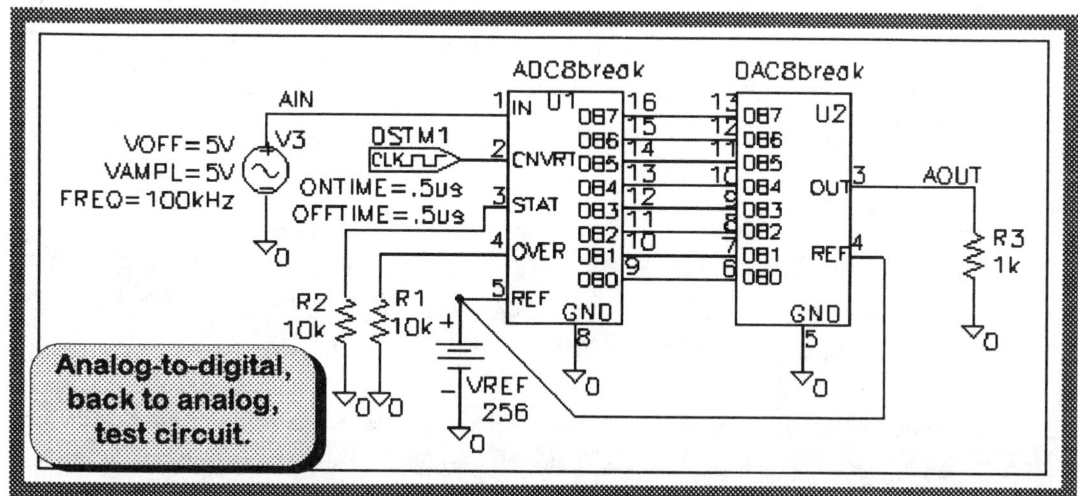

FIGURE 19.10
Adding a DAC

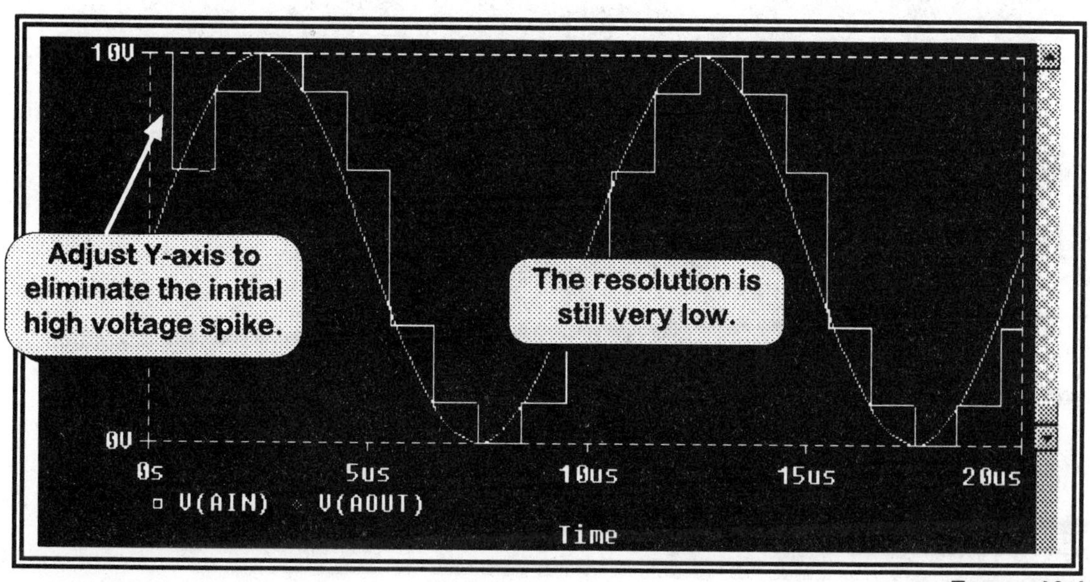

FIGURE 19.11
Integrated circuit test waveforms

As before, we could increase resolution by increasing clock speed—and we will. However, regardless of the clock speed, we would still be limited to 16 voltage levels. Therefore, to maximize resolution, we must use all 256 steps provided by this 8-bit device.

20. Based on the equation of Step 15, what reference voltage value will make the maximum input analog voltage (10V) equivalent to the maximum binary output (11111111)? (Hint: 11111111_2 = 255_{10}, giving an answer slightly above 10V.)

    V_{REF} = ___________

21. Change V_{REF} to the value determined in Step 20, increase the clock (DSTM1) frequency to 10MEGhz (ON and OFF times = .05μS) and generate the new waveform set of Figure 19.12.

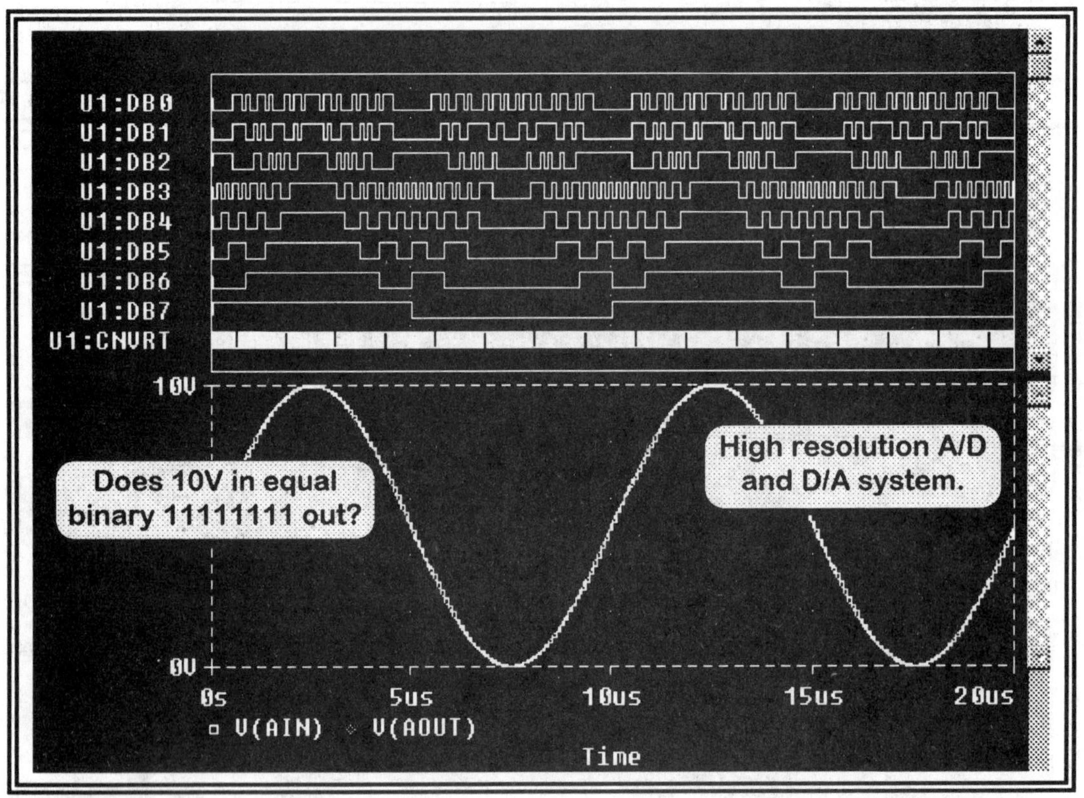

FIGURE 19.12
A/D and D/A precision waveforms

22. Zoom in on the AIN/AOUT waveforms of Figure 19.12.

    (a) Is *AOUT* a staircase of very small steps?

    Yes **No**

    (b) Do the steps range in size from approximately .1V to .3V?

    Yes **No**

Advanced activities

23. Redesign the DAC circuit of Figure 19.1 using a resistor ladder input to a VCVS buffer op amp. (Hint: A DAC resistor ladder is shown in Figure 19.5.)

24. By double-buffering the A/D ramp circuit of Figure 19.5 (using two latches), redesign the circuit so the conversion time is constant. (Hint: Let the conversion time be the time required for the counter to cycle fully from 0 to 15.)

EXERCISE

- Using basic circuit elements (such as those of Figure 19.5), design an A/D converter based on the successive-approximation technique.

QUESTIONS AND PROBLEMS

1. What two processes does the operational amplifier of Figure 19.1 perform?

2. Based on the equation of Step 7 for the integrated circuit D/A converter of Figure 19.3, what would the output voltage be if the binary input were 00001111?

3. Regarding the counter-ramp A/D converter of Figure 19.5, why do the conversion times vary?

4. Regarding the integrated circuit A/D converter of Figure 19.8, why is the reference voltage initially set at 256V?

5. What is the minimum (best) resolution of the 8-bit A/D of Figure 19.8?

6. A data acquisition system would sample analog voltages, convert them to digital, and store them for future reference. What circuit component would accomplish the storage feature? (Hint: See Chapter 20.)

CHAPTER 20

Random-Access Memory
Data Acquisition

OBJECTIVES

- To write data to and read data from a random-access memory (RAM) chip.
- To design a data acquisition system.

DISCUSSION

Random-access memory (RAM) is a *read/write volatile* storage device for digital data. Positive-feedback latching action maintains the digital states in internal latches—but all data is lost when power is removed. By following the proper timing requirements, data is *written* to and *read* from *addressable* locations.

A typical RAM chip is shown in Figure 20.1. It is an 8k × 8 device, meaning that it offers 8k (8,192) addressable memory locations, each storing 8 bits of data. Figure 20.2 shows the device timing specifications for the read and write operations.

DATA ACQUISITION

A *data acquisition system* samples and stores analog values for later playback. The most widely used storage medium is *random-access memory* (RAM).

In a typical case, a data acquisition system might be used to store the winter temperature readings at a remote arctic site. When spring comes, the data can be retrieved for analysis.

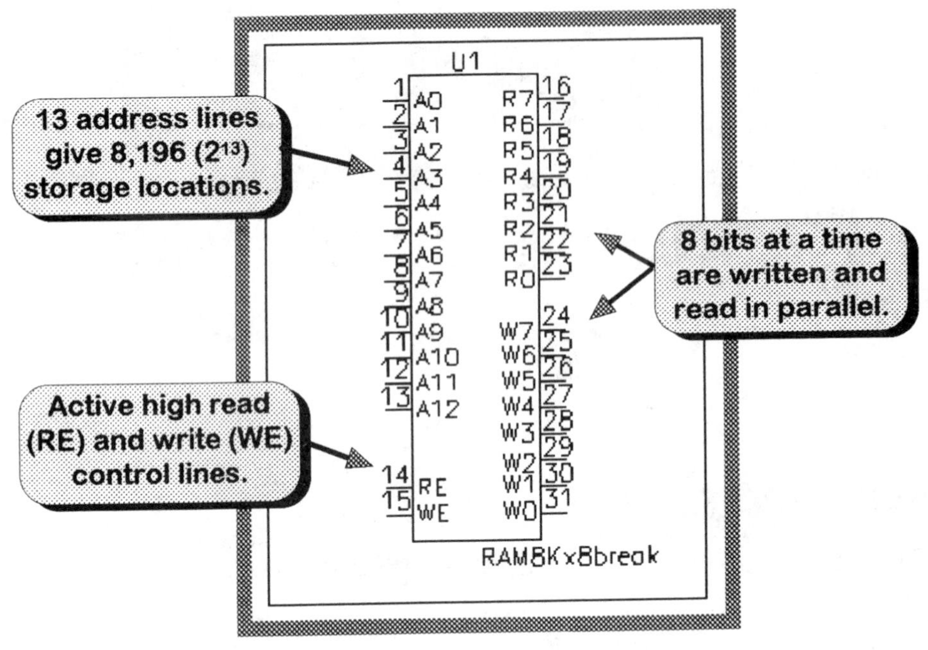

FIGURE 20.1

8k × 8 RAM device

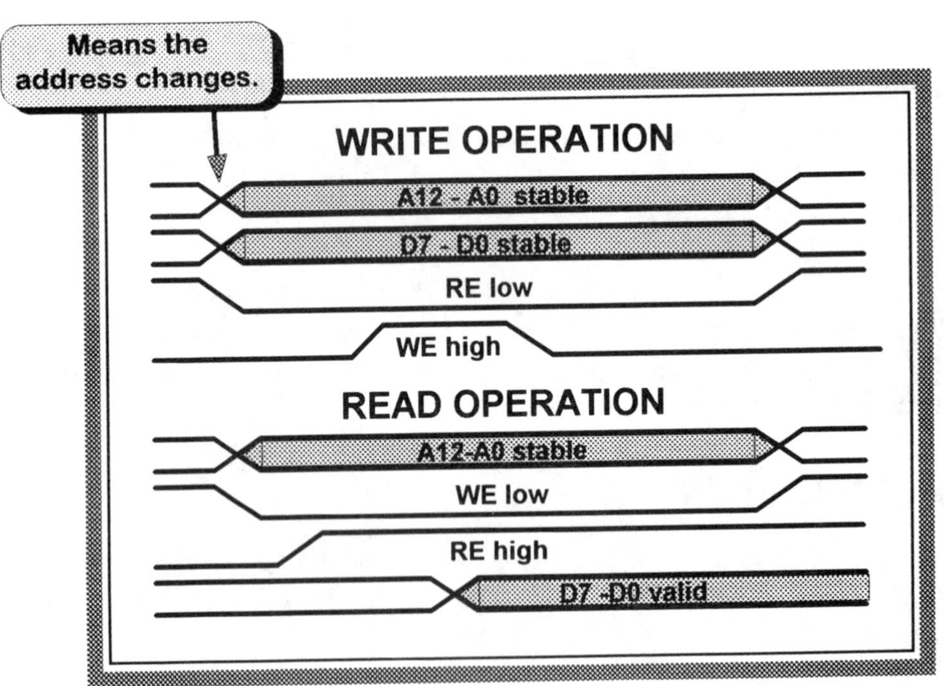

FIGURE 20.2

RAM read and write waveforms

Data acquisition and retrieval is performed as follows, where *n* is the number of RAM storage locations used:

Repeat *n* times to store:
1. Sample an analog signal.
2. Convert to digital.
3. Write to the next RAM location.

Repeat *n* times to retrieve:
1. Read from the next RAM location.
2. Convert to analog.
3. Display analog signal.

SIMULATION PRACTICE

RAM

Our first test of RAM is to write data byte C3H to location 2 and read it back.

1. Draw the binary-format RAM test circuit of Figure 20.3(a).

2. Run PSpice and generate the binary-format waveform set of Figure 20.4(a).

3. As first performed in Chapter 16, modify your circuit for hexadecimal-format display, as shown by Figure 20.3(b), and generate the hexadecimal-format waveform set of Figure 20.4(b).

4. Based on the results of Figure 20.4 (either format):

   (a) Do the read/write waveforms follow the standards of Figure 20.2?

   Yes **No**

   (b) Was data byte C3H (11000011B) written to location 002H (000000000010B), stored, and read back correctly?

   Yes **No**

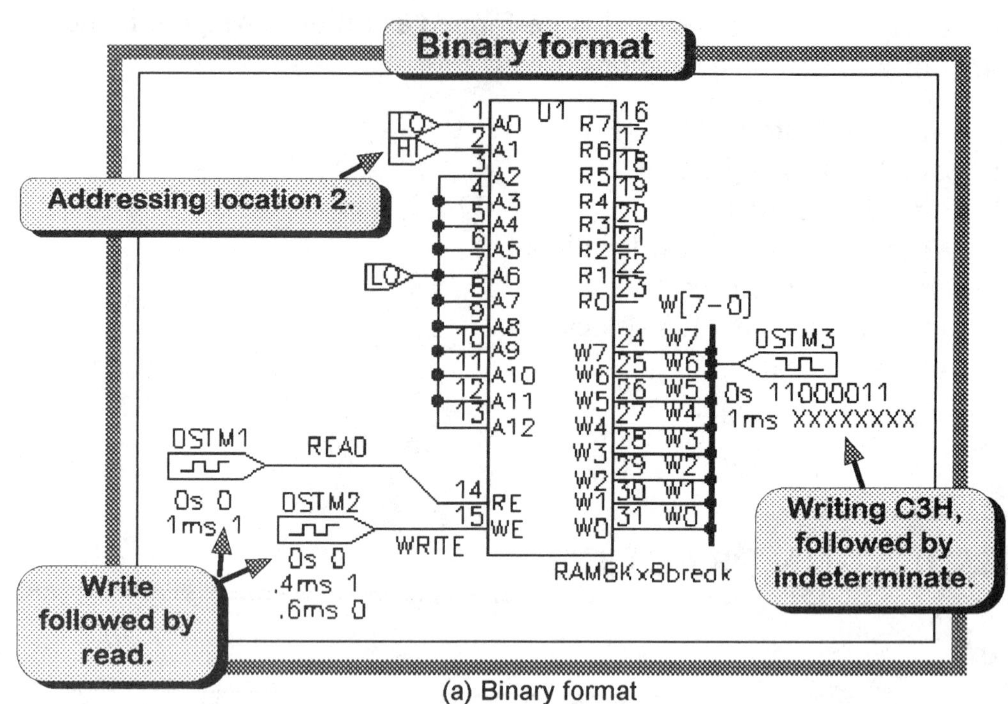

(a) Binary format

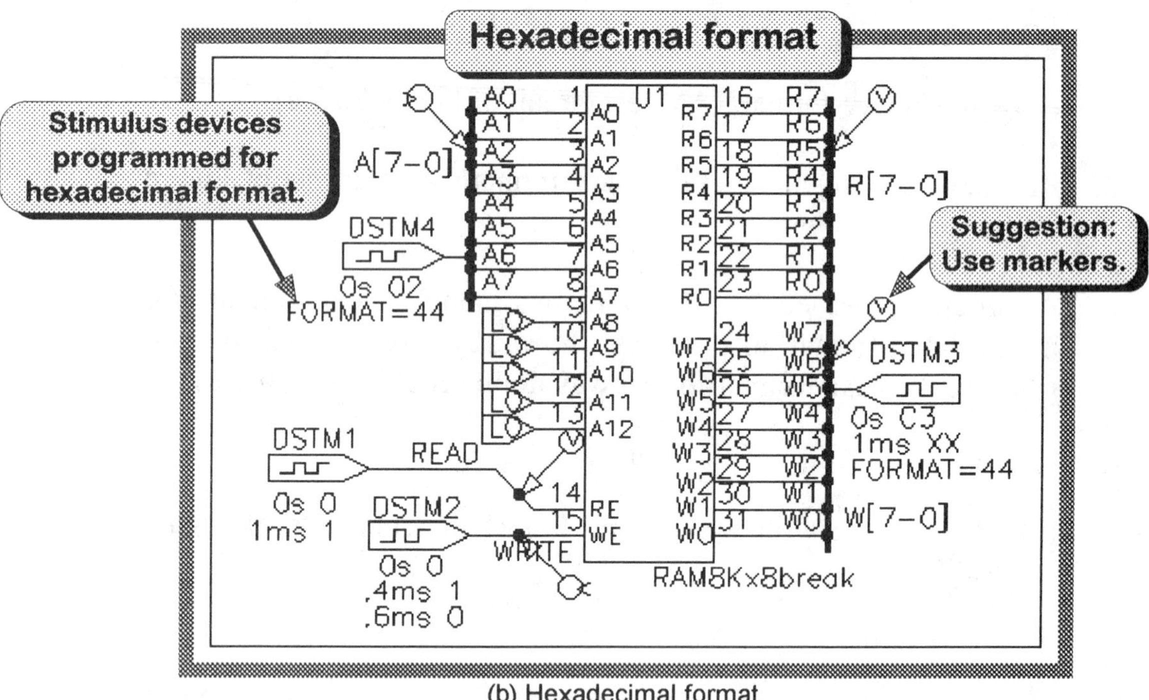

(b) Hexadecimal format

FIGURE 20.3

RAM test circuits
(a) Binary format
(b) Hexadecimal format

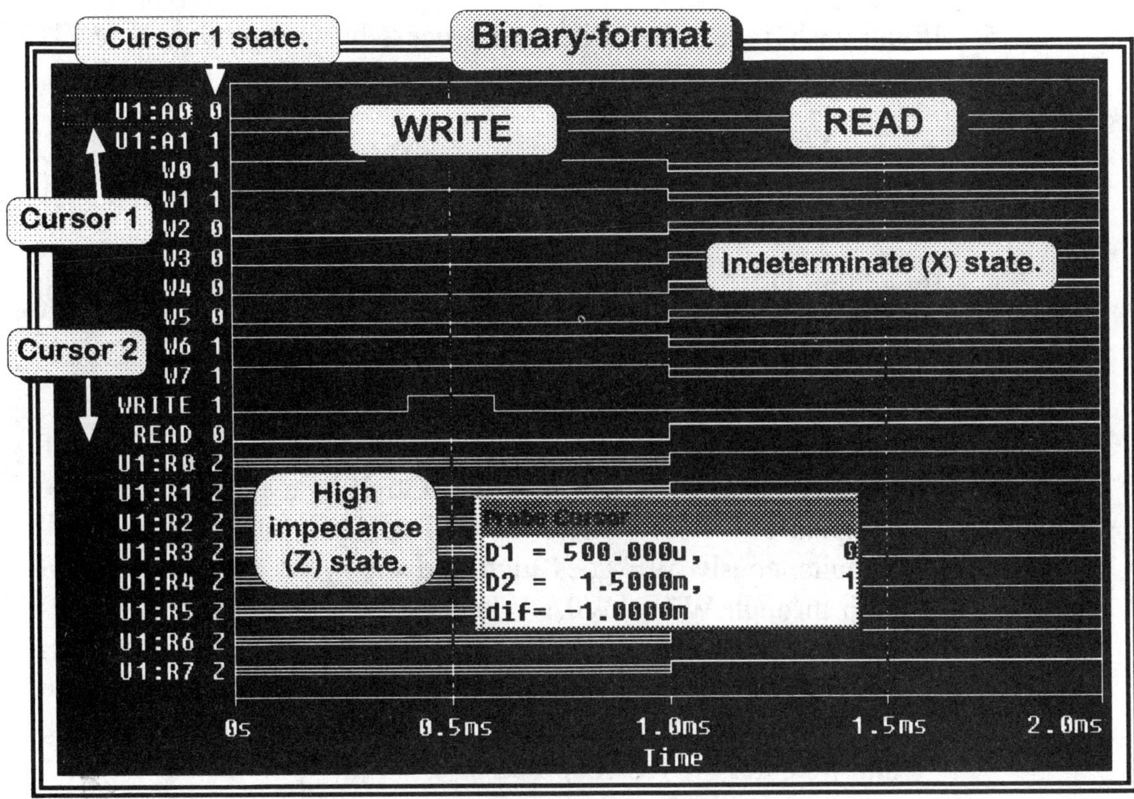

(a) Binary-format display

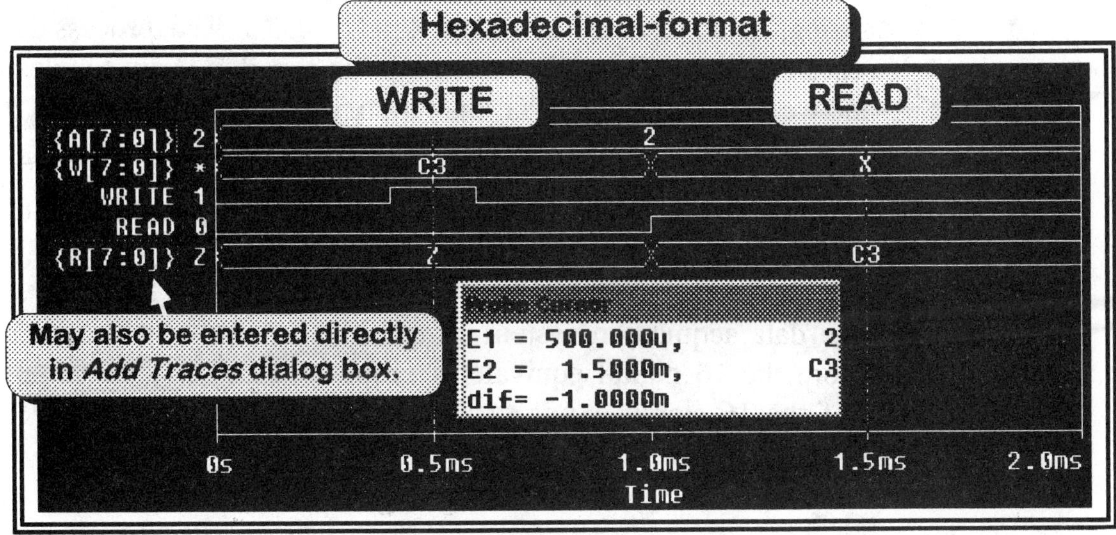

(b) Hexadecimal-format display

FIGURE 20.4
RAM test waveforms
(a) Binary-format
(b) Hexadecimal-format

5. If you wish, repeat the write/read process by writing data byte 71H to location A4H. Was the process successful?

   Yes **No**

Data acquisition

6. By adding DAC, ADC, and timing components to your test circuit, draw the data acquisition system of Figure 20.5.

- During the WRITE operation, a sinewave analog signal enters the system at line AIN. When CLOCK goes high, CNVRT (convert) also goes high, and the 0 to 10V analog waveform is sampled and converted to an 00H to FFH 8-bit digital word. Simultaneously WE goes high and the word enters the RAM chip through W7 to W0. When CLOCK goes low, WE goes low and the word is stored. Shortly thereafter, the address is incremented by the counter and the process repeats. After 16 WRITE cycles, the counter overflows, the flip-flip is toggled, and we enter the READ process.
- During each READ cycle, RE stays high (WE remains low), data is output to the DAC through R7 to R0, converted to analog, and presented at output line AOUT. The process is repeated for 16 CLOCK cycles until the flip-flop again toggles and we return to the WRITE operation.

7. Run PSpice and generate the input/output analog waveform set of Figure 20.6.

8. Did the data acquisition system sample the analog input 16 times and store the 16 digital equivalents properly? Did it then output each of the 16 stored words to the DAC and generate the proper waveform?

   Yes **No**

9. Is the system properly calibrated? (Does 10V in equal 10V out, with an error of less than .1%?)

   Yes **No**

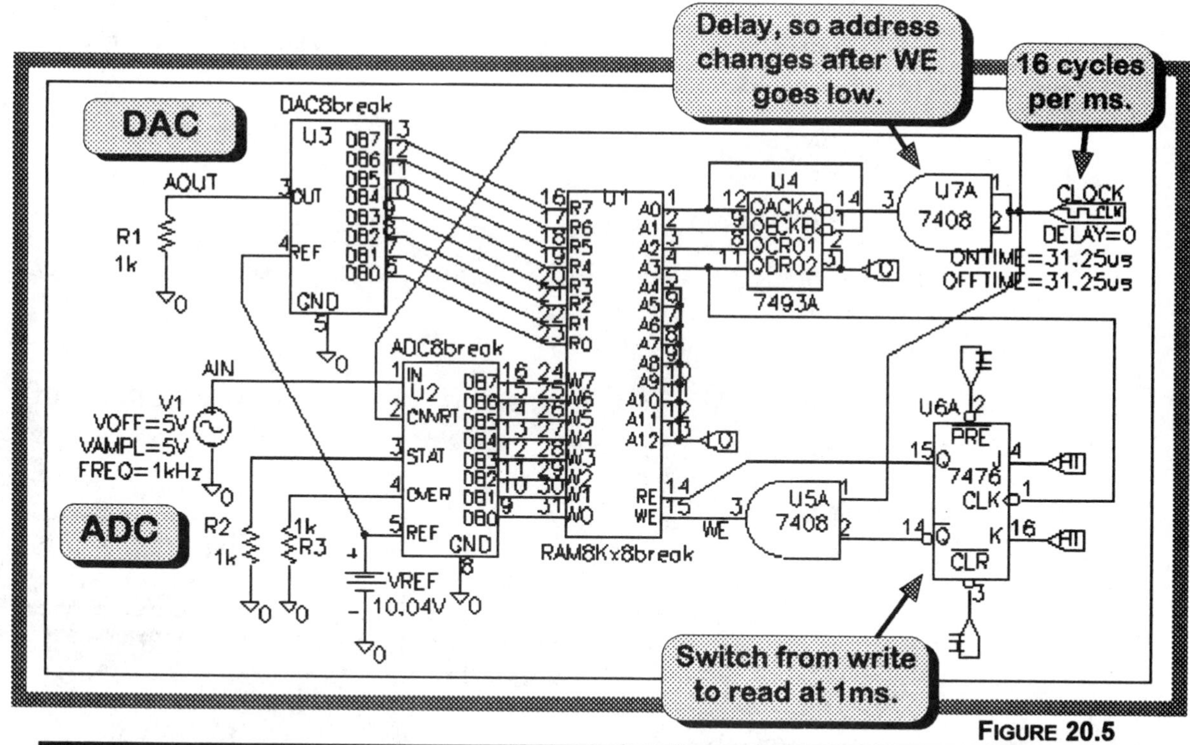

FIGURE 20.5
Data acquisition system

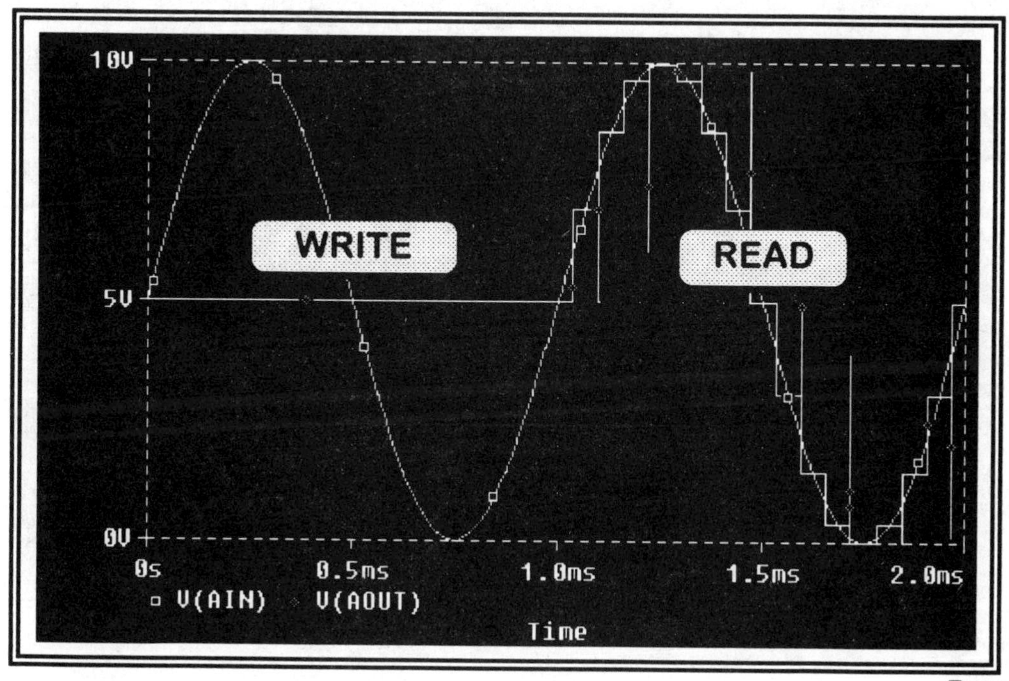

FIGURE 20.6
Analog data acquisition waveforms

10. Next, generate the binary-format digital waveform set of Figure 20.7.

    (a) Did the address lines (A3 to A0) cycle from 0 to 15?

    Yes **No**

    (b) Are the READ waveforms identical to the WRITE waveforms?

    Yes **No**

    (c) Do the READ/WRITE waveforms match the specifications of Figure 20.2?

    Yes **No**

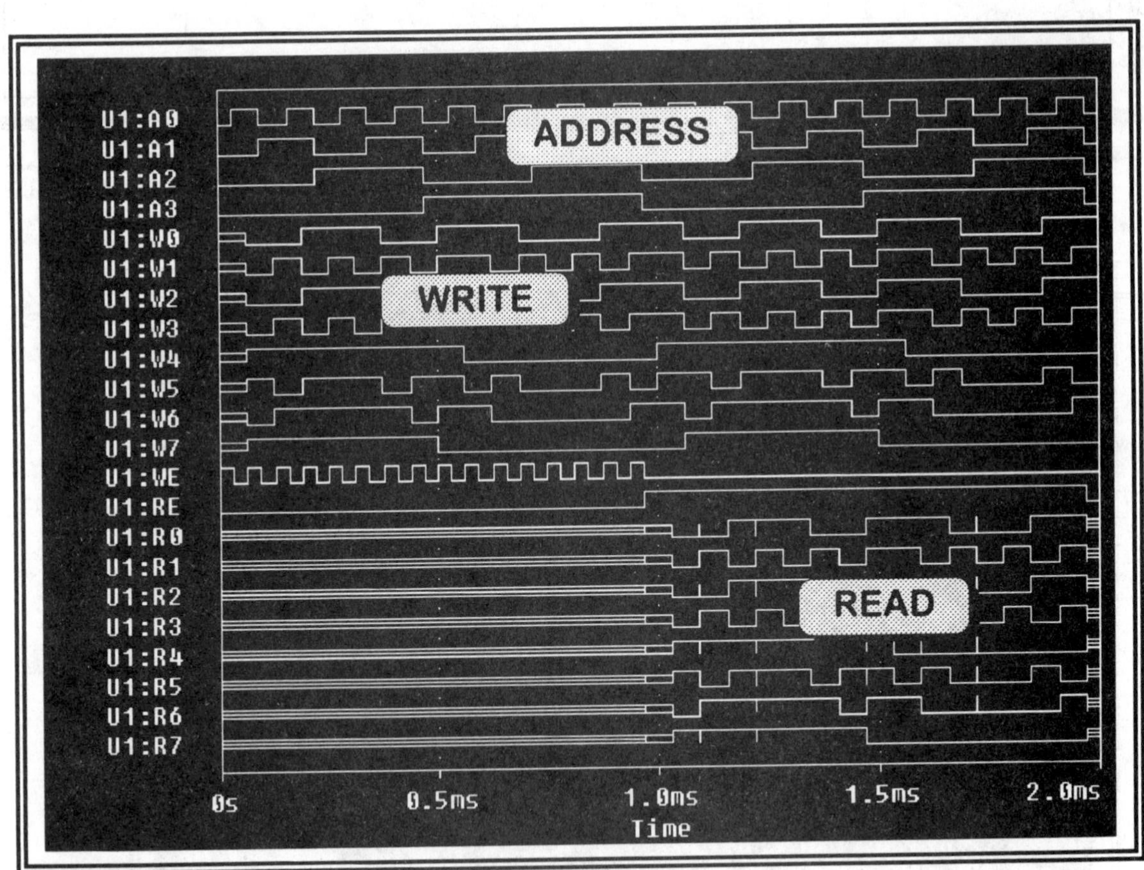

FIGURE 20.7
Binary-format digital data-acquisition waveforms

11. If you wish, add address, write, and read bus strips to the data acquisition circuit of Figure 20.5 and generate the hexadecimal-format waveform set of Figure 20.8.

    (a) The upper curve shows asterisks for most data states because there is insufficient space to write the numerical values.

    (b) By expanding the waveforms, as shown by the lower curve, sufficient room is created and the values are displayed.

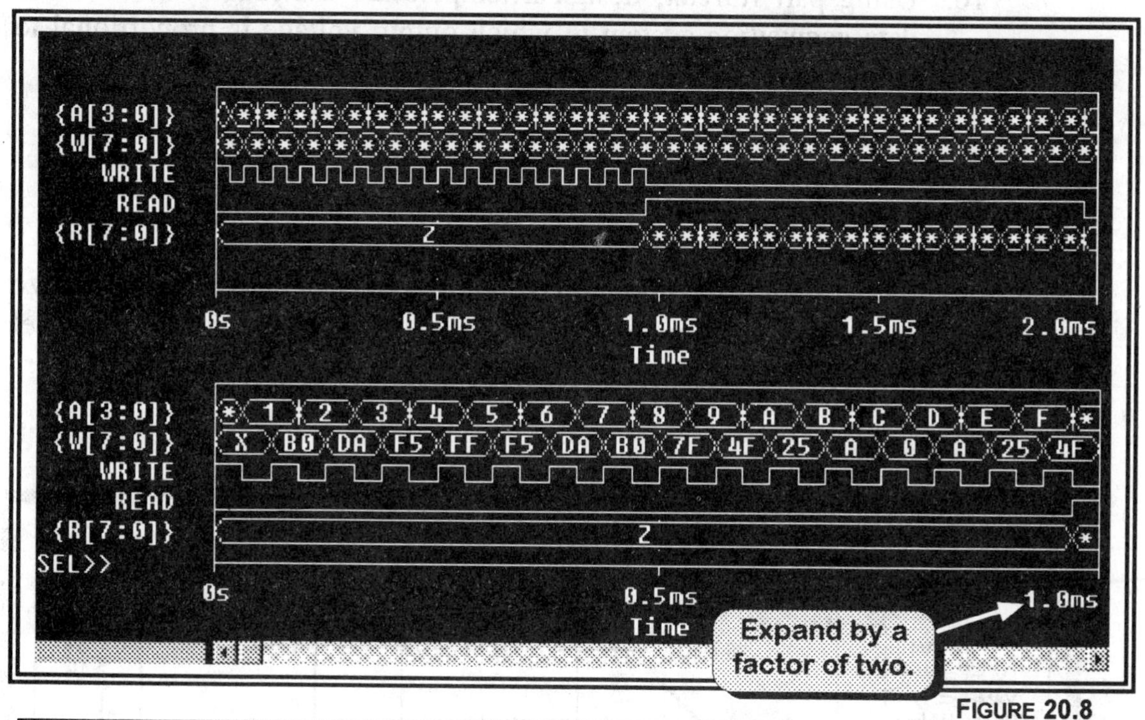

FIGURE 20.8
Hexadecimal-format display

12. Are the hexadecimal-format waveforms of Figure 20.8 equivalent to the binary-format waveforms of Figure 20.7?

    Yes **No**

Advanced activities

13. By increasing the number of storage locations, greatly increase the resolution of the data acquisition system of Figure 20.5. (Hint: Cascade two 7493A counters.)

14. Test the *Niquist theorem* on an input square wave. (To avoid losing information, the input analog signal must be sampled at a rate that is at least twice the highest input frequency of interest.)

15. By modifying the data acquisition system of Figure 20.5, design an advanced system that uses time multiplexing to alternately sample, store, and playback *two* analog inputs. (Hint: Look ahead to Chapter 24.)

16. Using part *Rbreak,* design a temperature transducer for use in a data acquisition system in which output voltage is proportional to temperature.

EXERCISE

- In preparation for the applications chapters of Part IV, modularize the data acquisition system of Figure 20.5. (A possible top-level block diagram is given in Figure 20.9.)

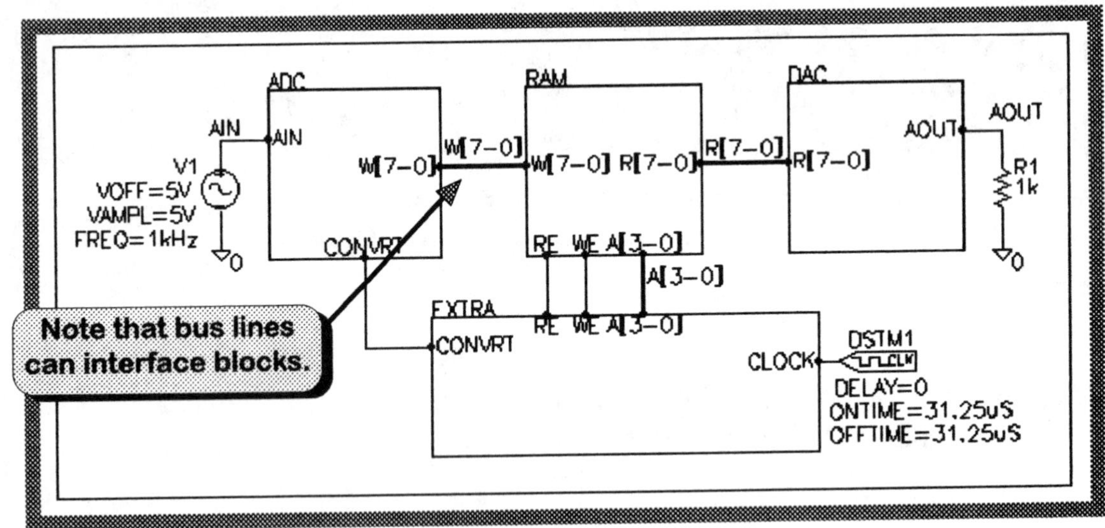

FIGURE 20.9
Top-level modular design

QUESTIONS AND PROBLEMS

1. How many bits of data can be stored in a 32k × 4 RAM?

2. During a write operation, why must WE be low when the address is changing?

3. Referring to the READ operation of Figure 20.2, why is there a lag between the time RE goes high and the data is valid?

4. Under what conditions would a data acquisition system not require a DAC or ADC?

5. How would the analog output waveform (AOUT) of Figure 20.6 change if VREF were doubled to 20.08V?

6. During a WRITE operation, when RE is low, in what state are the read outputs (R7 to R0)?

PART III

Filter Synthesis

In most cases throughout this text, we have first designed a circuit, then tested its characteristics with PSpice.

However, when designing filters, the process is reversed: we first specify the characteristics, then design the circuit to meet these specifications.

In the two chapters of Part III, we use MicroSims *Filter Synthesis* evaluation software (*fseval63*). It is identical to the professional version, except that all filters are limited to third order.

Obtain *fseval63* from the Instructor's Guide CD-ROM, or download from MicroSim's website.

CHAPTER 21

Passive Filters

Biquads

OBJECTIVE

- To design a second-order Butterworth LC low-pass filter using the special filter synthesis evaluation package.

DISCUSSION

A filter is a circuit that selectively passes certain ranges of frequencies and attenuates others. There are four major types of filters:

- Low-pass — passes low frequencies, blocks high.
- High-pass — passes high frequencies, blocks low.
- Bandpass — passes a frequency range, blocks frequencies above or below the range.
- Band-stop (reject) — blocks a frequency range, passes frequencies above or below the range.

The exact input/output characteristics of each type of filter are given by a *transfer function*, which is a complex-number mathematical description of the output divided by the input. Highly accurate transfer functions require very complex mathematical formulations for their description. However, transfer functions can be successfully *approximated* by limiting the mathematical description to quotients of polynomials:

$$F(s) = \frac{a_0 + a_1 s + a_2 s^2 + \ldots + a_n s^n}{b_0 + b_1 s + b_2 s^2 + \ldots b_n s^n} \quad \text{(where } s = j\omega = j2\pi f\text{)}$$

The order of the denominator (n) is called the filter order. The higher the order, the more closely the approximation fits the ideal. Loosely speaking, each order corresponds to an inductor or a capacitor in the resulting circuit.

Because the polynomial form of the transfer function is difficult to work with, the polynomials are usually factored to yield even and odd order linear and quadratic terms:

Even order functions

$$F(s) = \prod_{i=1}^{n} \frac{k_2 s^2 + k_1 s + k_0}{s^2 + s\frac{w_0}{Q} + w_0^2}$$

Odd order functions

$$F(s) = \left(\frac{k_1 s + k_0}{s + w_0}\right)\left(\prod_{i=2}^{n} \frac{k_2 s^2 + k_1 s + k_0}{s^2 + s\frac{w_0}{Q} + w_0^2}\right)$$

Circuits that use such functions are called biquads. The roots of the denominator (the complex frequencies that make the denominator zero) are termed the poles of the filter because the complex gain goes to infinity. The roots of the numerator are termed zeros because the complex gain goes to zero.

By selecting appropriate coefficients and terms in the biquad transfer function, we can design filter approximations that fall into five classic types:

- Butterworth
- Bessel
- Chebyshev
- Inverse Chebyshev
- Elliptic

In this chapter, we will select the popular Butterworth (maximally flat) type to generate frequency responses that are flat and smooth. To set a specific goal, we will design a *passive* LC filter that follows the ideal Bode characteristics of Figure 21.1. (In the following chapter, we will switch to an *active* filter.)

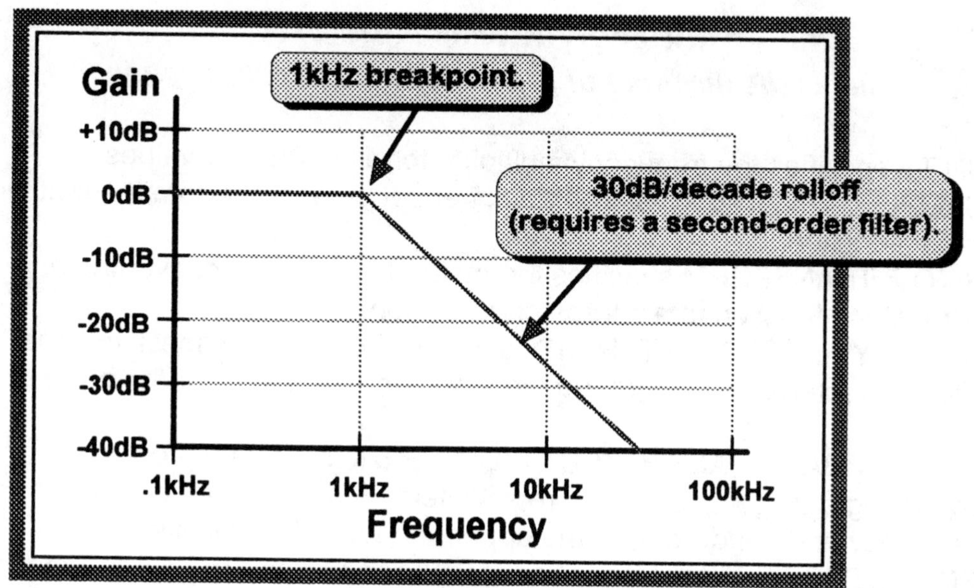

FIGURE 21.1
Desired frequency response

SIMULATION PRACTICE

Installation and execution

1. Install the Filter Synthesis Evaluation Package by following the directions in the *Introduction.*

2. Running under either DOS or Windows, run *filter.exe* in directory *fseval63*, press any key, and observe the main menu shown below:

   File Specify Coefficients Circuits Plots Preferences

Low Pass Filter design

3. From the main menu, select *Preferences* (press "r" or use the arrow keys to highlight Preferences, <**Enter**>). Review *Filter Synthesis Note 20.1* to learn the meaning of the preferences options available.

Filter Synthesis Note 21.1
What is the meaning of the various preferences options?

- UNITS specifies the frequency multiplier for all frequency values.
- SAMPLED DATA specifies the clock frequency when performing switched capacitor filtering.
- S TO Z TRANSFORM specifies the method used when converting an s-transform function to a z-transform function.
- SPECIFY ALL STD. FILTERS BY gives us the option of specifying the filter characteristics by entering the stop band frequency or the order of the filter.
- SPECIFY BAND PASS/REJECT FILTERS BY gives us the option of specifying bandpass and band-reject filters by entering the center frequency and bandwidth, or the upper and lower band limits.
- SCF NETLIST FORMAT specifies the type of analysis to be performed for switched capacitor filter (SCF) circuits. PSPICE (normal) makes use of conventional circuit elements, PSPICE (ideal switch) uses switches, and SWITCAP uses switching capacitors.

4. Using the appropriate function keys, enter the following into the *Preferences* screen:

   Units: *Khz* (Key F2.)
   Sampled data: Leave blank. (Erase if necessary using Key Del.)
   S to Z transforms: *mod bilinear* (Key F3.)
   Specify all std filters by: *stop band* (Key F4.)
   Specify band pass/reject filters by: *center freq & bw* (Key F5.)
   SCF netlist format *PSPICE(normal)* (Key F6.)

   Press **F1** to save settings and return to the main menu.

5. Next, open the *Specify* menu and select *Low Pass*.

6. Following the requirements of Figure 21.1, enter the *Low Pass Limits* and *Gain* specifications as follows. (Use arrow keys to move the cursor.)

Pass band cutoff	**1**	**kHz**	The 3dB down frequency
Stop band cutoff	**10**	**kHz**	The frequency where the stop band attenuation is reached
Pass band ripple	**3**	**dB**	3dB for Butterworth filter
Stop band atten.	**30**	**dB**	The minimum attenuation in the output at the stop band cutoff frequency
Gain	**0**	**dB**	The gain at low frequencies
Select	**Normal Approx**		

Press F2 to select the Butterworth approximation. Note that the number of orders required for each of the filter types is automatically generated. As shown, our Butterworth design requires a second-order circuit. (A rolloff of greater than 20dB requires a second-order filter.)

7. Press **ESC**, **ESC** to return to the main menu and save all settings.

Examine the Predicted Characteristics

8. Next, enter the *Plots* menu and select the *Bode* option.

9. Note the two horizontal red lines. Does the upper line correspond to the specified *pass band* cutoff values (–3dB and 1kHz), and does the lower line correspond to the specified *stop band* values (–30dB and 10kHz)?

Yes **No**

10. To compare our predicted design to the goal of Figure 21.1, use the left/right arrow keys to move the cursor and verify the following. (Note the cursor coordinates at the lower right of the screen.)

    (a) Is the gain approximately 0dB at low frequencies?

    Yes **No**

    (b) Is the amplitude –3dB at the pass band cutoff frequency (1k)?

    Yes **No**

    (c) Is the amplitude –30dB (*or better*) at the stop band cutoff frequency (10k)?

    Yes **No**

11. Press **ESC** and return to the *Plots* menu. Select *Time Domain* and draw the resulting curve on the graph of Figure 21.2. The curve is the filter's time-domain response to a step input.

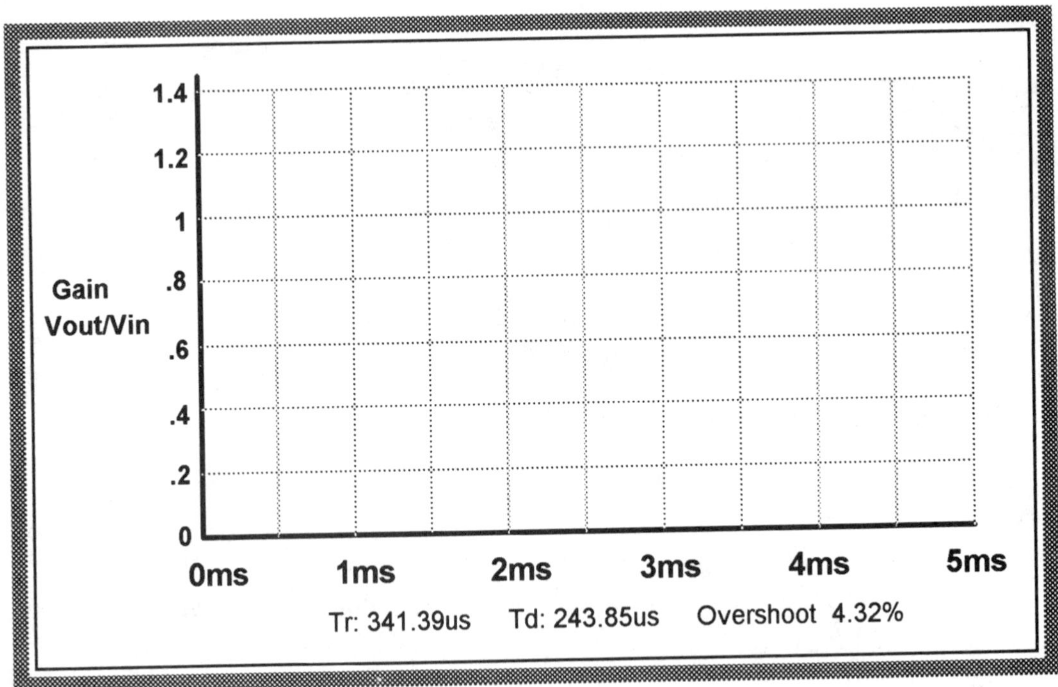

FIGURE 21.2
Time-domain response

Viewing the Circuit

12. Return to the main menu (press **ESC** twice). Select *Circuits* and bring up the circuits menu.

13. Select *LC Ladders*, followed by *Singly Terminated.* The table on the screen is a description of the circuit in terms of branches, types, and roles (series or shunt).

14. To interpret this table better, bring up the circuit schematic of Figure 21.3 (press F2). This, by the way, is the solution to our problem!

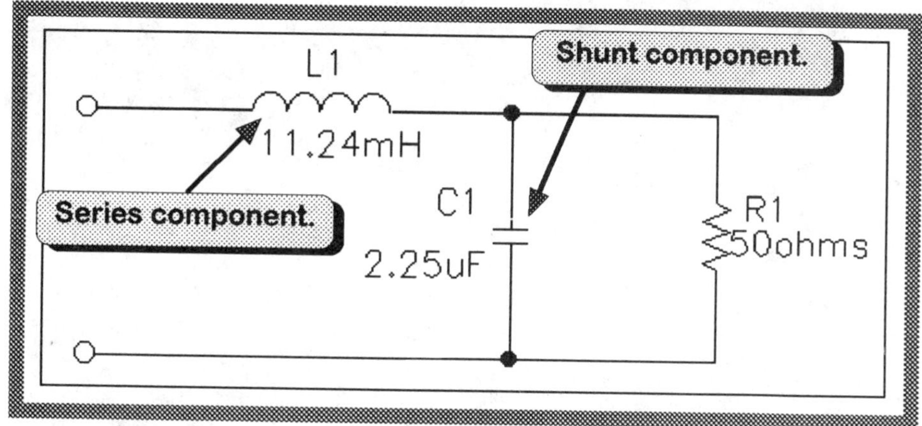

FIGURE 21.3
The finished design

15. Press **ESC** to return to the table.

    (a) Is the first branch listed as a series inductor of 11.24mH, and the second branch listed as a shunt (parallel) capacitor of 2.25μF?

    Yes **No**

    (b) Is the default value of the singly terminating resistor 50Ω, and can it be changed from the bottom of the table?

    Yes **No**

16. Return to the main menu (**ESC**, **ESC**, **ESC**).

Verify the solution using PSpice

17. Exit the filter program (**File**, **Quit**), enter Schematics, and draw the circuit of Figure 21.3. Using the appropriate analysis and voltage source, generate the *AC sweep* graph of Figure 21.4.

    (a) Is the breakpoint approximately 1kHz?

    Yes **No**

    (b) Is the rolloff 30dB/decade *or greater*?

    Yes **No**

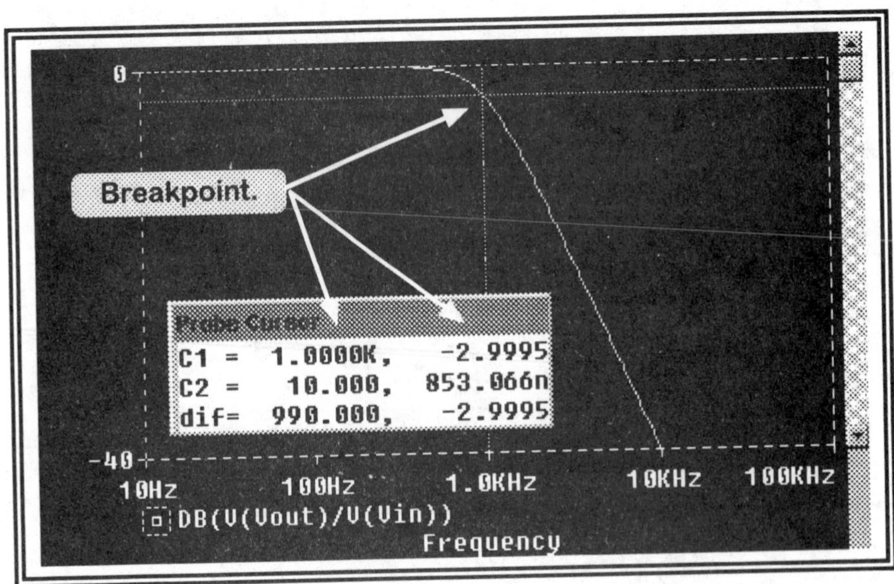

FIGURE 21.4
Bode plot test

18. Switching to the appropriate analysis and voltage source, generate the time domain response plot of Figure 21.5. Use the cursors to determine each of the following. Compare your answers to the predicted values of Figure 21.2.

    (a) *Tr* (Rise time from 10% to 90%) = ________

    (b) *Td* (Delay from 0s to 50%) = ________

    (c) *Overshoot* (% rise above steady state) = ________

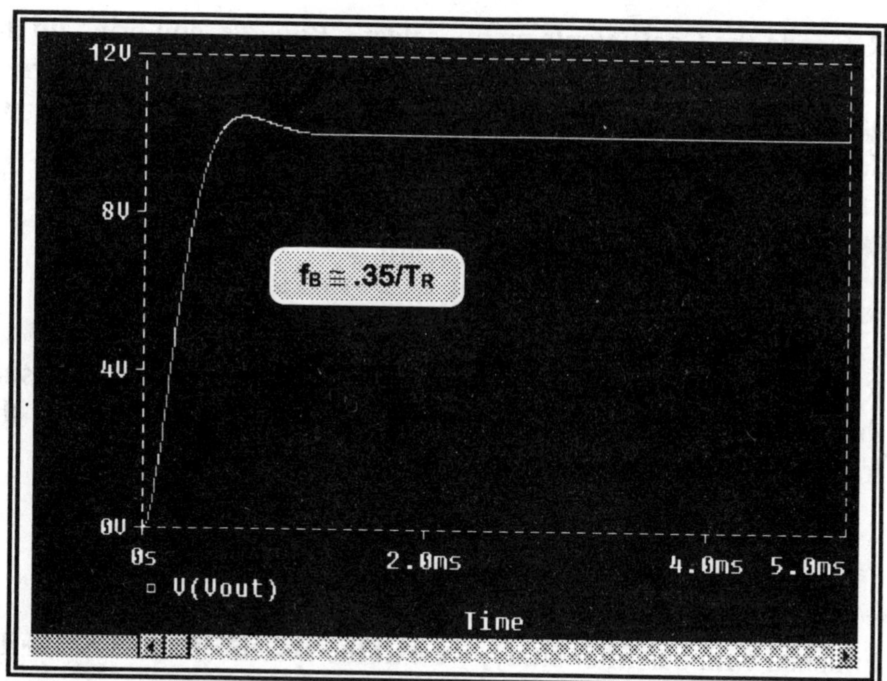

FIGURE 21.5
Time response plot

Advanced activities

Re-enter the filter design program and respecify all values.

19. From the main menu, select *Plots*, *Pole-Zero*, and note the graph. The two Xs show the location of the two poles (roots of the denominator of the biquad s-plane equation). Using complex notation, record the approximate coordinates of each pole:

	Real	Imaginary
Pole 1	______________	______________
Pole 2	______________	______________

20. Exit to the main menu, bring up the *Coefficients* menu, and select *s Poles & Zeros*. Press F3 and note the location of the poles. Does this match the pole/zero plot of step 19?

21. Return to the Coefficients menu, select *s Biquads*, and bring up the *S Domain* table. Verify that the equation below is the correct biquad transfer function for the single-stage first-order filter we designed earlier:

$$F(s) = \frac{39.6M}{s^2 + s\frac{6.29k}{.707} + 6.29k^2}$$

22. Return to the *Coefficients* menu and select *s Polynomial.* Using the information given, we generate the polynomial form of the transfer function (are the two forms the same?):

$$F(s) = \frac{39.6M}{39.6M + 8900s + s^2}$$

23. Finished with our filter study, we exit the system (**ESC**, **ESC**, **File**, **Quit**.)

EXERCISE

- Using the sketch of Figure 21.6 to help select the proper specifications, design a bandpass filter using a third-order LC ladder.

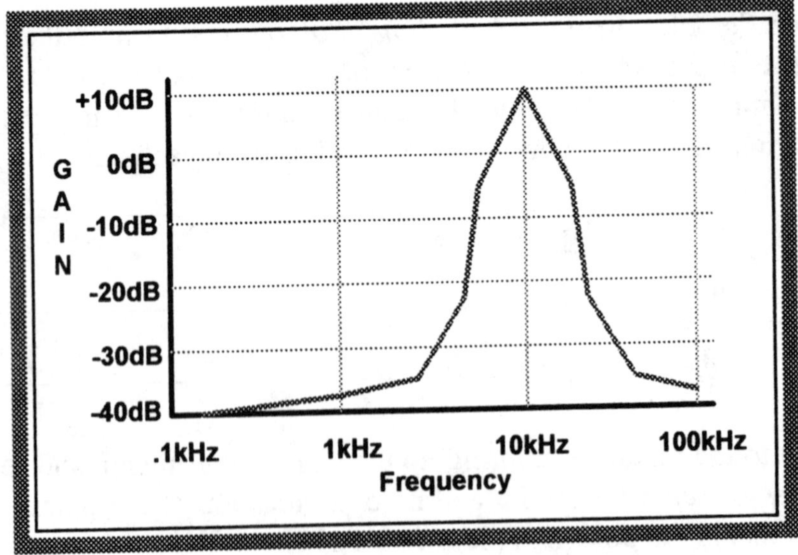

FIGURE 21.6
Bandpass design goal

QUESTIONS AND PROBLEMS

1. What is the difference between a *passive* and an *active* filter?

2. What is a *transfer function*?

3. Using your calculator, prove that the *s Biquad* form of the tansfer function (Step 21) is equivalent to the *s Polynomial* form (Step 22).

4. In Figure 21.1, we asked for a rolloff of 30dB/decade. Why then did the system generate a filter with a 40dB/decade rolloff? (Hint: What is the inherent rolloff of a second-order filter?)

5. What are *poles* and *zeros*?

6. Based on this chapter, design a third-order passive LC high-pass filter (components only; no values.).

7. What is the difference between a *singly terminated* and a *doubly terminated* circuit?

8. Based on the equation highlighted in Figure 21.5 (and shown below), the break frequency of a filter is inversely related to the risetime. Based on the value of T_R from Figure 21.2, use the relationship to predict the breakpoint frequency (f_B) of the low-pass filter designed in this chapter. Is the result close to 1kHz?

$$f_B = \frac{.35}{T_R} = \underline{\hspace{3cm}}$$

CHAPTER 22

Active Filters

The Switched-Capacitor Filter

OBJECTIVES

- To design a bandpass filter based on an active filter design.
- To introduce the *switched-capacitor* filter.

DISCUSSION

The *passive* filters of the last chapter contain only resistors, capacitors, and inductors. An *active* filter, on the other hand, is one that contains active elements—such as op amps. Compared to the passive filter, an active filter has sharper characteristics, can generate an overall gain, and does not require expensive inductors.

The switched-capacitor filter (SCF)

Another way to implement a biquad filter section is by way of switched capacitor circuits. To create such a filter, we start with a standard RC active biquad circuit and replace each resistor by a charge pump (a clocked switch and capacitor). Switched capacitor filters are the most common way to create a complete filter on a single chip.

The bandpass filter

The bandpass filter passes only a single frequency and blocks all others. In this chapter, we will design an active bandpass filter that passes only the musical tone of A (880Hz).

SIMULATION PRACTICE

1. Enter the filter design main menu, use the left/right arrows to select the *Preferences* menu, press **Enter** to open, and select the following:

    - **Units:** Hz
    - **Sampled data:** (Not applicable; leave blank; erase any numbers)
    - **S to Z transforms:** mod bilinear
    - **Specify all std filters by:** stop band
    - **Specify band pass/reject filters by:** center freq & bandwidth
    - **SCF netlist format :** PSPICE(normal)

    Press **F1** to save preferences and return to the main menu.

2. Open the *specify* menu, select *bandpass*, and enter the design parameters shown in Figure 22.1 in the appropriate sections (but do not press **ESC** yet).

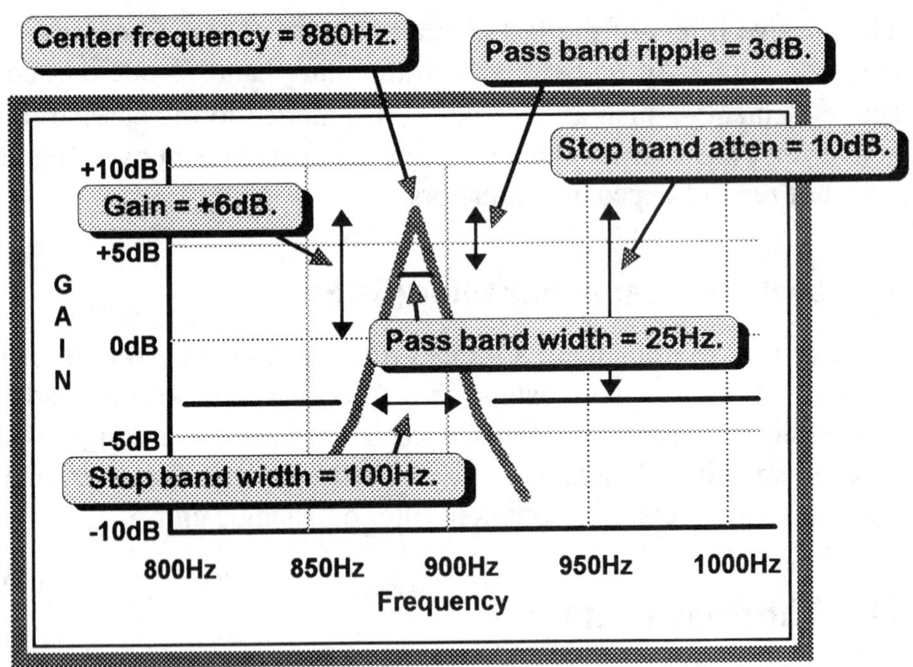

FIGURE 22.1
Desired frequency response

3. Press **F2** to select the Butterworth approximation. Note that all filter types are second order.

4. Press **ESC** twice and return to the main menu.

5. Next, enter the *Plot*s menu, select the *Bode* option, and view the plot.

6. Use F2 to change the horizontal plot type to linear and press **Enter**.

> If you wish, change the horizontal scale as follows: Use the up/down arrow keys to select *Min* and *Max*, change the *Min* and *Max* limits to 800 and 1000, and press **Enter**.

7. Using the left/right arrow keys, move the cursor and verify that the plot meets the requirements of Figure 22.1:

   (a) Is the *center frequency* approximately 880Hz?

   Yes **No**

   (b) Is the *gain* approximately 6dB at the center frequency?

   Yes **No**

   (c) Is the *pass band ripple* approximately 3dB?

   Yes **No**

   (d) Is the *pass band width* approximately 25Hz or less?

   Yes **No**

   (e) Is the *stop pass width* approximately 100Hz or less?

   Yes **No**

   (f) Is the *stop band atten* approximately 10dB (or better)? (Remember: We have a 6dB gain.)

   Yes **No**

8. Return to the main menu (press **ESC** twice). Open the *Circuits* menu, select *RC Biquads*, *Basic MLF* (multiloop feedback), and display the schematic circuit (**F4**).

9. Looking at the circuit, are the resistors set to conventional values?

Yes **No**

10. To set the resistors to conventional values, press **ESC** to return to the RC biquads screen. Press **F1** to begin the *round* process. Press **F2** to set the resistors to values commonly found in the 1% tolerance category; press **F3** to set the capacitors to values commonly found in the 5% tolerance category.

    Press **F1** to set, and notice the new (conventional) values. Press **F4** to select the schematic, and draw the circuit below:

11. Return to the *RC Biquads* submenu (**ESC**, **ESC**), view the various circuit options listed below, and place an asterisk beside those that are *state* filters. (A *state* filter uses integrators and differentiators to solve the filter's time-domain differential equation.)

    Deliyannis-Friend
    Akerberg-Mossberg
    Tow-Thomas
    KHN

12. Exit the filter program, analyze the circuit under PSpice, generate the *AC sweep* graph of Figure 22.2, and fill in the table below:

	Design goals	Test results
Center frequency	880Hz	
Pass band width (at 3dB down from peak)	25Hz	
Gain	6dB	
Pass band width (at 10dB down from peak)	100Hz	

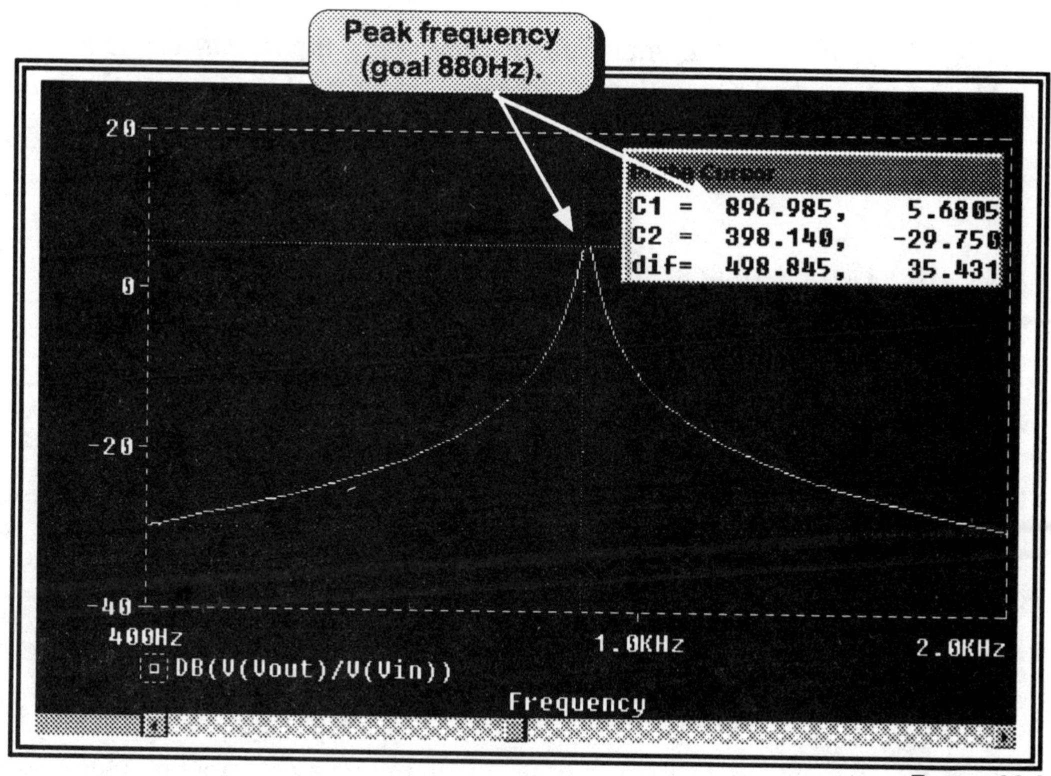

FIGURE 22.2
PSpice test results for bandpass filter

13. Based on the results, does the circuit recorded in step 10 approximately match our original design of Figure 22.1?

    Yes **No**

Switched Capacitor Filters

14. Restart the filter program and enter the *preferences* menu. Set the clock frequency to 100kHz and *SCF netlist format* to SWITCAP. Press **F1** to set values and exit.

15. Enter the *specify* menu, select *band pass*, enter the specifications of Figure 22.1, select *Butterworth*, and return to the main menu.

16. Enter the *plots* menu, select *bode*, and display the filter characteristics. Does the filter equal or better the specifications?

    Yes **No**

17. Return to the main menu and enter the *circuits* menu. Select *Switched Cap Biquad*, **Enter**, and press **F4** to view the circuit. Does the circuit contain only op amps, capacitors, and switches?

    Yes **No**

18. Exit the filter synthesis program.

Advanced activities

19. Add tolerances to the design of Step 10, and generate bandwidth and peak frequency histograms. Comment on all results.

20. Based on the design of Figure 22.1, how low can you take the *stop band width* (presently 100Hz) and still stay within the second order Butterworth limitations?

21. Repeat the design using *Bessel, Chebyshev, or elliptic* types.

22. Repeat the test using other available PSpice op amps (such as the LM 324).

23. Design and test a bandstop filter by cascading a low-pass and high-pass filter.

EXERCISE

- Design an active bandpass filter for a radio station of center frequency 1MEGHz and a pass width of 6K. To prevent overlap into neighboring stations, the stop width should be 12K at a stop attenuation of 10dB.

QUESTIONS AND PROBLEMS

1. What are some of the advantages of an active filter design over a passive filter design?

2. Why is an operational amplifier used to implement analog filters? (Hint: Does a filter make use of feedback?)

3. Why are switched capacitor filters normally implemented using commercial integrated circuits?

PART IV
Applications

In the four chapters of Part IV, we have an opportunity to put our skills to the test.

The four application projects consist of analog/digital combination circuits that push the component limits of the evaluation version.

In all cases, we take full advantage of the modularization techniques introduced in Volume I.

Special Note

All the projects of Part IV make use of hierarchical blocks. Because of a possible design flaw in version 7.1 related to nested blocks, it may be necessary to place a wire segment between all components and block edges.

If simulation generates error messages like "less than 2 connections at node" and "node floating," and no obvious reason exists for these errors, then the wire segments may be necessary.

CHAPTER 23

Touch-Tone Decoding
The Bandpass Filter

OBJECTIVE

- To design, test, and modify a touch-tone decoder circuit.

DISCUSSION

Our first application is representative of those found in the modern-day *public switched telephone network* (PSTN). To provide the dialing function, they use *dual-tone multifrequency* (DTMF)—more commonly known as *touch-tone*.

Using a keypad similar to that of Figure 23.1, each keypress generates a *pair* of tones. The two tones are summed and sent to the central office (CO), where they pass through a series of filter-decoders to determine which key was pressed.

In this chapter, we will follow the action of a single key—key 6. When key 6 is pressed, tones 770Hz and 1477Hz are generated.

SIMULATION PRACTICE

1. Draw the top-level block design of Figure 23.2.

> The two frequencies of key 6 are summed by block PHONE and transmitted to the CO via ports LOOP and LOOPBAK over wire COPPER. Within block CO, the combined signal passes through filters corresponding to key 6. The filter outputs are ANDed together (decoded) and presented at port KEY6.

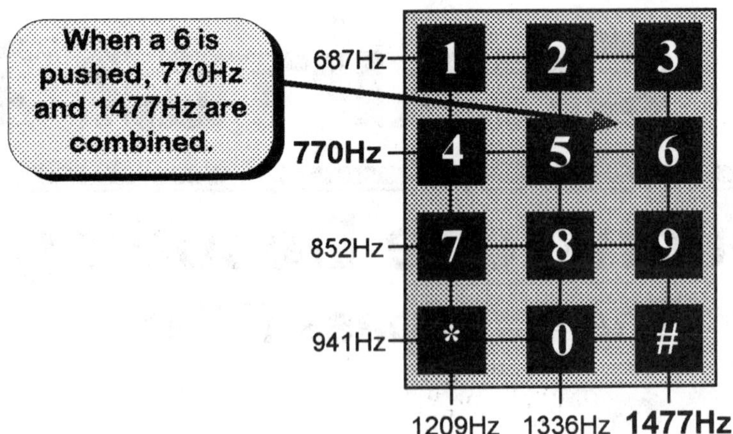

FIGURE 23.1
Touch-tone keypad

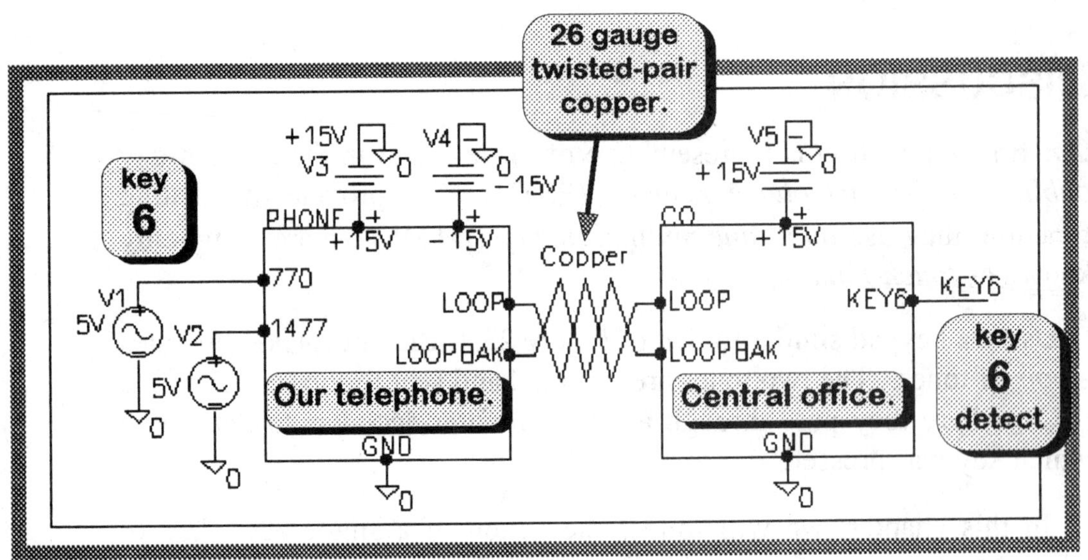

FIGURE 23.2
Touch-tone top-level design

2. First, we PUSH into block PHONE (**DCLICKL** within block PHONE), enter an appropriate file name (such as *sum.sch*), and create the design of Figure 23.3.

> The op amp summing circuit combines the two inputs in a linear manner, and transmits them by way of interface ports LOOP and LOOPBAK.

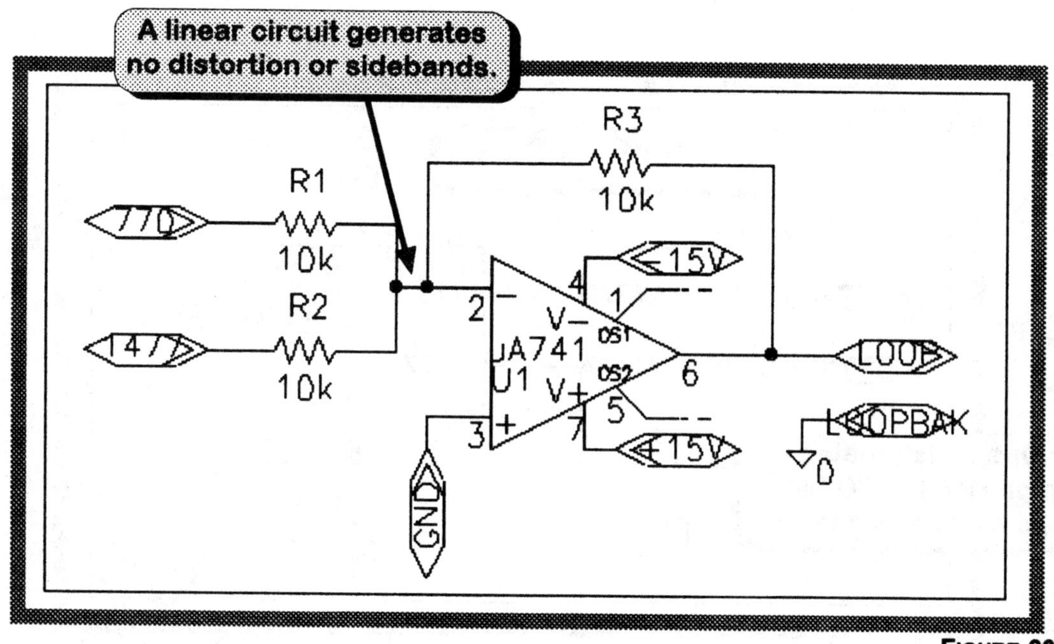

FIGURE 23.3
Circuit SUM

3. When the SEND block is done, we POP back to the top level (**Navigate**, **Pop**) and PUSH into block CO. This time, because the CO process is more complex, we create the two-block midlevel diagram of Figure 23.4.

> The received signal from port LOOP is bandpass filtered by block FILTER to separate its 770Hz and 1477Hz components. These components are sent by way of ports F770 and F1477 to block AND, where they are decoded and output from port KEY6.

4. We next PUSH into midlevel block FILTER and create the bottom-level circuit of Figure 23.5.

> Behavioral modeled bandpass filters E1 and E2—which are tuned to corresponding frequencies 770Hz and 1477Hz—receive the combined signal via port LOOP and pass on the 770Hz and 1477Hz components to block AND via ports F770 and F1477.

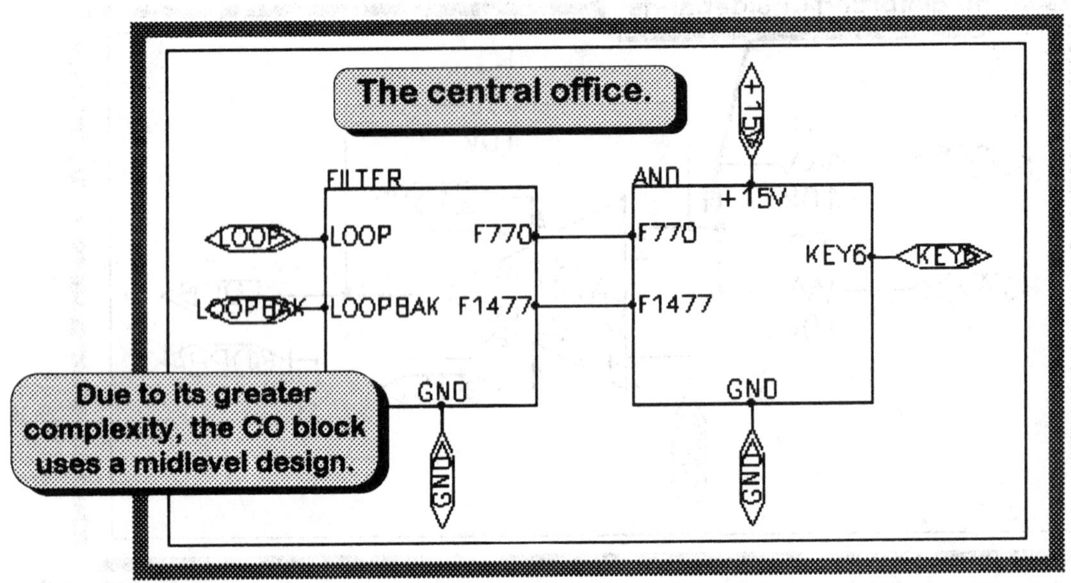

FIGURE 23.4
RECEIVE midlevel block diagram

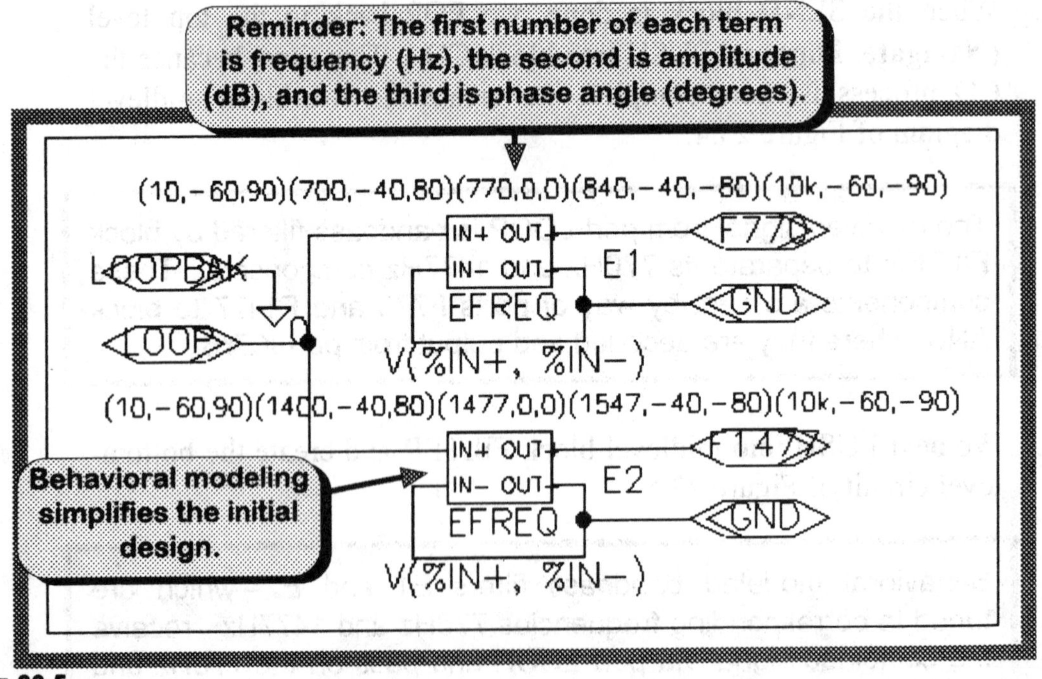

FIGURE 23.5
Block FILTER low-level design

Special Note

The filters of Figure 23.5 are designed using the behavioral modeling techniques of Chapter 30 in Volume I. This not only simplifies the initial design, but reduces the component count to within the evaluation version limits.

To program each filter we use the characteristics of the bandpass filter of Figure 23.6 (based on the bandpass circuit from Chapter 9).

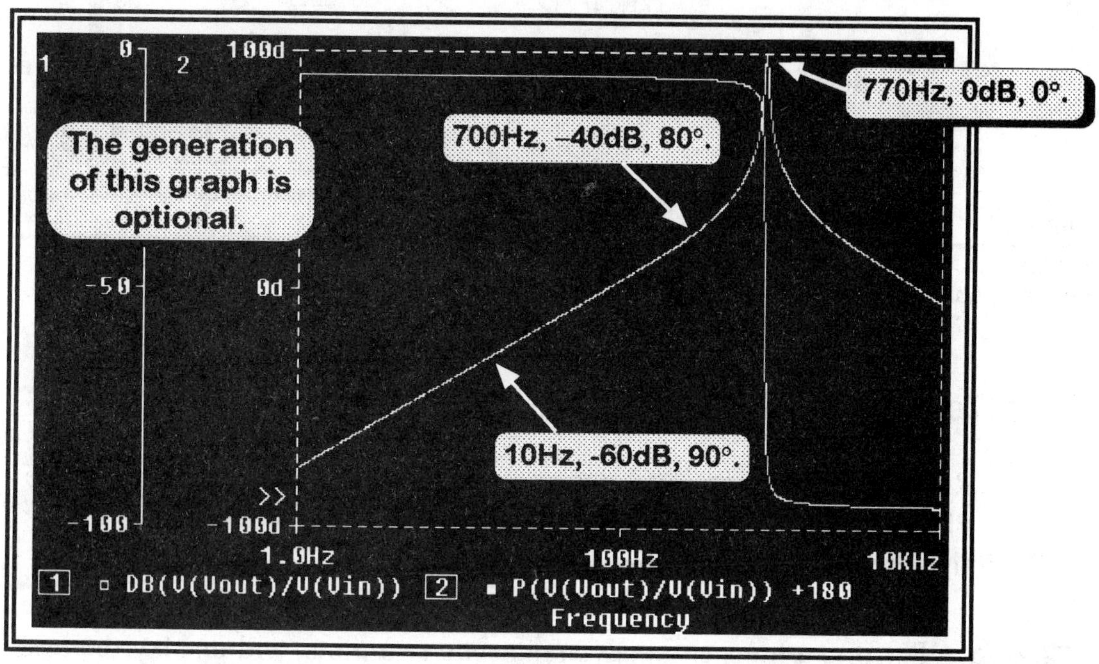

FIGURE 23.6
Bandpass filter characteristics

5. Finally, we POP back to the midlevel, PUSH into block AND, and create the circuit of Figure 23.7.

The two filtered signals enter the diode AND gate and pass to the sample-and-hold circuit where the output is presented at port KEY6. If the combined transmitted signal matches the filters, the AND gate output is high and we have a hit. Otherwise, the output is low and we have a miss.

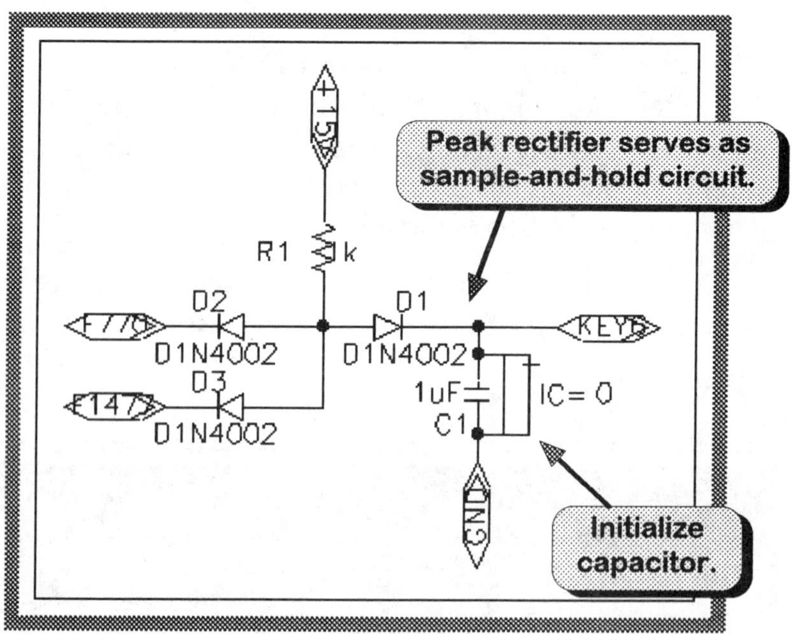

FIGURE 23.7

AND gate and sample-and-hold

Test the touch-tone circuit

Returning to the top-level design of Figure 23.2, we will first test the PHONE block. If it is operating properly, the 770Hz and 1477Hz signals should be combined linearly and presented at output LOOP.

Fast Fourier transform

6. Perform a transient analysis, unsync two plots, and display the time and FFT output waveforms of the PHONE block (Figure 23.8).

Note: For smoother waveforms, be sure to set *Step Ceiling* to approximately period/50, or about 15μs in this case [1/(1477×50)].

(a) Does the time-domain signal appear to be a composite of two frequencies?

Yes **No**

(b) Do the frequency components peak at approximately (within 5% of) 770Hz and 1477Hz, and are they of nearly equal amplitude?

Yes **No**

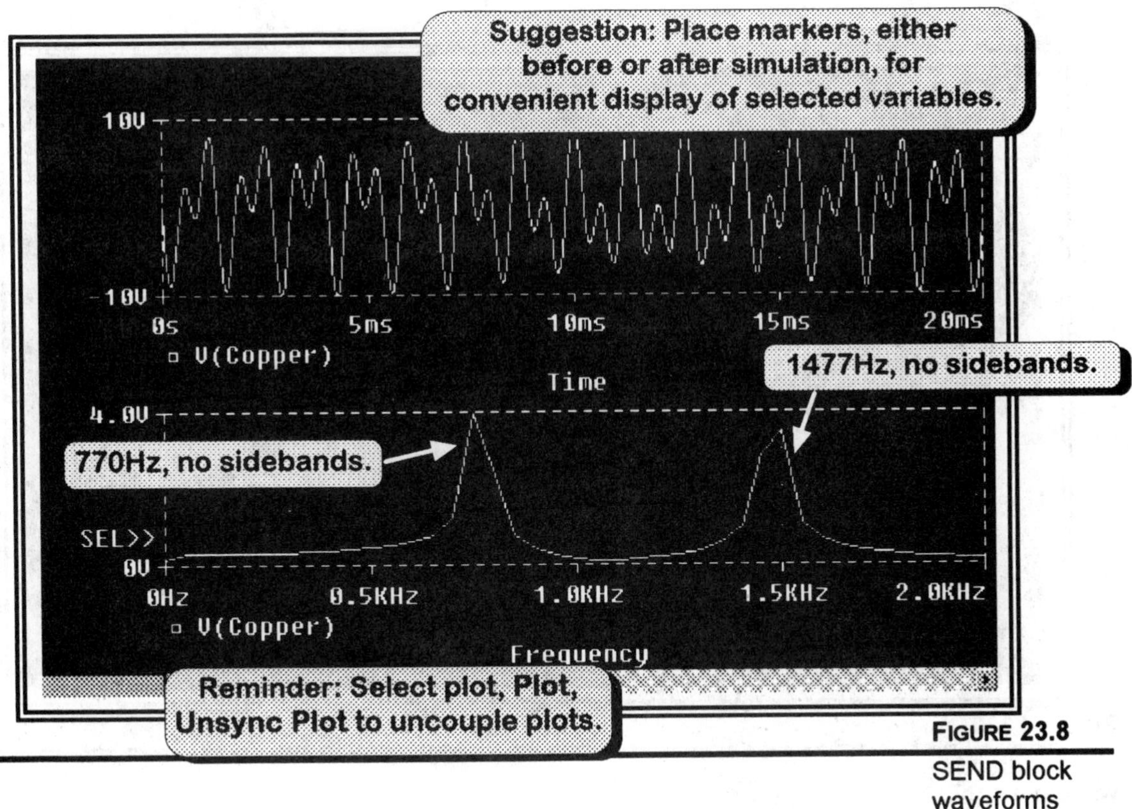

FIGURE 23.8
SEND block waveforms

Referring to the top level design of Figure 23.2, we will next test the CO (central office) block. If it is operating properly, the KEY6 output should be high when key 6 is pressed, and it should be low for any other key.

7. Return to the midlevel state of Figure 23.4, and display the output waveform of the CO block, as well as the outputs of the two filters (Figure 23.9).

   Are the two filter outputs high and of nearly equal amplitude, and is the final CO output a hit (greater than 1.5V at steady state)?

   Yes **No**

8. Change the input frequencies to match the key of 3 (change 770Hz to 687Hz), and again display the output waveforms (Figure 23.10).

   Is the F770 output less than the F1477 output, and is the final CO output a miss (less than 1.5V)?

   Yes **No**

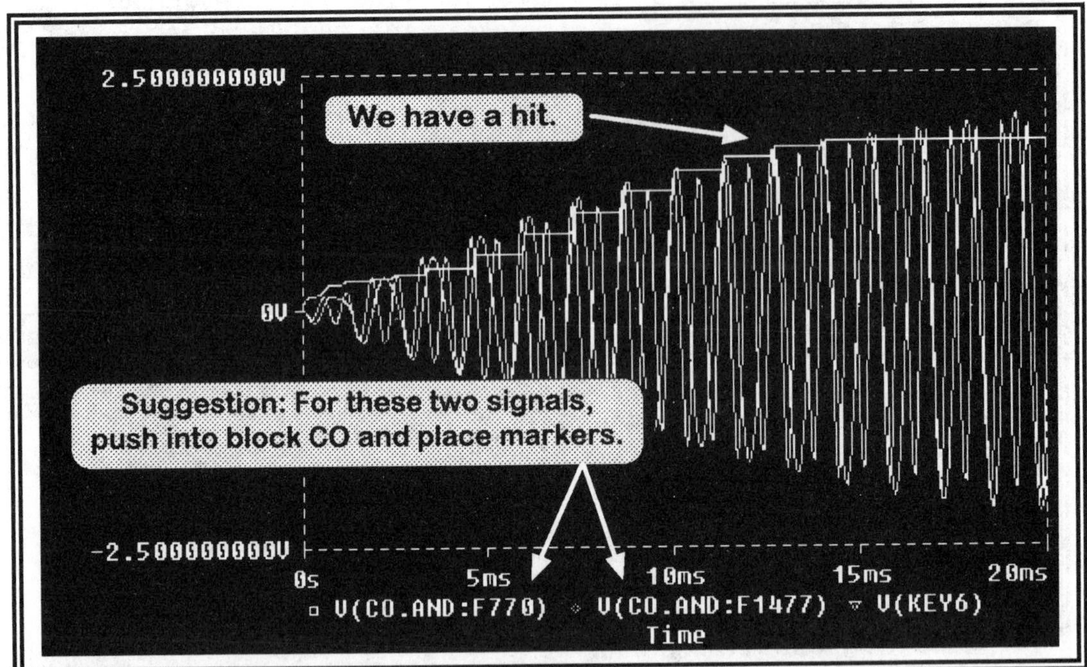

FIGURE 23.9

Key 6 output waveforms (hit)

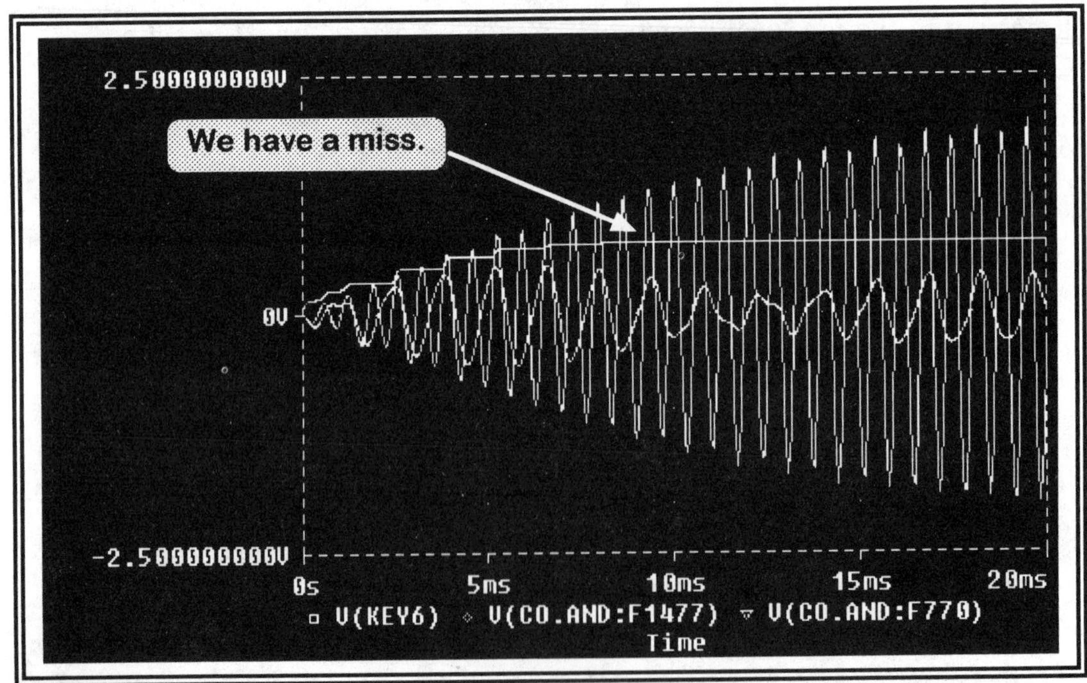

FIGURE 23.10

Key 3 output waveforms (miss)

Test and modification suggestions

> Note: Due to the limitations of the evaluation version, most of the modification suggestions in this and future applications chapters will have to be tested separately, evaluated with the full-fledged version, or designed on paper. Report all findings on a separate sheet.

9. With the help of Figure 23.11, modify the output *sample-and-hold* circuit to include the ability to clear the output when the key is released.

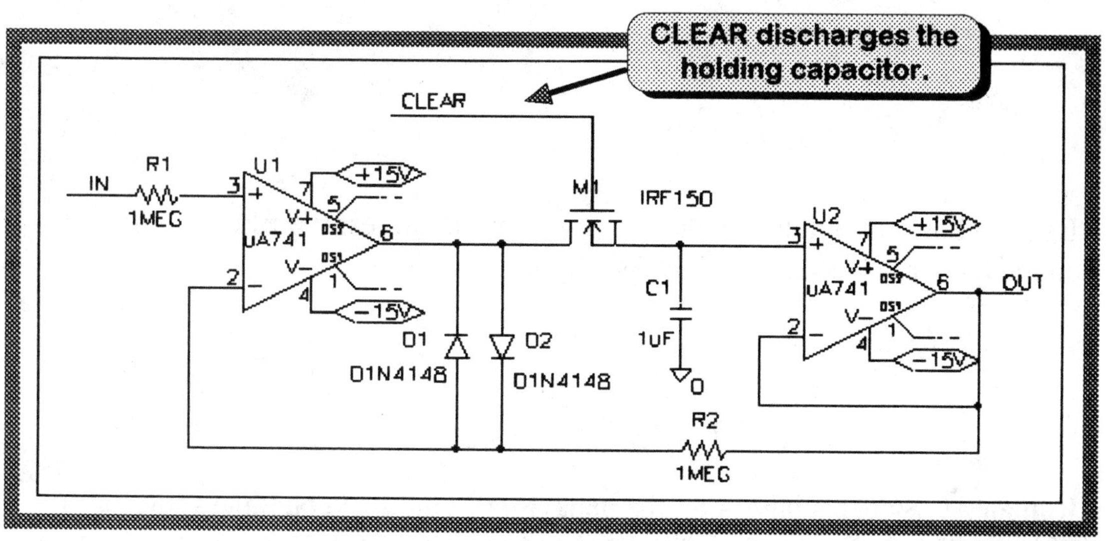

FIGURE 23.11
Sample-and-hold with CLEAR

QUESTIONS AND PROBLEMS

1. Which tones are generated when a 0 is pressed?

2. Were it not for the limitations of the evaluation version, could we have used a conventional digital AND gate in the circuit of Figure 23.7, rather than the diode-based AND gate?

3. What function is performed by output components D1 and C1 of Figure 23.7?

4. Why is the bandwidth of the two filters an important factor in the design?

5. Explain how you might use the AC Sweep mode to test the filter characteristics of the two bandpass filters.

6. Run an AC Sweep analysis on the behaviorally modeled bandpass filter array of Figure 23.5. Does it have the expected characteristics?

CHAPTER 24

Pulse-Code Modulation
Time-Division Multiplexing

OBJECTIVE

- To design, test, and modify a pulse-code modulation (PCM) system.

DISCUSSION

In the last chapter we investigated the dialing function between home telephone and central office. In this chapter, we examine the *trunk* lines that interconnect the central offices and carry the thousands of phone signals between major cities.

For efficiency, each individual wire making up these trunk lines must carry many telephone calls simultaneously. This is accomplished by a process called *multiplexing*, the subject of this chapter.

> *Multiplexing* is the process of combining two or more signals onto a single transmission line.

FDM versus TDM

There are two major types of multiplexing:

- FDM (frequency-division multiplexing), in which two or more signals are placed in different *frequency* slots.
- TDM (time-division multiplexing), in which two or more signals are placed into different *time* slots.

The older FDM system is now rapidly being replaced by the newer TDM system. When TDM includes conversion of each time-sampled analog signal to a serial digital stream, we have a PCM (pulse-code modulation) system—the subject of this chapter.

In our simplified PCM system, two input analog voice channels are continuously sampled in sequence. Each analog signal is converted to a 4-bit digital word and transmitted in serial by time-multiplexing.

At the destination, the process is reversed and the serial digital words are demultiplexed, converted back to analog, and output on two analog lines.

SIMULATION PRACTICE

1. The circuit of Figure 24.1 shows a complete PCM system, including transmission and reception. However, due to the limitations of the evaluation version, we restrict our discussion to the transmission block.

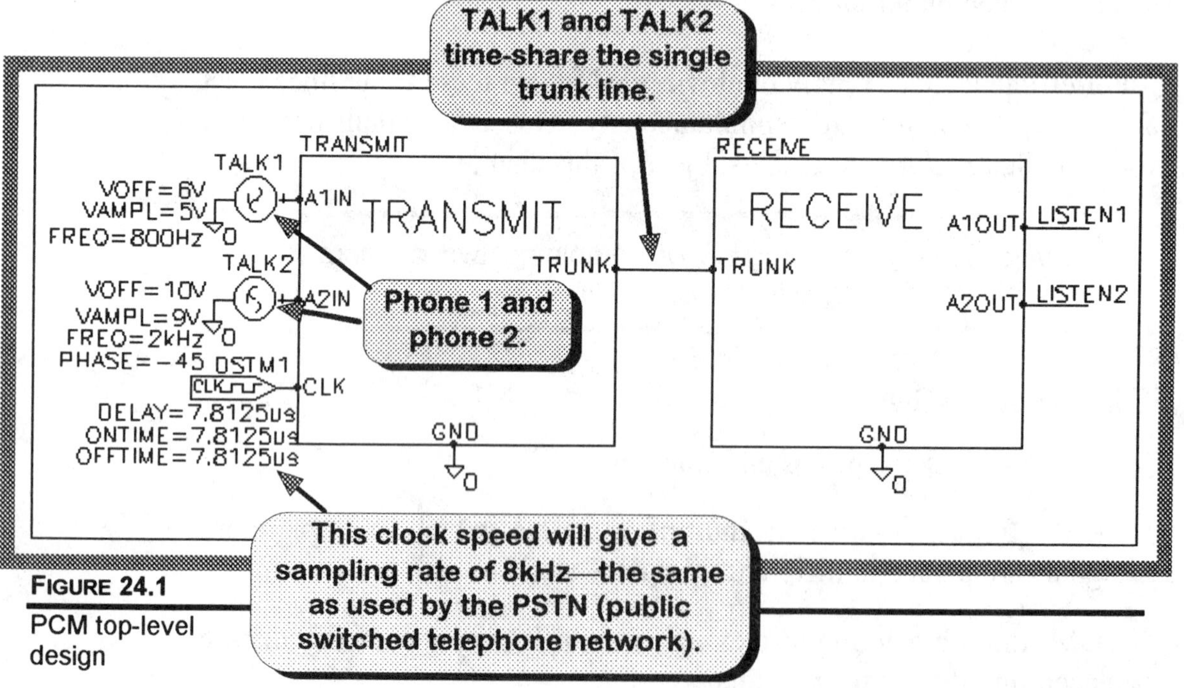

FIGURE 24.1

PCM top-level design

2. Draw the PCM transmission block circuit of Figure 24.2.

> Inputs TALK1 and TALK2 represent two phones simultaneously transmitting analog signals. The two analog inputs are converted to a PCM digital serial stream by block TRANSMIT and sent out on line TRUNK.
>
> Digital stimulus DSTM1 provides the basic clock signal for timing all operations.

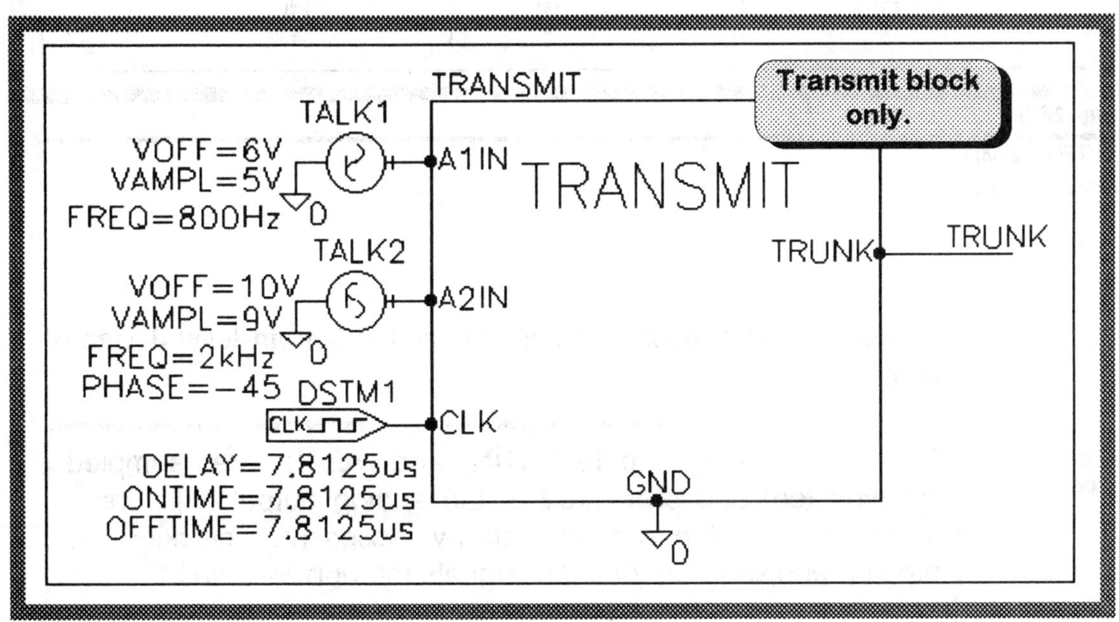

FIGURE 24.2
PCM TRANSMIT block design

3. We PUSH into the TRANSMIT block and create the midlevel block design of Figure 24.3.

> Block PAM (pulse-amplitude-modulation) samples the analog input signals, block PCM (pulse-code-modulation) converts each analog signal to digital, and block P2S (parallel-to-serial) converts each 4-bit digital word to serial and transmits it to output line TRUNK.

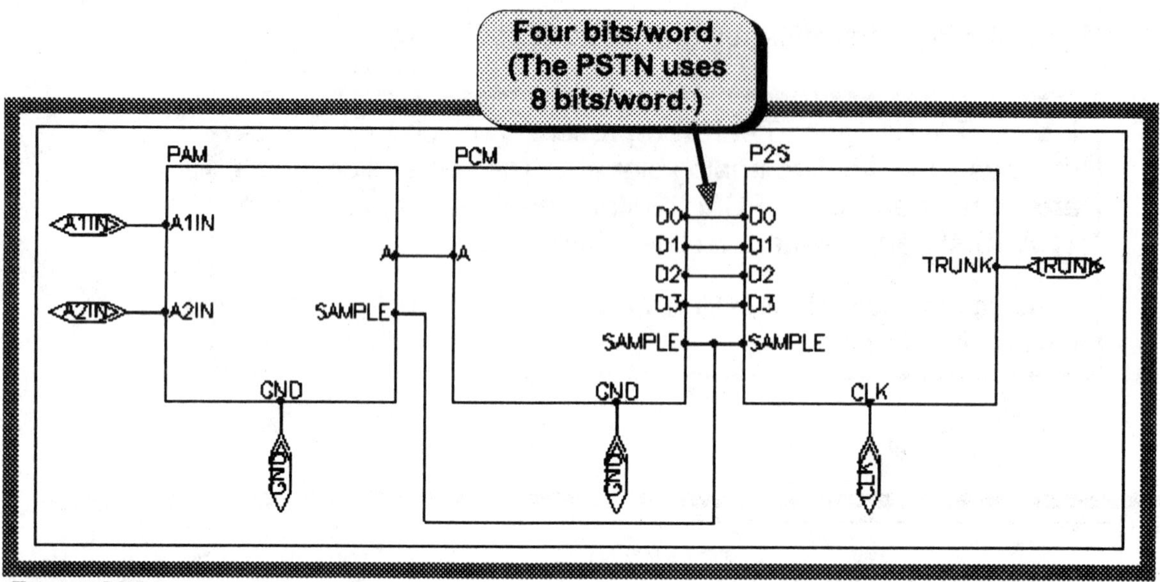

FIGURE 24.3
PCM TRANSMIT
midlevel design

4. We push first into block PAM and create the bottom-level design of Figure 24.4.

> The two analog inputs (A1IN and A2IN) are sampled (multiplexed) and presented at the analog output (A). Input TOGGLE times the multiplex rate by successively clocking the flip-flop and shorting the input signals through M1 and M2.

5. Next, we return to the midlevel, push into the PCM block, and create the circuit of Figure 24.5.

> Each time SAMPLE goes active high, the multiplexed analog voltage (A) is converted to 8-bit digital by IC ADC8break and presented at output lines DB0 to DB7.
>
> Setting VREF to 256V calibrates the system for direct binary conversion (10V in equals 1010B out, etc.).

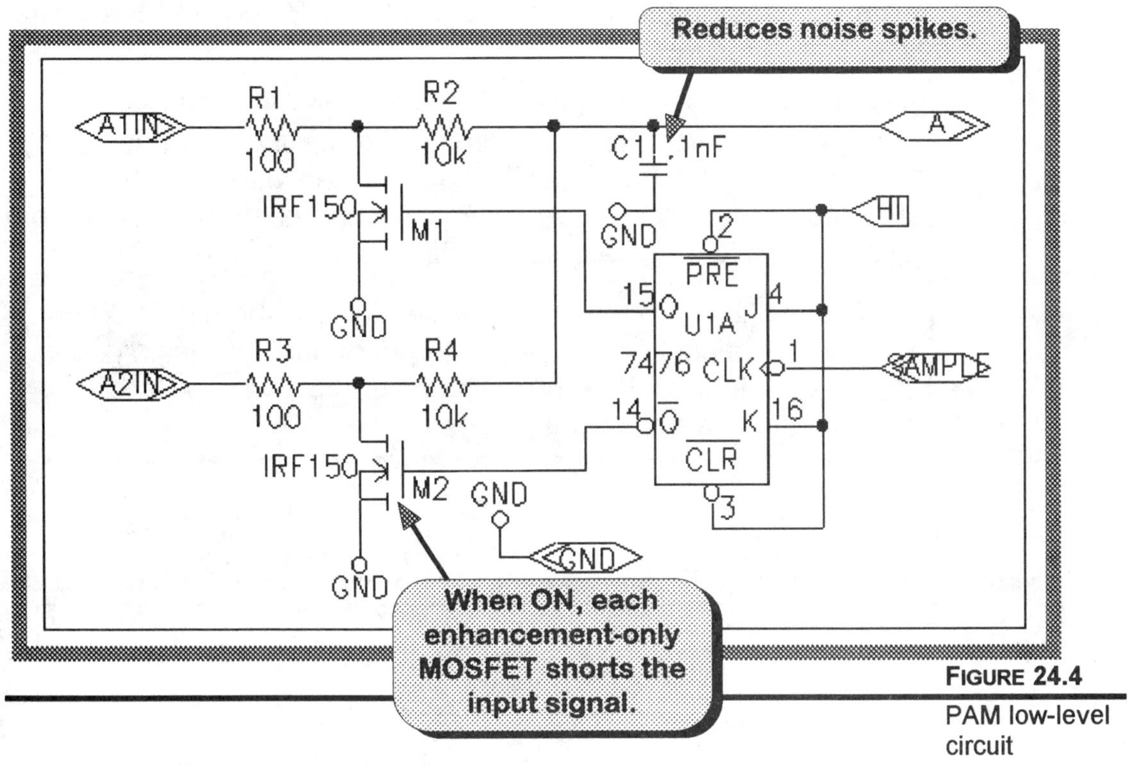

FIGURE 24.4
PAM low-level circuit

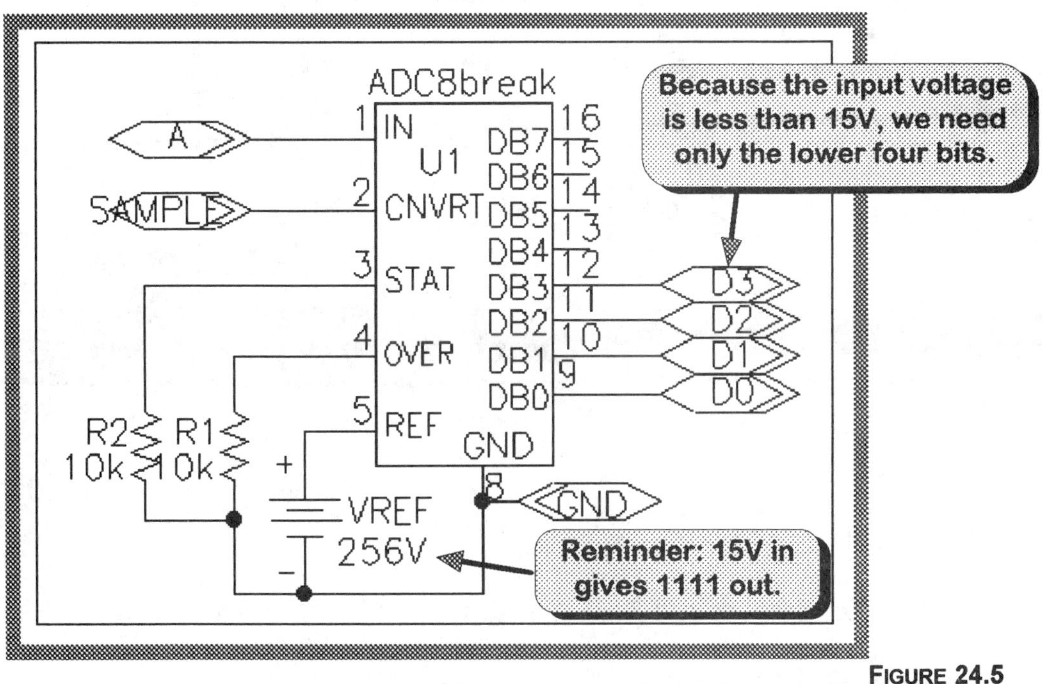

FIGURE 24.5
PCM low-level circuit

6. Again we return to the midlevel, push into the P2S block, and create the circuit of Figure 24.6.

> The input parallel bits are presented to the four input lines of the 74153 dual 4-input multiplexer. In step with CLK, the QA and QB output lines of the 74163 synchronous binary counter cycle from 0 to 3 and are sent to the multiplexer's select lines.
>
> The multiplexer steers the parallel data to the single output line as the select lines are activated in order. QB, which goes high every fourth clock pulse, is returned to the rest of the circuit to time the sampling process.

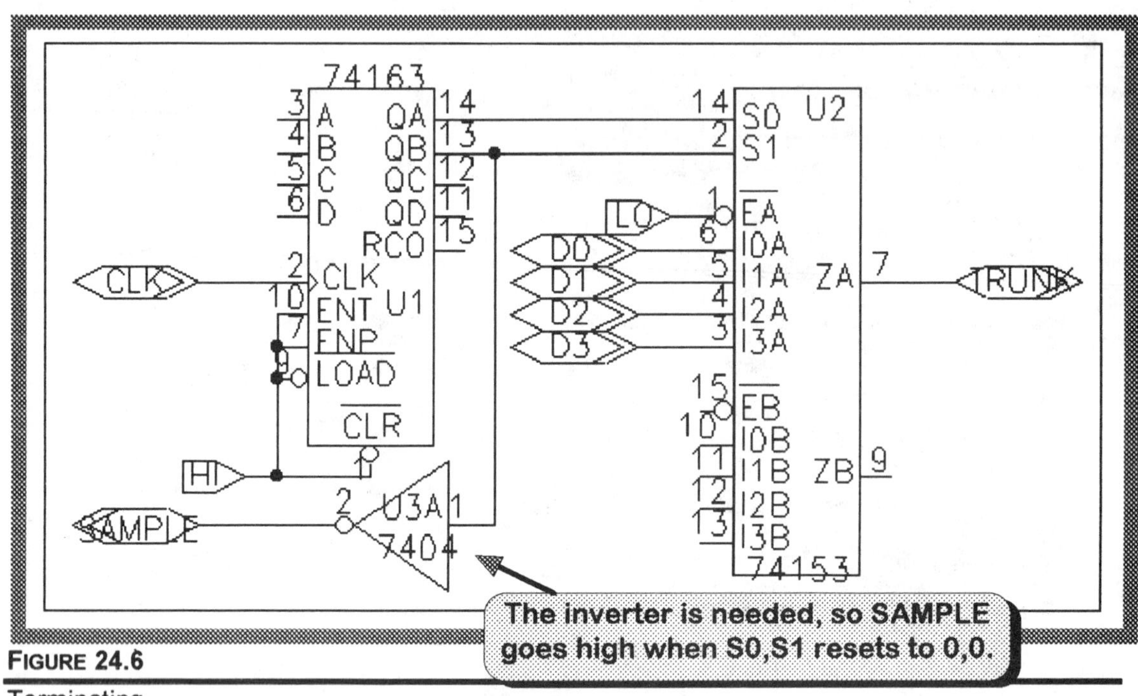

FIGURE 24.6
Terminating block P2S

Testing our design

If necessary, initialize all flip-flops to 0 by *Sets up the simulation analysis for active* toolbar button, **Digital Setup**, **All 0**, **OK**. (Timing should already be **Typical**.)

7. Analyze the circuit and enter Probe. Ignore for now (**Cancel**) any digital simulation errors that may occur.

8. First show the input analog waveforms (A1IN and A2IN) of Figure 24.7. Do the waveforms match the attributes of TALK1 and TALK2 from Figure 24.2?

   Yes **No**

> Suggestion: To generate waveforms, consider pushing into blocks as necessary and placing markers.

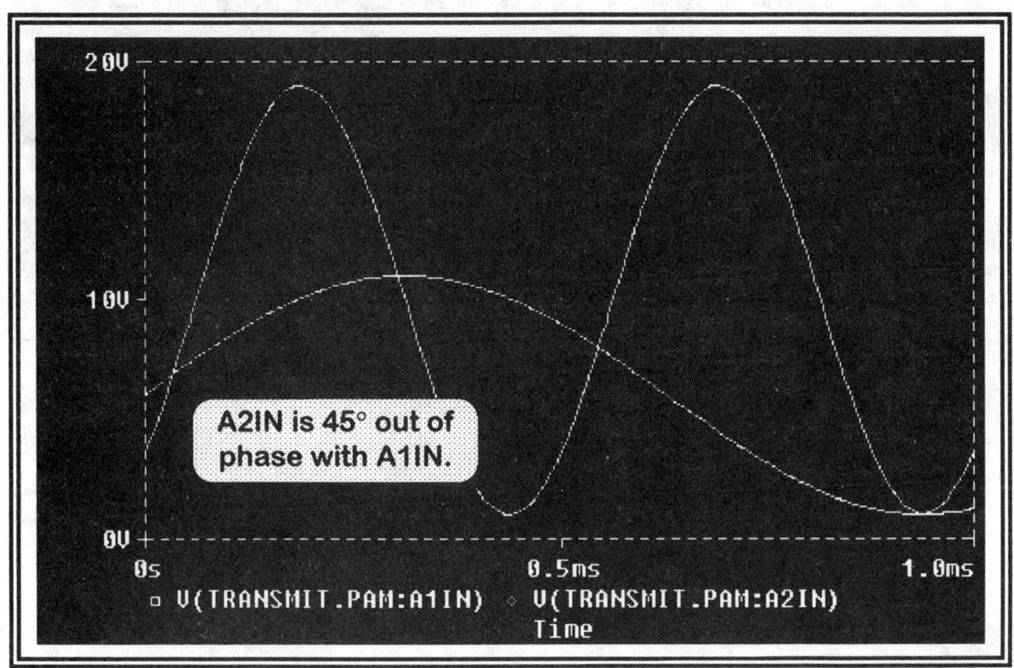

FIGURE 24.7
Input analog waveforms

9. Erase the two input analog waveforms of Figure 24.7 (**Trace**, **Delete All**). Add the one analog and three digital trace variables of Figure 24.8. Each sampling process begins with a rising SAMPLE signal.

Zoom Area

> If you wish, expand the curves for easier viewing with *Zoom Area*, **CLICKLH** and drag expansion bars. *Zoom All* to return.

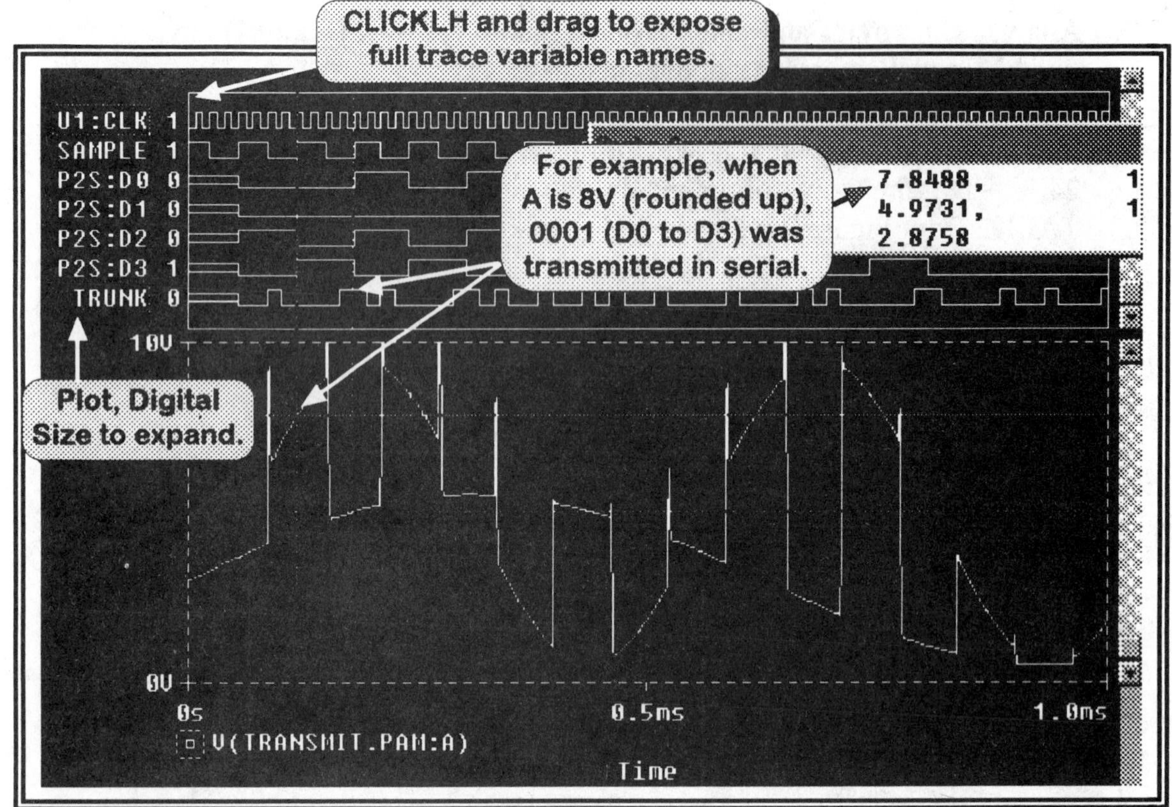

FIGURE 24.7

Internal PCM waveforms

10. Referring to Figure 24.7:

    (a) Is the PAM circuit working properly? (Are the input signals alternately sampled?)

    Yes **No**

    (b) Is the PCM circuit working properly? (Do the parallel digital output signals match their analog counterparts?)

    Yes **No**

    (c) Is the P2S circuit working properly? (Does each complete cycle transmit 4 bits of serial data corresponding to the analog input voltage?)

    Yes **No**

    (d) Is each phone sampled at 8000Hz (8 times in 1 ms)?

    Yes **No**

11. Does the system obey the *Nyquist* theorem? (Is the sampling rate for both A1IN and A2IN at least twice the input frequency?)

    Yes **No**

12. Perform a *worst case* analysis on the circuit. Does reducing the clock speed solve the problem? (If not, is it reasonable that the ADC device may contain internal loops that are invalid under worst case analysis?)

Test and modification suggestions

Reminder: Because we have already reached the maximum component count for the evaluation version, most of these modifications will have to be made separately, on paper, or with a full-fledged version of PSpice. Report all findings on a separate sheet.

13. Design and test the RECEIVE block of our PCM system. (Hint: Reverse the PAM, PCM, and P2S processes.)
14. Expand the system to multiplex four phones.
15. Redesign the system for 8-bit operation.
16. Use a 74194 shift register, instead of the multiplex, to carry out the P2S block.
17. Redesign the PAM block of Figure 24.3 to eliminate crosstalk. (Hint: Make use of an ICVS op amp.)

QUESTIONS AND PROBLEMS

1. Why are we guaranteed that switches M1 and M2 of block PAM can never be ON at the same time?

2. According to the Nyquist rule, to avoid losing data, we must sample the input analog signals at twice the highest frequency of interest. Based on this, what is the highest analog input frequency that the circuit of Figure 24.2 can sample without losing data?

3. The basic PCM multiplexing unit for the phone system consists of 24 phones. If each phone is sampled at the 8kHz rate, and each analog voltage is converted to 8-bit digital, what is the serial output bit rate from this basic unit? (Why do you think it's slightly less than the telephone system's 1.544MHz T1 rate?)

4. With our system, the transmitter and receiver were synchronized with a separate clock line. In the real phone system, there is no separate clock line. How, then, does the phone system remain synchronized? (Hint: What is a *phased-locked-loop*?)

5. Why does TDM (time-division multiplexing) tend to produce noise spikes?

6. If the transmission block consists of a multiplexer, analog-to-digital converter, and parallel-to-serial converter, what should the receive block consist of?

CHAPTER 25

Serial Communication

The UART

OBJECTIVE

- To design and test a UART-based RS232C serial communication system.

DISCUSSION

Our third application performs a classic duty: computer-to-computer serial communication using the asynchronous RS-232C protocol standard of Figure 25.1.

FIGURE 25.1
RS-232 asynchronous serial format

In order to synchronize the transmitter and receiver, each 8 bits of data are preceded by an active-low start bit and two active-high stop bits. When the receiver detects each high-to-low transition of the start bit, all clocks are reset.

At the transmitting end, each 8 bits of parallel data are converted to serial, the start and stop bits are factored in, and the serial data are transmitted. At the receiving end, the process is reversed. The serial data are received, the start and stop bits are stripped off, and the resulting serial data are converted to parallel.

For efficiency, both the transmitting and receiving processes are normally carried out by a UART (universal asynchronous receiver transmitter) chip. In this chapter, we use a multiplexer pair as an unusual way of simulating the action of the UART.

SIMULATION PRACTICE

1. The circuit of Figure 25.2 shows a complete RS-232C system, including transmission and reception. However, due to the limitations of the evaluation version, we restrict our initial design to the transmission block.

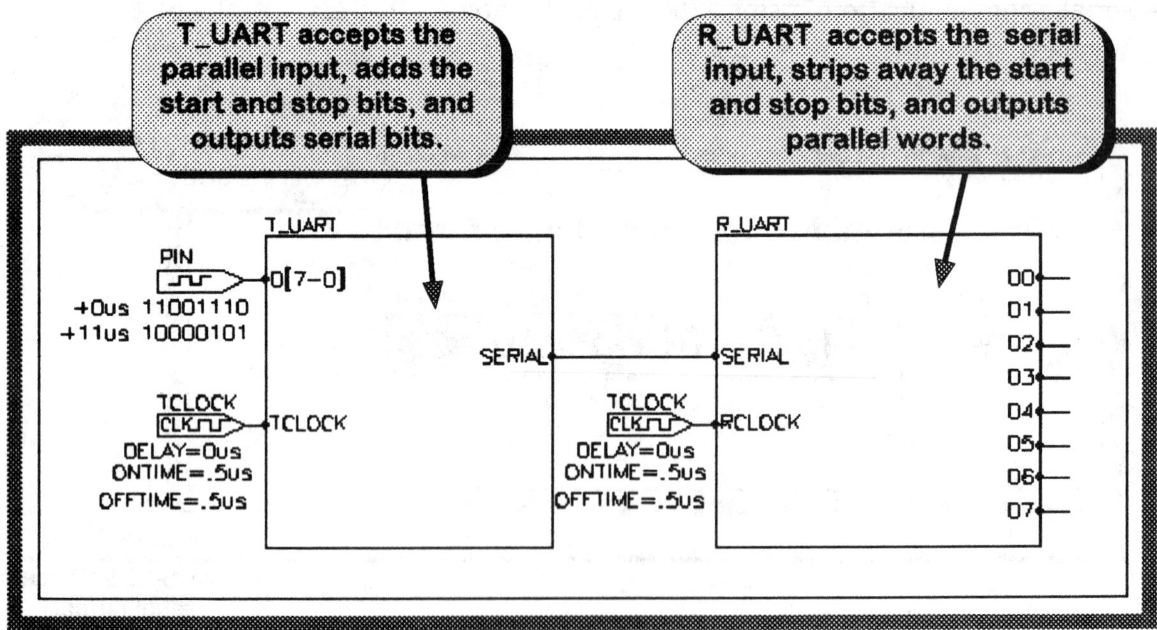

FIGURE 25.2

RS-232 top-level design

2. Draw the top-level block design of Figure 25.3.

> 8-bit parallel data enters block T_UART by way of digital source PIN (parallel-in). START and STOP bits are added and the resulting serial data are transmitted from pin SERIAL in the following order: one start bit, eight data bits D0 to D7, two stop bits—for a total of 11 bits.

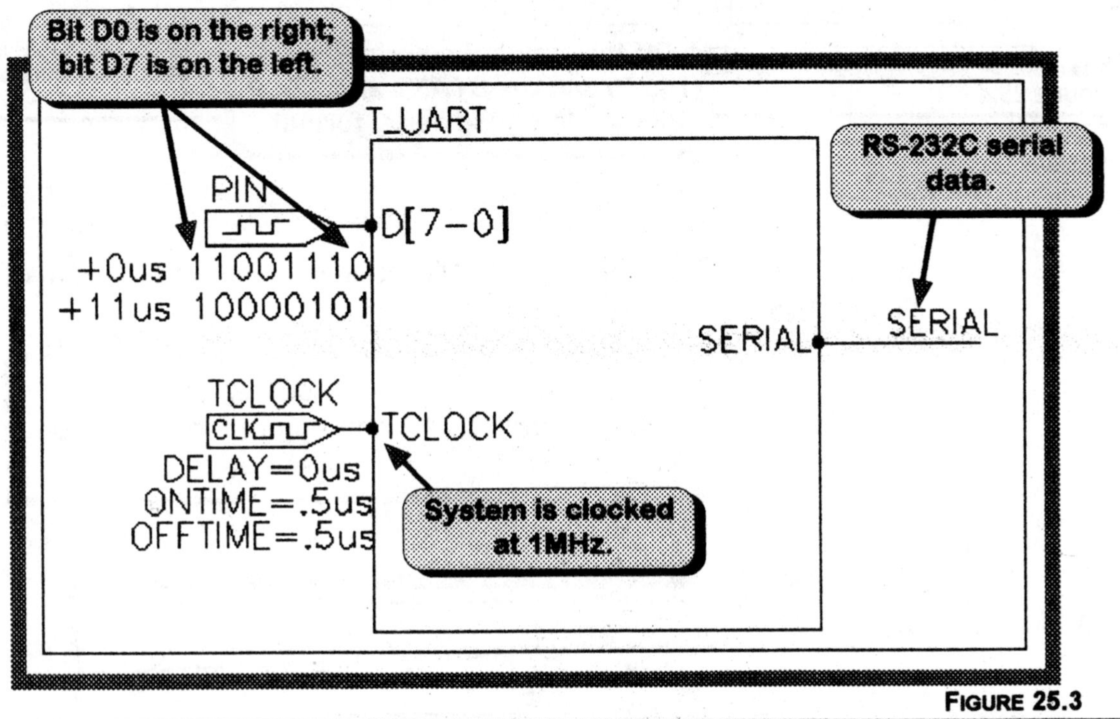

FIGURE 25.3
TRANSMIT top-level design

3. We PUSH into block T_UART and create the midlevel block design of Figure 25.4.

> Block COUNTER counts from 0 to 10. Block MUX accepts these counts, as well as the original parallel input data (D[7-0]), and outputs the serial data and the START and STOP bits in coded serial form on four output lines (G0 to G3). Block GATES inputs the coded serial data and combines them into a single RS-232C serial output.

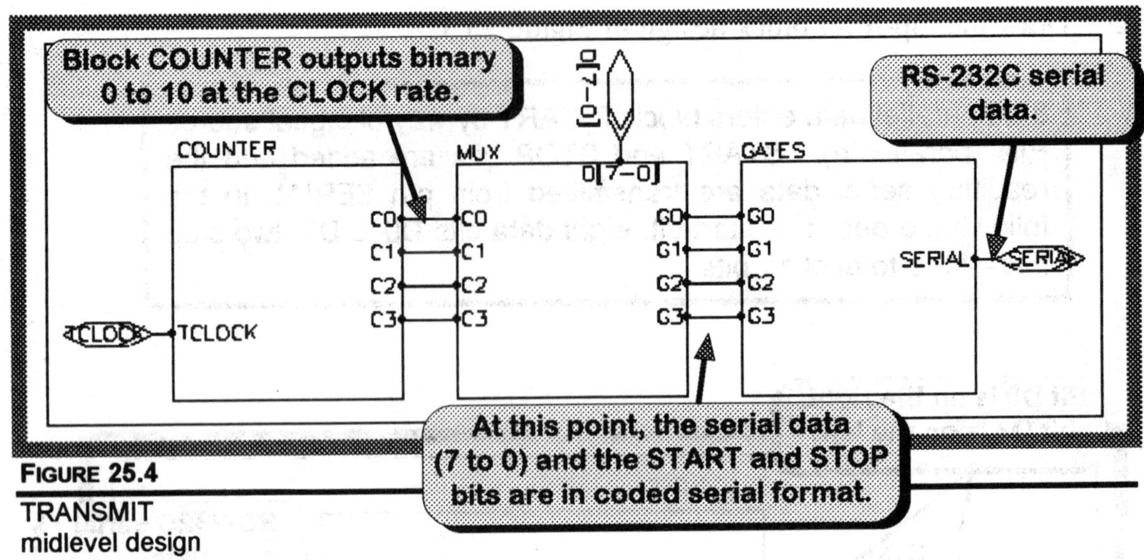

FIGURE 25.4
TRANSMIT midlevel design

4. We PUSH into the block COUNTER and create the design of Figure 25.5.

> Block COUNTER uses a synchronous counter chip (74161) and feedback to count from 0 to 10 and output the count from C0 to C3.

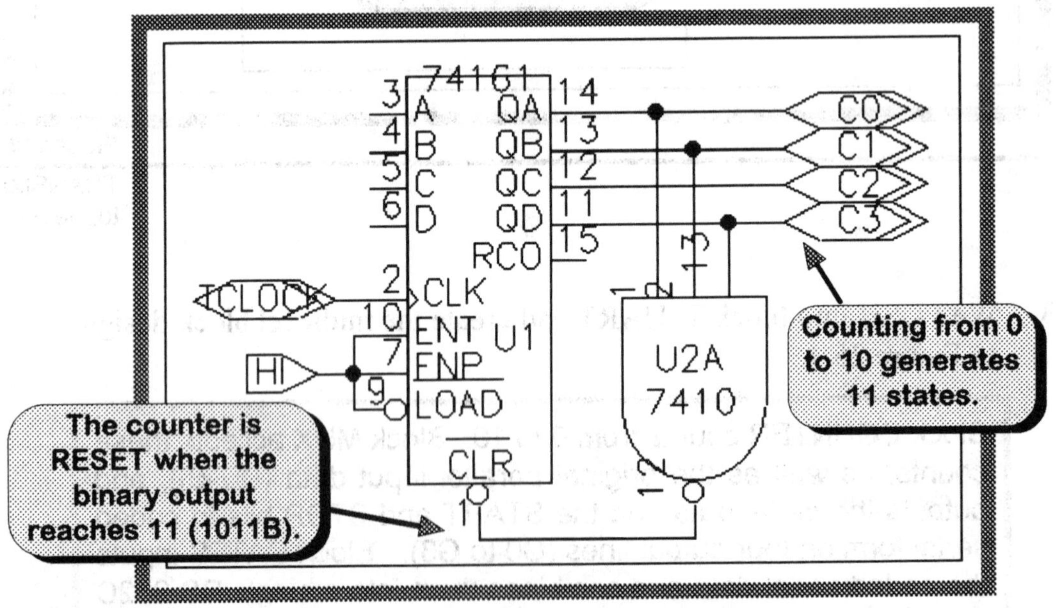

FIGURE 25.5
Block COUNTER design

5. With the block COUNTER done, we POP back to the midlevel, PUSH into block MUX, and create the design of Figure 25.6.

> To simulate a UART, the 0 to 10 binary sequence from block COUNTER is sent to a pair of 74151A 8-in/1-out multiplexers. Starting at the top, we first output the START bit, then the 8 data bits, and finally the two STOP bits. The serial outputs from each multiplexer (G0 and G3), as well as two enabling signals (G1 and G2), are output to block GATES.

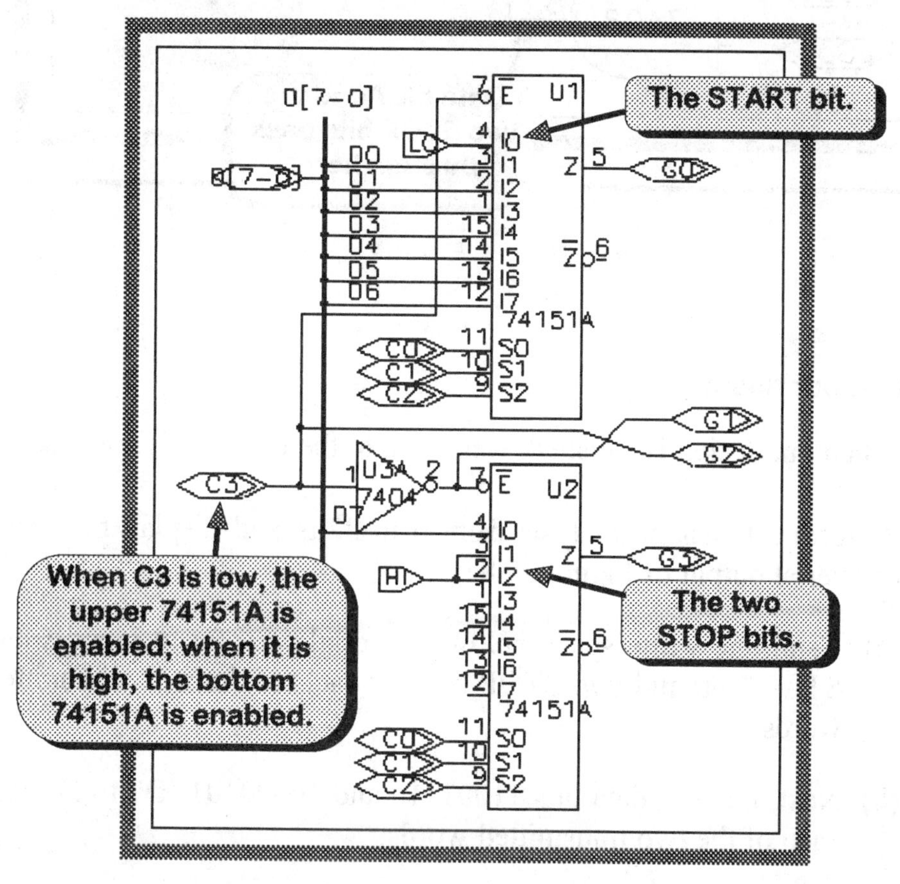

FIGURE 25.6
Block MUX design

6. Finally, we repeat the process to create the block GATES of Figure 25.7.

> Block GATES combines the serial outputs from the two multiplexers. Steering logic G1 and G2 switches the serial flow as we move from the upper multiplexer to the lower.

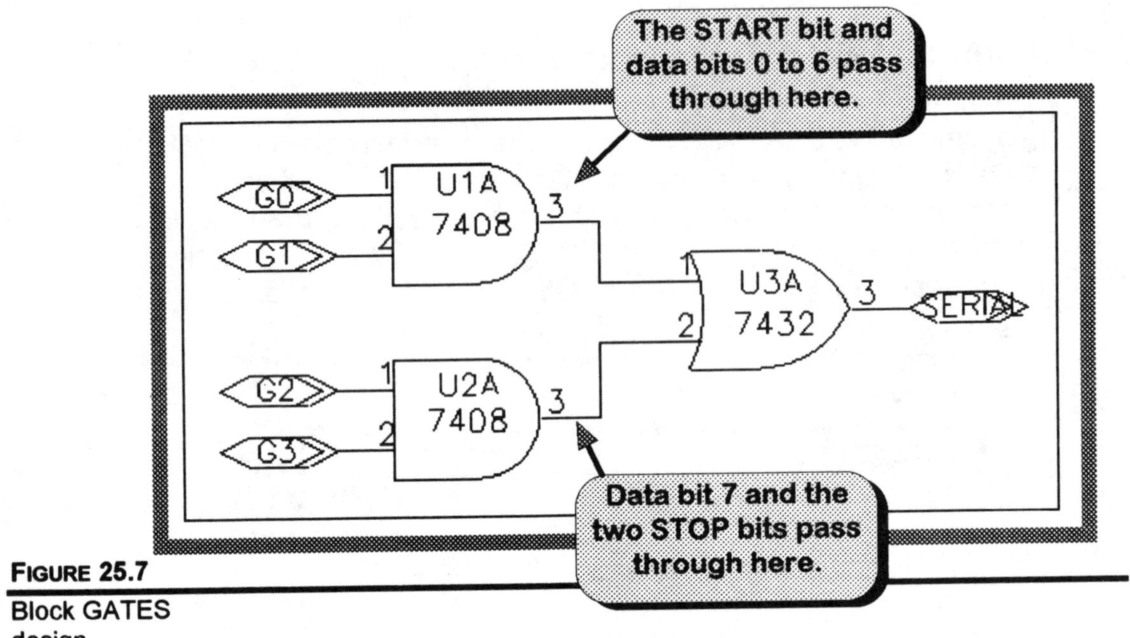

FIGURE 25.7
Block GATES design

Testing our design

7. When our design is finished, we return to the top (**Navigate, Top**).

8. Perform a transient analysis from 0 to 22μs and display the clock and serial output of Figure 25.8.

   (a) Using the white space below the serial waveform, identify the START bit and two STOP bits for each of the two transmitted words.

   (b) Next, identify data bits 11001110 and 10000101 (D7 to D0) for each of the two transmitted words.

   (c) Was D0 or D7 transmitted first? (Circle your answer.)

   D0 **D7**

   (d) Did the transition from one bit to the next take place on a rising clock pulse or a falling clock pulse?

   Rising **Falling**

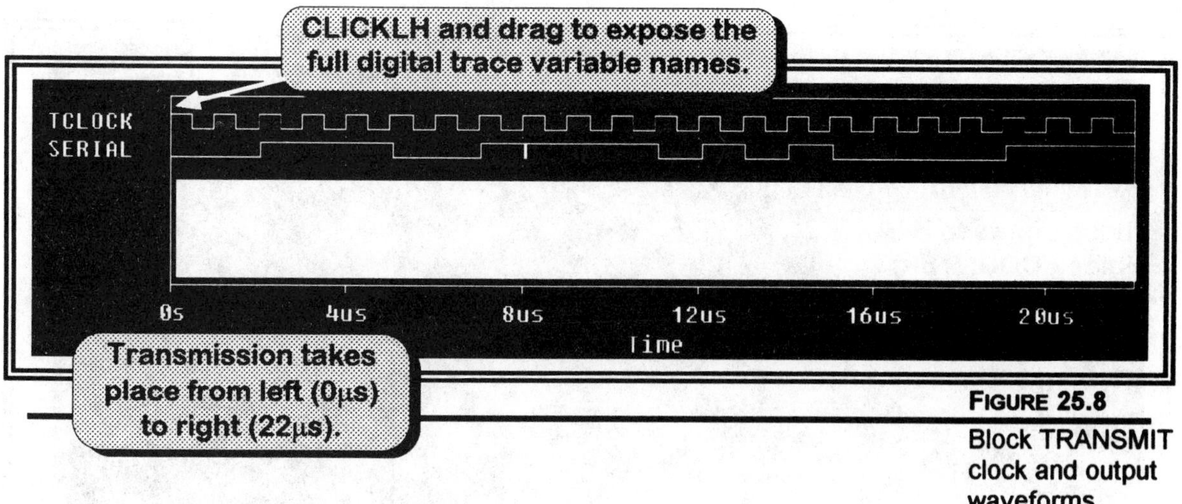

FIGURE 25.8
Block TRANSMIT clock and output waveforms

9. View additional signals of your choice (for example, the counter output to check for 0 to 10 operation).

Test and modification suggestions

10. Show how to perform both the transmit and receive UART processes more directly using three 74194 universal shift registers (three for transmission and three for reception).

11. Perform a worst case digital analysis of the TRANSMIT circuit and generate the output shown in Figure 25.9.

    (a) What caused the hazard?

    (b) What action can you take to eliminate the hazard?

12. Modify the TRANSMIT circuit to include only a single STOP bit.

13. Modify the TRANSMIT circuit so that if no parallel input is available, a HIGH (MARK) is transmitted continuously until the next parallel data byte is available.

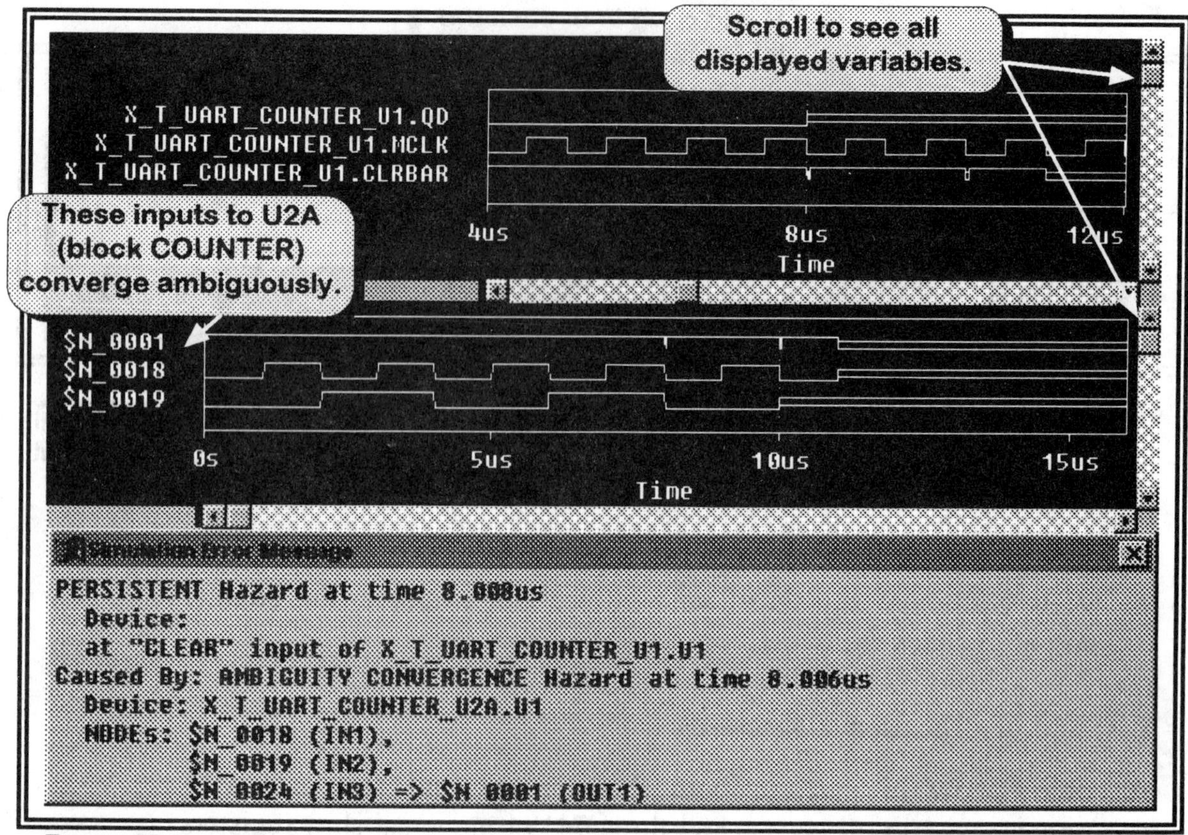

FIGURE 25.9

Block TRANSMIT worst case hazard

14. Design a circuit to carry out the RECEIVE block.

QUESTIONS AND PROBLEMS

1. During RS-232 asynchronous communication, what is the purpose of the START bit? Of the two STOP bits?

2. What is the major advantage of serial communication over parallel communication?

3. As each serial byte is transmitted, what is the percentage of overhead (START and STOP bits) required?

4. What is the difference between *synchronous* and *asynchronous* serial transmission? (Hint: Which one requires a phased-locked-loop?)

CHAPTER 26

Big-Ben Chime Control
The Comparator

OBJECTIVE

- To design and test a digital circuit that controls clock chimes.

DISCUSSION

Many tower clocks (such as London's Big Ben) sound the time each hour by generating the corresponding number of chimes. A main clock counts continuously in hours from 1 to 12. At each hour transition, the corresponding number of chime pulses is sounded.

SIMULATION PRACTICE

1. Draw the top-level block design of Figure 26.1.

> Pulse trains HOUR and SECOND enter block **BIGBEN**, and the output to the chime generator appears at line BELL. Every hour, a series of chime signals equal to the hour (from 1 to 12) appears from output line BELL at the rate of one chime every second.
>
> *In order to shorten computation times during development, all clock sources have been greatly speeded up. For example, the 1 hour counter clocks at 10μs, and the 1 second counter clocks at .2μs (200ns). After the design has been confirmed, all clocks can be reduced to normal speed for final testing.*

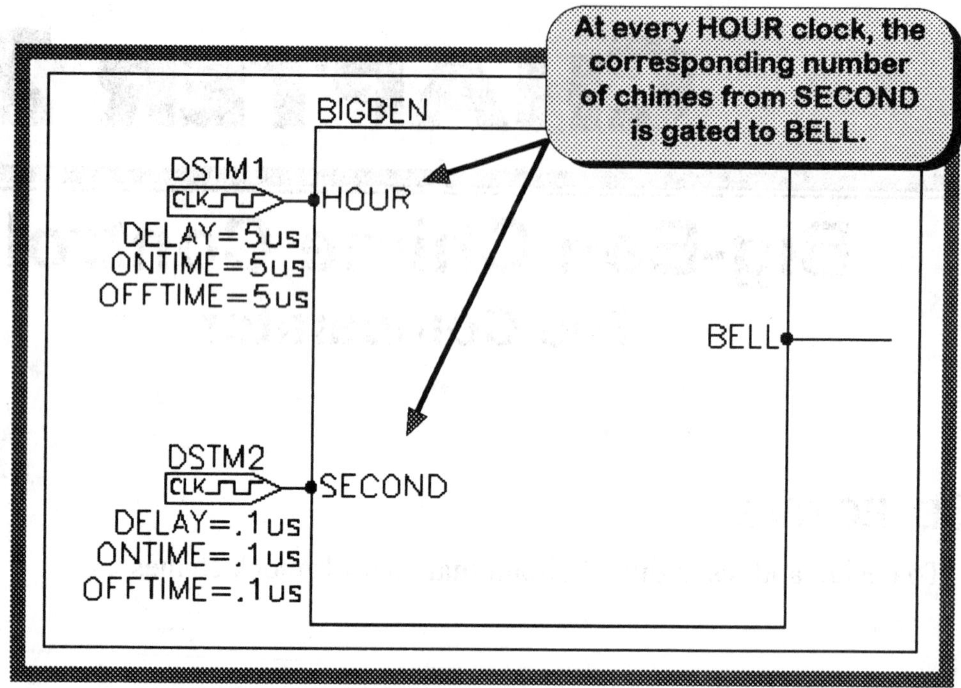

FIGURE 26.1
Big-Ben top-level design

2. We PUSH into block BIGBEN and create the midlevel design of Figure 26.2.

> **BLOCK HOURCTR** counts up at the normal rate of once per hour from 1 to 12 and outputs the time as a digital word on H0 to H3. The hour clock is passed on to block SECONDCTR via line START.
>
> **BLOCK COMPARE** takes digital inputs from HOURCTR and SECONDCTR and outputs a CLEAR pulse when the two are equal.
>
> **BLOCK SECONDCTR** gates the clock pulses of source SECOND to output BELL. The gate circuit is enabled upon each START pulse from HOURCTR. The count value is sent to block COMPARE. When the count reaches the proper value, a CLEAR is issued and the gate circuit is disabled.

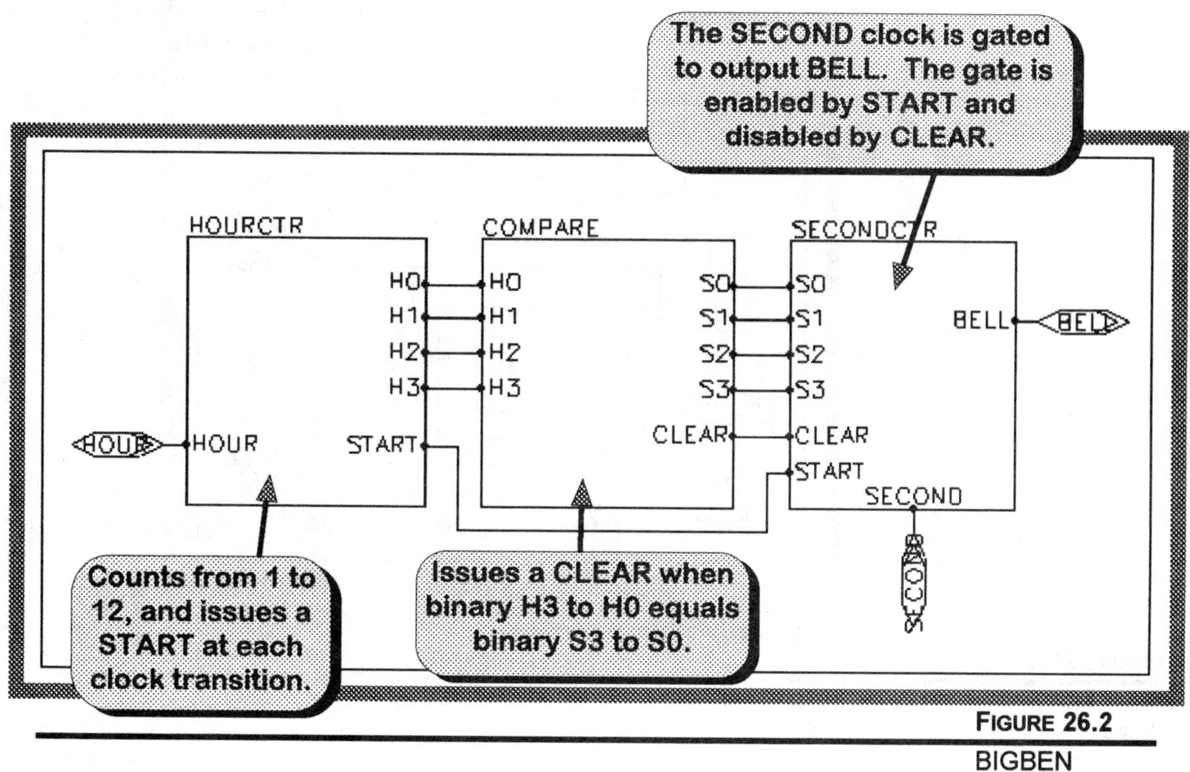

FIGURE 26.2
BIGBEN midlevel design

3. We PUSH into the HOURCTR block and create the design of Figure 26.3.

> Block **HOURCTR** uses a synchronous counter chip (74161) and feedback to count from 1 to 12 and output the count from lines H0 to H3. Output START is sent to block SECONDCTR to begin the chime process.

4. When the HOURCTR block is done, we POP back to the midlevel, PUSH into block COMPARE, and create the design of Figure 26.4.

> Block **COMPARE** accepts binary values from blocks HOURCTR (H0 to H3) and block SECONDCTR (S0 to S3). When the two counts are equal, a CLEAR pulse is sent to block SECONDCTR.

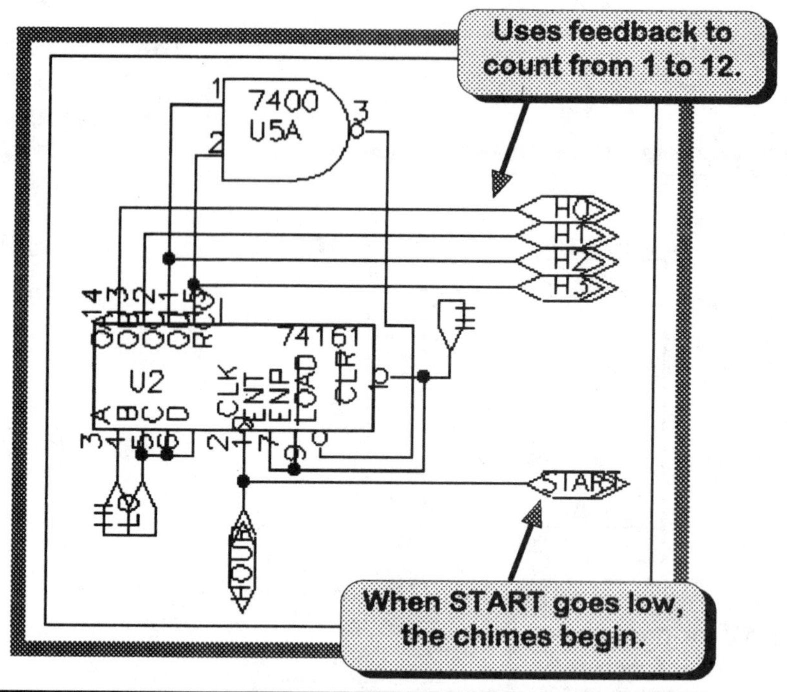

FIGURE 26.3

Block HOURCTR design

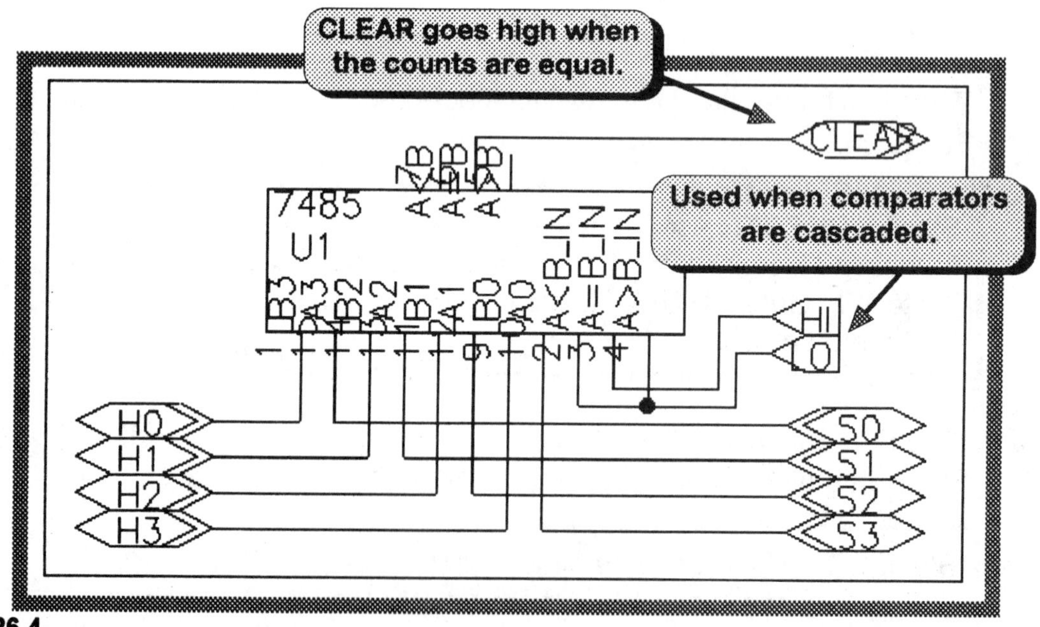

FIGURE 26.4

Block COMPARE design

5. Finally, we repeat the process to create block SECONDCTR of Figure 26.5.

> Block **SECONDCTR** receives the START signal, which sets the flip-flop and gates the SECOND count out to BELL. Simultaneously, the 7493A counter counts up and sends the count to block COMPARE. When the second count equals the hour count, CLEAR is issued and the second pulses are blocked.

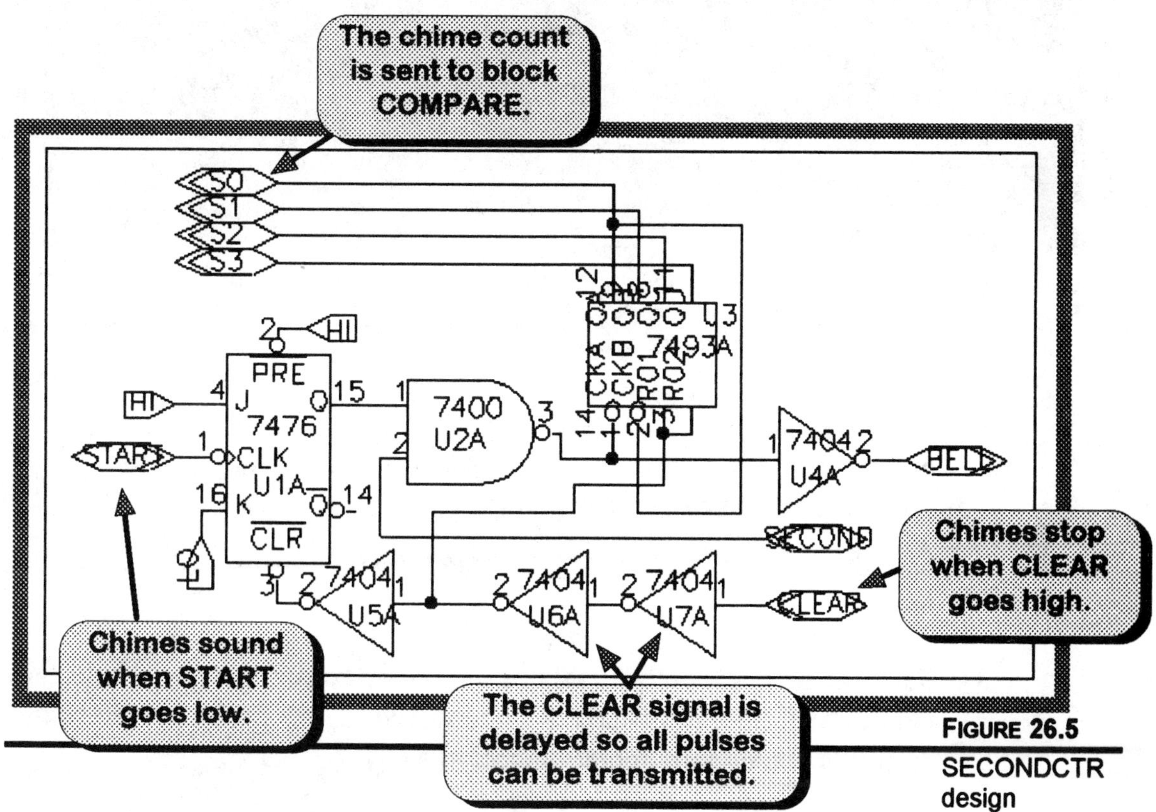

FIGURE 26.5
SECONDCTR design

Testing our design

6. When our design is complete, we perform a transient analysis from 0 to 200μs and display the output (BELL) curves of Figure 26.6. (The bottom curve is an unsynced blowup of hours 3 and 4.)

   Is the circuit functioning as expected?

   Yes **No**

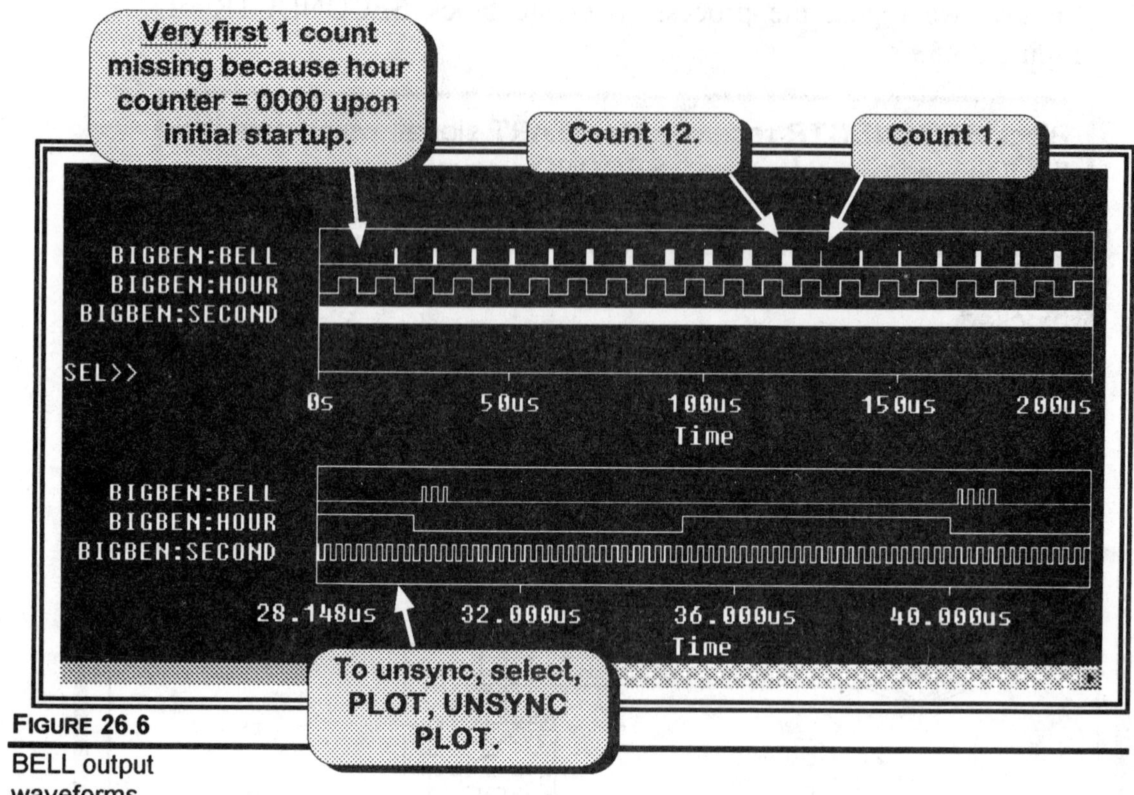

FIGURE 26.6
BELL output waveforms

Test and modification suggestions

7. Modify the design so only a single (second) clock source is required.

8. Perform a worst case analysis of the present circuit. Why would the cumulative ambiguity hazards uncovered be no problem when the circuit is placed into actual operation?

9. Modify the circuit to include an AM/PM indicator.

10. Following up on step 9, further modify the circuit so the chimes are inhibited during the 1A.M. to 5A.M. period (when most people are sleeping).

11. Modify the circuit to sound a single chime at either the half-hour or quarter-hour.

QUESTIONS AND PROBLEMS

1. Although the hour counter counts up very slowly (in real life), why did we choose a high-speed synchronous 74161 counter?

2. When is the CLEAR signal issued?

3. What is accomplished by the R01 and R02 inputs to the 7493A of Figure 26.5?

PART V
PC Boards

The final step in electronic design is to create a printed circuit board (PCB).

MicroSim has made the process especially easy by permitting a MicroSim Schematics design to be read directly into PCBoards.

In the single chapter of Part V, we will create a PCB from a circuit we have designed in a previous chapter.

CHAPTER 27

Printed Circuit Boards
Layouts

OBJECTIVE

- To use MicroSim's *PCBoards* to lay out circuits.

DISCUSSION

The design and development of a circuit is one objective—and by now we are quite adept at this process. However, the production and manufacturing of the circuit is quite another. The parts, components, and traces must be laid out for a *printed circuit board* (PCB) and the control files and artwork made available for downloading. Furthermore, we must have knowledge of the electrical and physical characteristics of the PCB layout. This is the purpose of *MicroSim PCBoards*.

Layers and Layout Objects

The central feature of *MicroSim PCBoards* is the use of *layers*. All PCB designs are presented as a collection of 19 or more layers of functionally related *layout objects*, each classified as either type *signal* or type *graphic*. Loosely speaking, these layers correspond to the final artwork images required to complete the manufacturing of a PCB.

Two of the most important layers are *component* and *solder*. The component layer shows all the metalized pads and traces on the top (component) side of the PCB; the solder layer shows all metalized pads and traces on the bottom (solder) side of the PCB.

During PCB design, we refer to various layers as we perform the following steps:

- Package and netlist
- Create the layout
- Define the border
- Arrange the components
- Define the signal area
- Route the traces
- Add power and ground
- Finalize the design

To emphasize the start-to-finish feature of DesignLab, we will produce a PCB layout for a familiar circuit designed in a previous chapter.

SIMULATION PRACTICE

1. Draw or bring back the op amp clamper circuit of Figure 27.1.

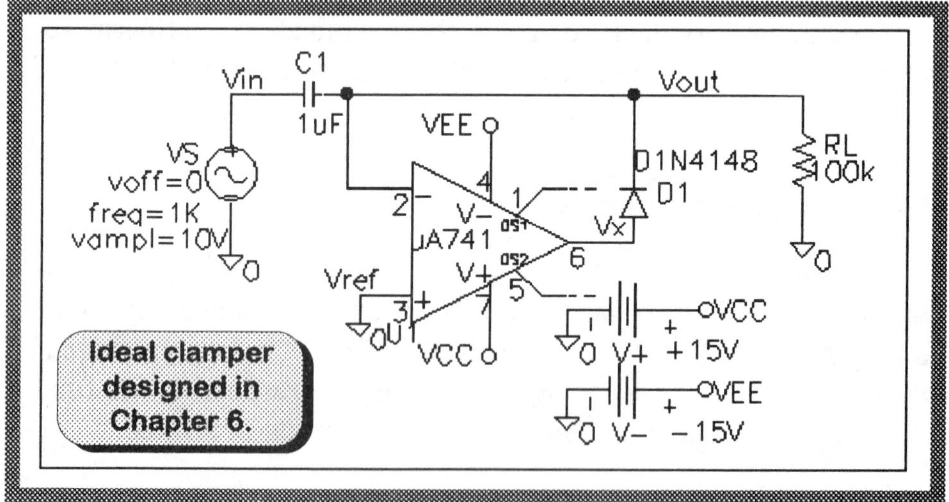

FIGURE 27.1
Op amp clamper design

Package and netlist

2. The first step is to *package* the circuit to create the list of gates and pin assignments. This is accomplished by **Tools**, **Package** to bring up the *Package* dialog box of Figure 27.2. **OK** to select all default assignments.

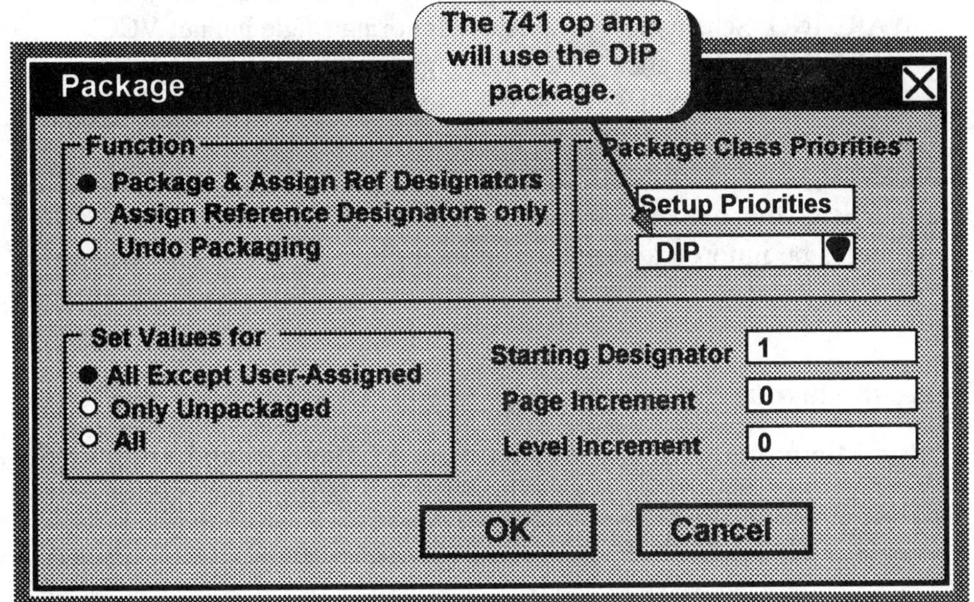

FIGURE 27.2
Package dialog box

3. Note that during packaging, a warning was issued. Bring up the *MicroSim Message Viewer* dialog box and note the following:

INFO Schematics 01:18AM Creating PCBOARDS netlist...
WARNING Schematics 01:18AM Simulation only part will not be packaged VS
INFO Schematics 01:18AM Netlist/ERC warnings occurred

Based on this warning, will the voltage source (VS) be used for simulation purposes only and not be assigned a place on the layout?

Yes **No**

4. We then produce the special netlist for layout design by **Tools, Create Layout Netlist**. Again, a number of warnings were issued by the MicroSim Message Viewer.

INFO	Schematics	10:20AM	Creating PCBOARDS netlist...
WARNING	Schematics	10:20AM	Ignoring simulation only part VS
WARNING	Schematics	10:20AM	Ignoring simulation only part V+
WARNING	Schematics	10:20AM	Ignoring simulation only part V−
WARNING	Schematics	10:20AM	Ignoring single pin net VCC
WARNING	Schematics	10:20AM	Ignoring single pin net VEE
WARNING	Schematics	10:20AM	Ignoring single pin net Vin
INFO	Schematics	10:20AM	Netlist/ERC warnings occurred

Scan the list of warnings. Will all power supply and source voltages be ignored and not assigned a place on the layout design?

Yes **No**

Create the layout

5. Run MicroSim PCBoards (**Tools**, **Run PCBoards**) and bring up the *PCBoards Layout Editor* of Figure 27.3. Within the layout grid is the default layout design of Figure 27.4.

> Special note: *Run PCBOARDS* automatically creates the netlist. We performed the netlist process separately in step 4 simply to concentrate on the warnings.

6. The default layout design of Figure 27.4 consists of up to 19 system defined layers, each represented by a default color code. Review *MicroSim PCBoards Note 27.1* for a brief description of each layer.

7. To display the present combination of layers: **Configure**, **Layer Display**, to bring up the *layer display* dialog box. The blackened layers are those presently selected and displayed by default. The layer tagged with an asterisk is the default *current* layer.

   (a) Are all but the paste and mask layers presently selected?

   Yes **No**

   (b) Is the component layer the current layer?

   Yes **No**

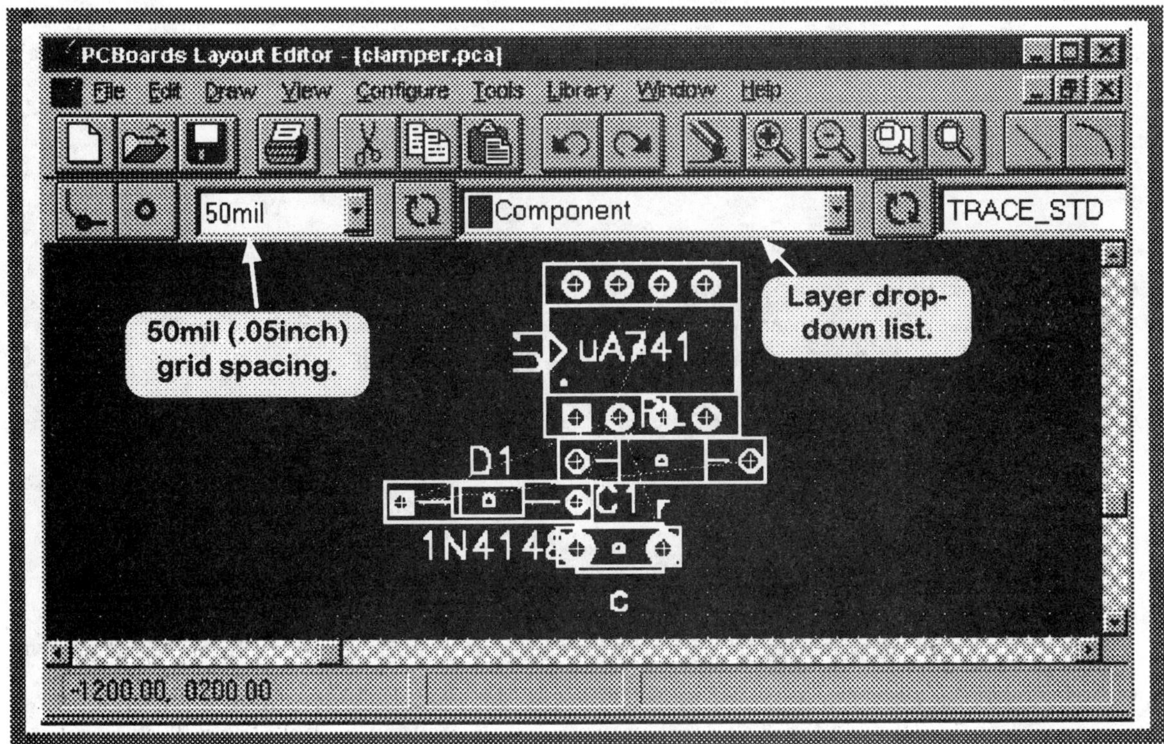

FIGURE 27.3
PCBoards window

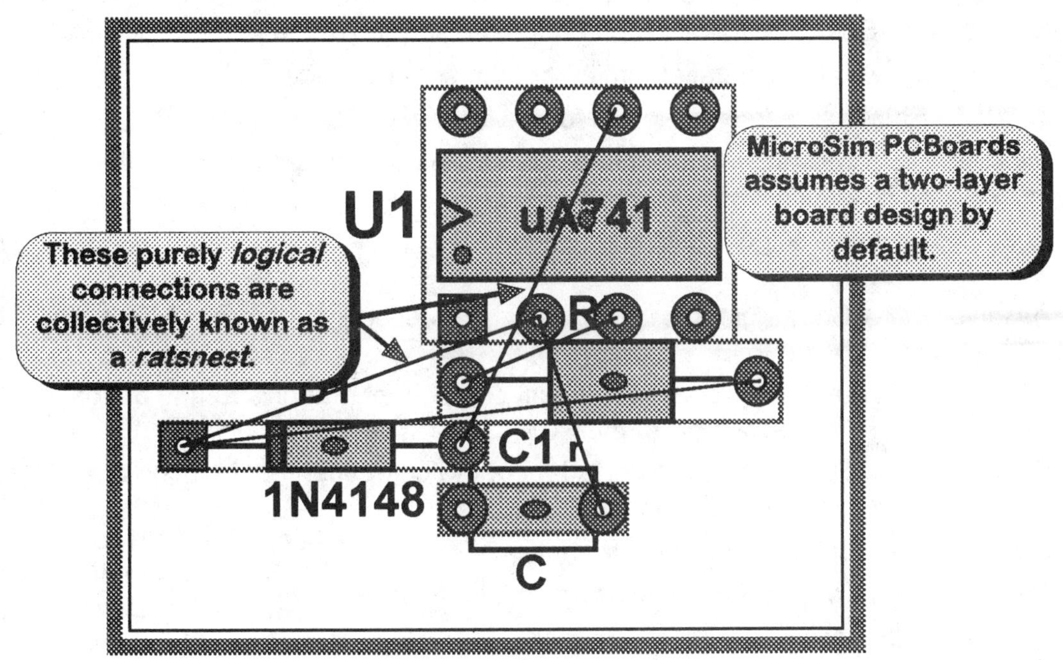

FIGURE 27.4
Clamper circuit default layout

MicroSim PCBoards Note 27.1

What is the purpose of each of the 19 system-defined PCB layers?

Each of the 19 predefined (system-defined) layers is of type *graphic* or type *signal*. Graphic layers accept graphic objects related to text, silkscreen, paste, and mask processes. Signal layers accept electrical objects related to metalized pads and traces. Each of the 19 system-defined layers is briefly described as follows:

- *BoardOutline*—Displays the perimeter of the board.
- *BoardSigKeepin*—Displays the boundary within which electrical objects are allowed.
- *BoundaryTop*—Shows the physical extent of the component footprint on the top surface of the PCB.
- *AssemblyTop*—Shows graphic layout objects representing the top assembly of the PCB.
- *SMTAssemblyTop*—Shows surface mount component graphics for the top assembly of the PCB.
- *PasteTop*—Shows solder paste shapes related to pads on the top of the PCB.
- *MaskTop*—Shows solder mask shapes related to pads on the top of the PCB.
- *SilkTop*—Shows silkscreen graphics for the top surface of the PCB.
- *Component*—Shows all metalization on the top surface of the PCB.
- *Solder*—Shows all metalization on the bottom surface of the PCB
- *SilkBot*—Shows silkscreen graphics for the bottom surface of the PCB.
- *MaskBot*—Shows solder mask shapes related to pads on the bottom of the PCB.
- *PasteBot*—Shows solder paste shapes related to pads on the bottom of the PCB.
- *AssemblyBot*—Shows graphic layout objects representing the bottom assembly of the PCB.
- *SMTAssemblyBot*—Shows surface mount component graphics for the bottom assembly of the PCB.
- *BoundaryBot*—Shows the physical extent of the component footprint on the bottom surface of the PCB.
- *Drill*—Shows drill coding symbols for holes.
- *Ratsnest*—Shows logical electrical connections (rats).
- *DRC*—Shows design rule check violations.

MicroSim PCBoards allows up to 45 additional *user-defined* layers.

8. Select the *component* and *silktop* layers and make the component layer current, **OK**.

> To select or deselect any layer, **CLICKL** on that layer name. To make any layer the current layer, **DCLICKL** on that layer name (or make use of the *Select All*, *Unselect All*, or *Make Current* buttons).

Does the work area display two layers (two colors), and is the current layer (component) displayed in the *layer-dropdown-list*?

Yes **No**

> The *layer-dropdown-list* gives us an alternative way of adding layers and selecting the current layer. (Selecting a layer always makes that layer current. If the layer is not presently displayed, it is added.)
>
> Layers cannot be deleted by way of the *layer-dropdown-list*. To delete layers, we must return to the layer display box (**Configure**, **Layer Display**).

9. Open (**CLICKL**) the *layer-dropdown-list* and scroll to the top of the list. Starting at the top, select one layer at a time until all 19 are displayed. As each layer is selected, note the changes (if any). Are the graphics associated with the current layer (the last one selected or added) always displayed on top?

Yes **No**

10. If necessary, practice using the *layer-display* dialog box and *layer-dropdown-list* until you are comfortable with their display features. When you are done, return to the layout editor with all layers (except DRC) displayed.

11. Make the *ratsnest* layer current, and note the straight-line "as the crow flies" connections between electrical pads.

Is it reasonable that these purely *logical* connections (known as a *ratsnest*) will later be routed and turned into metalized traces to be etched on the top and bottom of the PCB?

Yes **No**

Placing a border

The next design step is to define the outside border of the PCB. This design from the outside in is in keeping with the real world where space is a major design constraint.

12. We are told by the structural engineer that our PCB must be a square, 1in. on a side, and it must plug into an edge connector. If the grid on your layout editor shows 50mil spacing, and if 1000mils equals 1inch, what is the allowed size in terms of grid units?

    1in. by 1in. = 1000mils by 1000mils = ________ by ________

Draw/ Polyline

13. To draw the board outline of Figure 27.5, open the *Layer-dropdown-list* and select the *BoardOutline* layer (making this layer current). (If necessary, **CLICKL** on the *Zoom In* toolbar button to make room for the border, and *View/Fit* when done.)

    To draw the border in one continuous step: **CLICKL** on the *Draw/Polyline* toolbar button, place crosshairs at any convenient corner, **CLICKL** to place, **DRAG** to draw the first line, **CLICKL** to anchor, **DRAG** to draw the second line, **CLICKL** to anchor, etc. **DCLICKR** to abort. (To draw one line segment at a time: Follow the same steps as above, but **CLICKR** after each segment has been placed.)

    Your layout design should now resemble Figure 27.5.

14. The lower left corner of the border is typically located at (0,0), the location of the crosshairs on the grid. To accomplish this, select the entire layout (**CLICKLH** and draw a box around layout to turn all components white), **DRAG** the mouse to the lower left corner, and **CLICKLH** to move the mouse and layout to the crosshairs at coordinate location 0,0 (as shown in the lower left of the screen).

15. Referring to Figure 27.5, we are told to set aside the left side of the PCB for add-on hardware. Will the diode have to be repositioned?

    Yes **No**

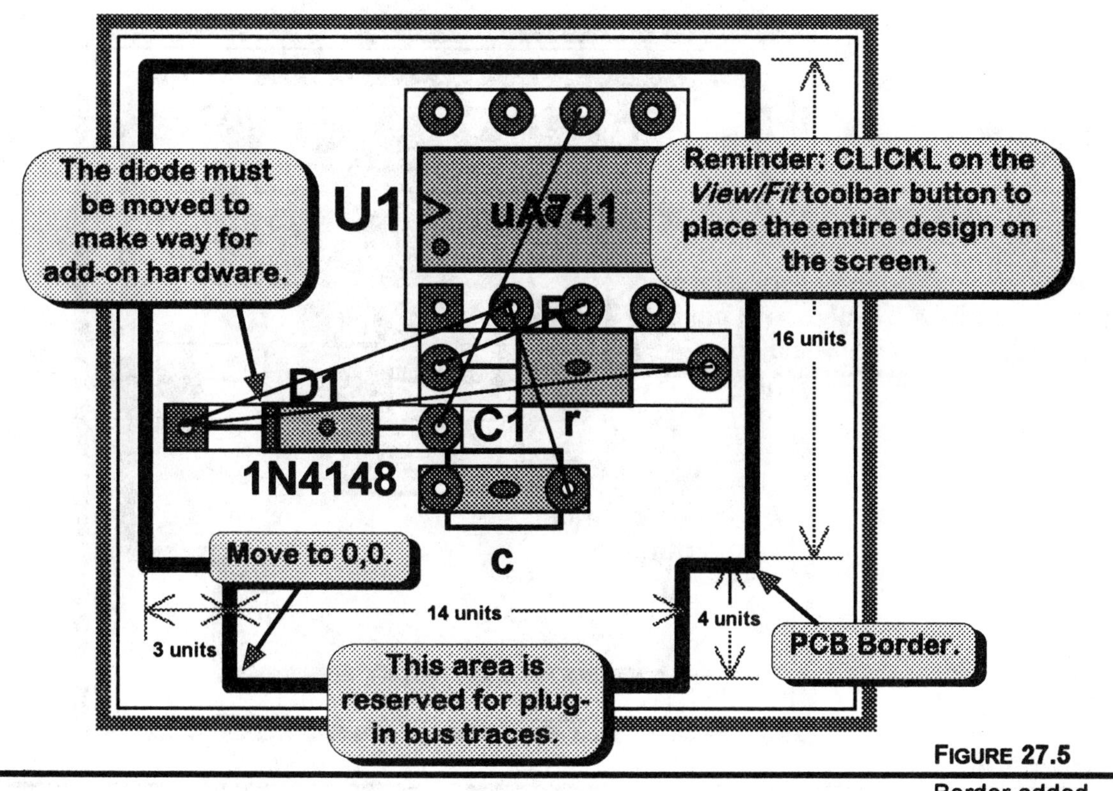

FIGURE 27.5
Border added

Arranging components

> The parts layout initially generated by MicroSim PCBoards should not be taken as a suggested placement pattern. Proper placement of all parts is the responsibility of the designer.

16. In this step, we will move RL, C1, and D1 to achieve a squared off design that allows for add-on hardware and provides for efficient wiring. The results are shown in Figure 27.6.

    (a) First, rotate and move D1. **CLICKL** on component D1 (to select and turn it white). Using **CLICKLH** and **DRAG** to move and Ctrl/R to rotate as necessary, reposition D1 as shown.

    (b) Next, select and move C1 so it fits closely to R1.

    (c) Finally, flip RL and C1 180° (Ctrl/F) to improve the connectivity.

    When you are done, your design should resemble Figure 27.6.

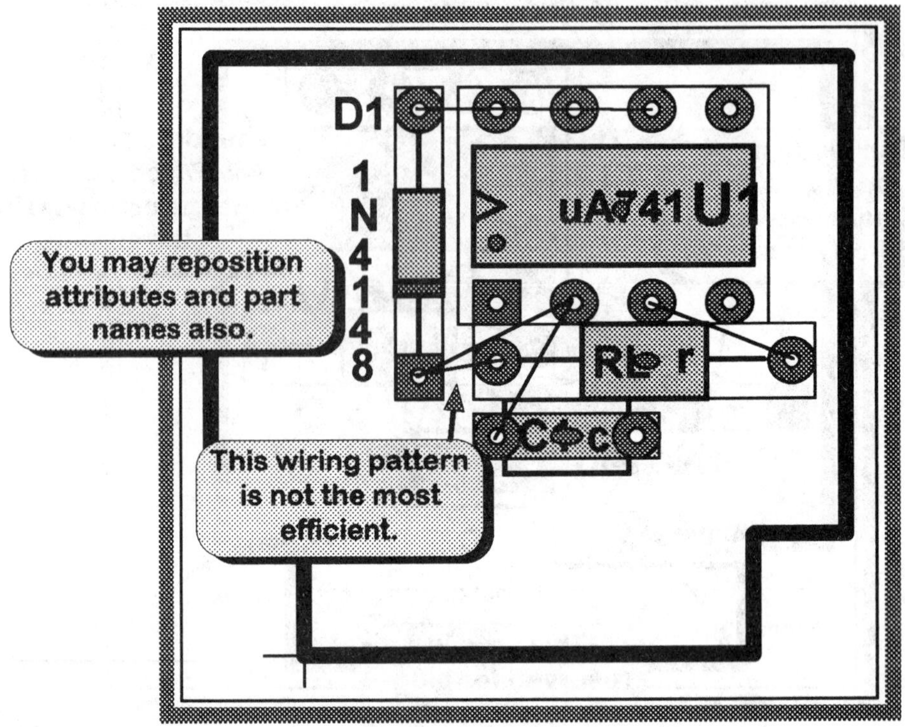

FIGURE 27.6
After rearranging D1 and C1

17. When you moved the components, did the "rats" (logical traces) automatically move as rubberbands?

    Yes **No**

18. **Tools**, **Optimize Rats**, and use a pen or pencil to note the changes on Figure 27.6. Do the new optimized traces appear to be a more efficient wiring scheme?

    Yes **No**

Define the trace area (Keepin)

Before actual metalized trace paths can be determined (routed), we must define the region that the traces must stay within. This is known as *Signal Keepin*. The signal keepin is typically .050inches (50mils) inside the board outline to prevent the etch from being too close to the edge of the board.

19. From the layout editor, **Draw**, **Board Signal Keepin** to bring up the keepin pencil. Using a series of **CLICKL** and **DRAG** steps, enclose the PCB design with a keepin border (and **CLICKR** when done). The result should resemble Figure 27.7.

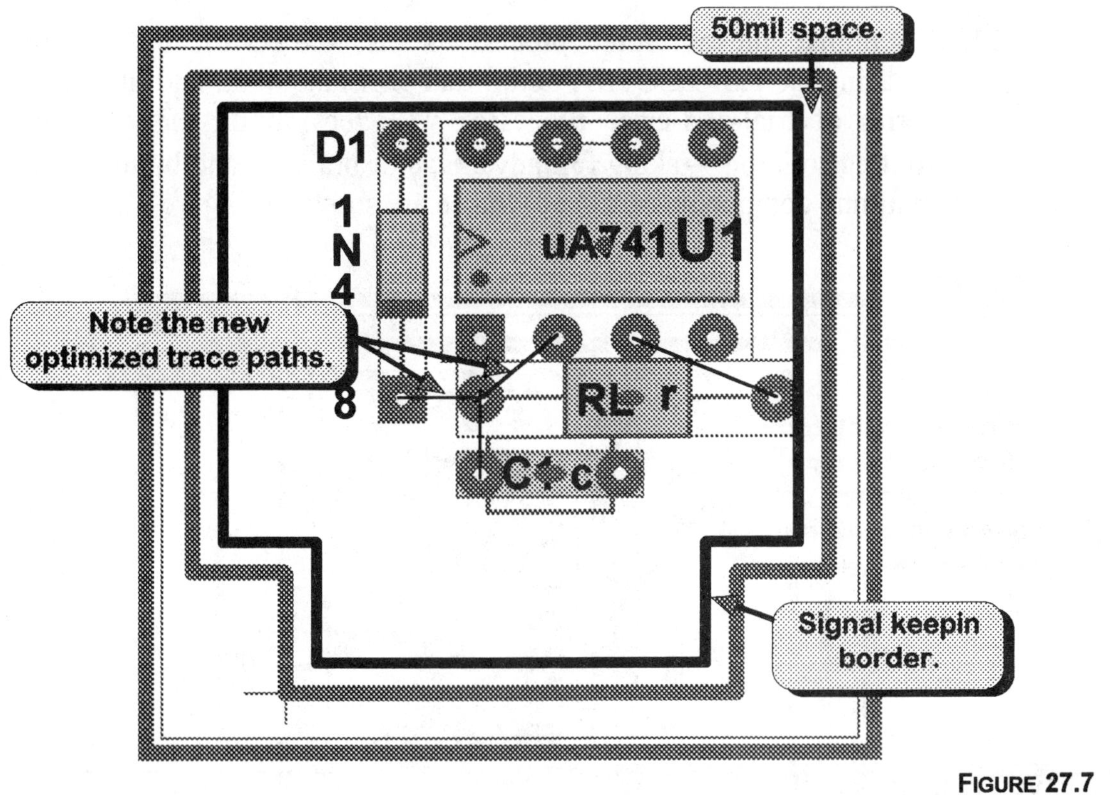

FIGURE 27.7
The keepin border

Defining padstacks

Pads are the square or round metal areas surrounding a component pin or *via* (connection between layers). Because pins or vias can extend through multiple layers, pads must be defined for each relevant layer. Collectively, these pads form a padstack definition—which must be set before routing can begin.

20. To set the padstack definition: **Configure**, **Styles**, **Trace** to bring up the *Configure Trace Styles* dialog box. **OK** to accept default values (standard 8mil trace widths, 3mil DRC error margin, and rnd-040-020 via specification).

Routing

The next step is to replace the purely logical (ratsnest) paths with metalized traces. Given the choice of the do-it-yourself (interactive) method or the powerful *autorouter*, we opt for the autorouter.

21. To initiate the autoroute process, **Tools**, **CCT:Autoroute**. The amazing CCT:SPECCTRA autorouter will take over, and through a series of trial-and-error processes will zero in on the optimum routing solution—taking full advantage of both top and bottom. The resulting component-side (red) traces are shown in Figure 27.8.

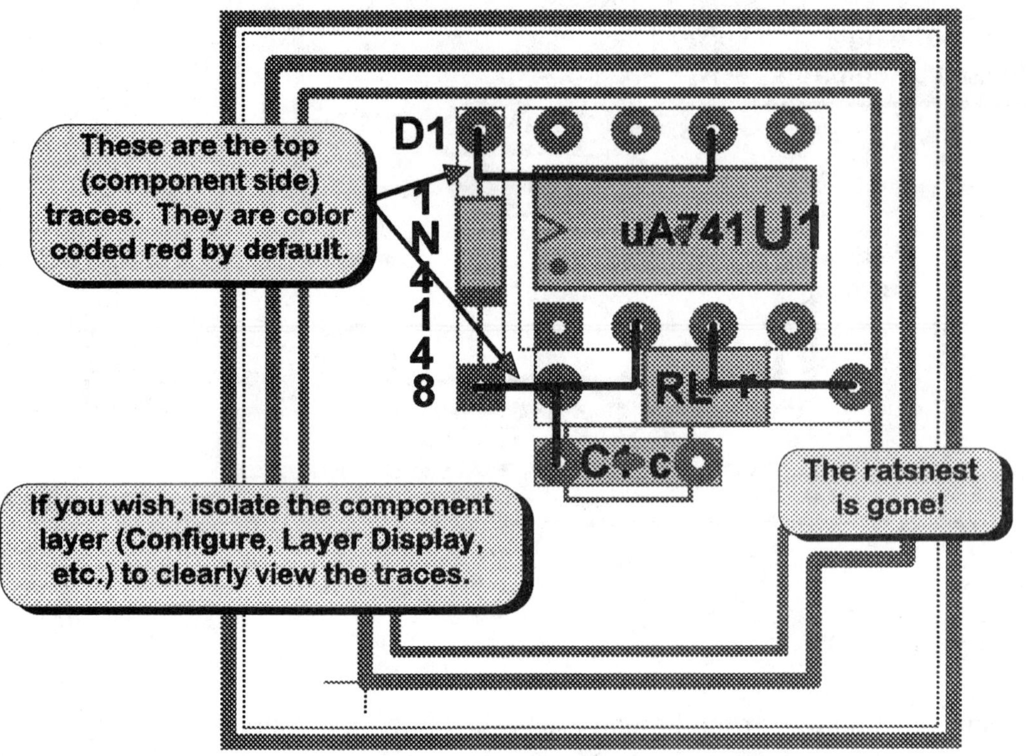

FIGURE 27.8
The top (component side) traces

22. Using the *layer-dropdown-list*, make the solder layer current. Were any solder-side (blue) traces generated by the autorouter? (To be sure of your answer, you may wish to isolate the solder layer by **Configure**, **Layer Display**, etc.)

    Yes **No**

Adding power, I/O, and bus-system traces

To complete the electrical design, we must connect the following five signals to edge-connector traces: *ground*, *+15V*, *–15V*, *Vin*, and *Vout*. To accomplish this, we add both via and edge-connector traces to the top of the board. The vias are placed to allow convenient connection to either side and to avoid DRC (design-rule-check) errors.

Draw/Add Via

23. If necessary, bring bank all layers but *paste* and *mask* (**Configure, Layer Display**, etc.). To add the five vias shown in Figure 27.9: **CLICKL** on the *Draw/Add Via* toolbar button, and use a series of **DRAG**, **CLICKL** steps. **CLICKR** to abort.

gle layer splay

Note that when adding vias we automatically enter the *Layer/Pairs* mode (as displayed by the *layer dropdown list*) because vias, by definition, are placed on bode sides (component and solder) at the same time. To toggle between the *Layer* and *Layer/Pairs* modes, **CLICKL** on the *Toggle Layer Display* toolbar button.

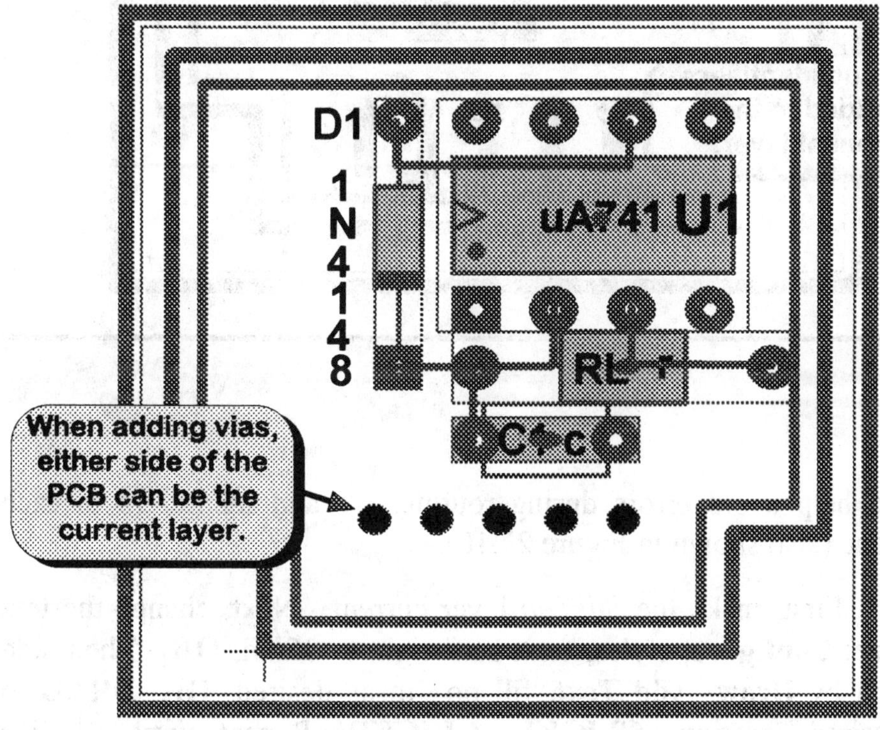

FIGURE 27.9
Vias added

24. To add the five thick edge-connector traces shown in Figure 27.10:

- First select the component side. Next, select thick traces: **Configure, Styles, Trace, TRACE_FAT, Change, OK**.

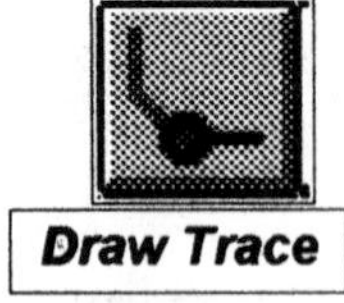

- To add the five traces: **CLICKL** on the *Draw/Trace* toolbar button, and use a series of **DRAG**, **CLICKL**, **CLICKR** steps. **DCLICKR** to abort.

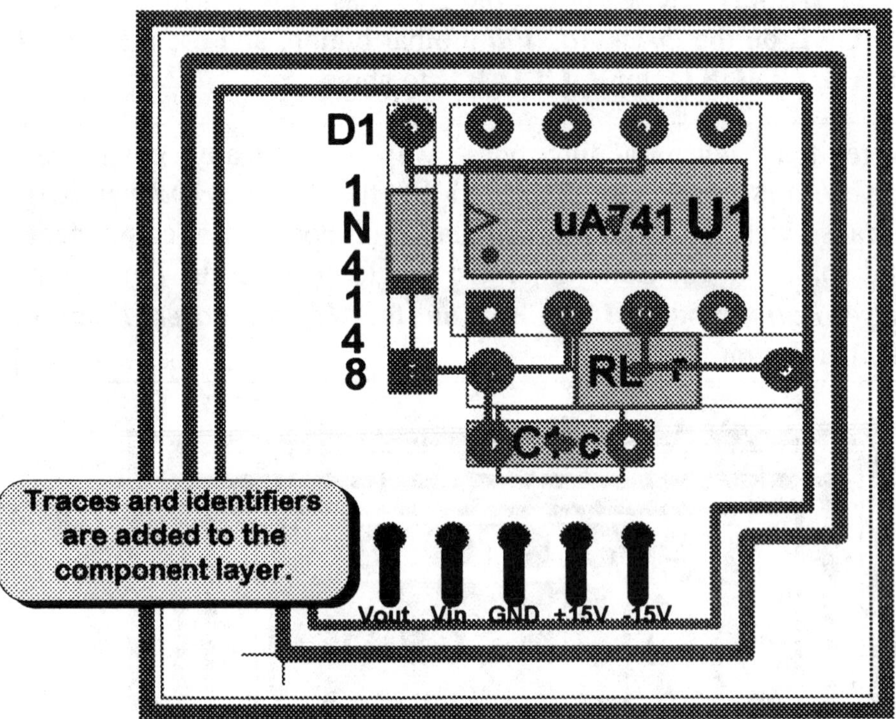

FIGURE 27.10
Traces and identifiers added

25. To help avoid errors during routing, we add identifiers to each trace (also shown in Figure 27.10).

    First, make the *SilkTop* layer current. Next, change the text size: **Configure**, **styles**, **text**, set height to 25mil, **OK**. Then, add text by **Draw**, **Add Text**, fill box in as desired, **OK**, **DRAG** to desired location, **CLICKL**, **CLICKR**. Repeat until all five identifiers are added.

26. In preparation for adding traces, return to a standard trace size by: **CLICKL** on the *Configure select* toolbar button and select the desired trace (or **Configure, Styles, Trace, TRACE_STD, OK**).

27. Examining our PCB layout design so far, we determine the following:

    - Traces *Vin*, *Vout*, and *Gnd* can best be drawn on the component side.
    - Traces *+15V* and *–15V* can best be drawn on the solder side.

28. First, to draw the top-side traces, make the component-side active. Referring to Figure 27.11, first place the *Vout* trace: **CLICKL** on the *Draw/Trace* toolbar button and move the bull's eye to a starting point. By using a series of **CLICKL, DRAG** steps, draw a trace. **CLICKR** to abort. Repeat for the other two traces.

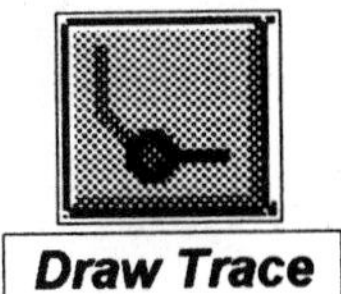

29. Make the solder layer current, and follow the same steps to add the final two traces to the solder side—as shown by Figure 27.12. *And the design is done!*

> ### Special Design Note
>
> The preferred method for adding power and ground traces involves adding a *connector* part to the original schematic of Figure 27.1. We then define a *footprint* for the connector part and add it to a footprint library. When the layout is first created (Figure 27.4), ratsnest traces appear for these power and ground connections. During autorouting, they are automatically converted to metalized traces.
>
> However, the evaluation version of MicroSim PCBoards does not allow additional footprints to be created, so we must revert to the hand-placement of the power and ground traces in this chapter.

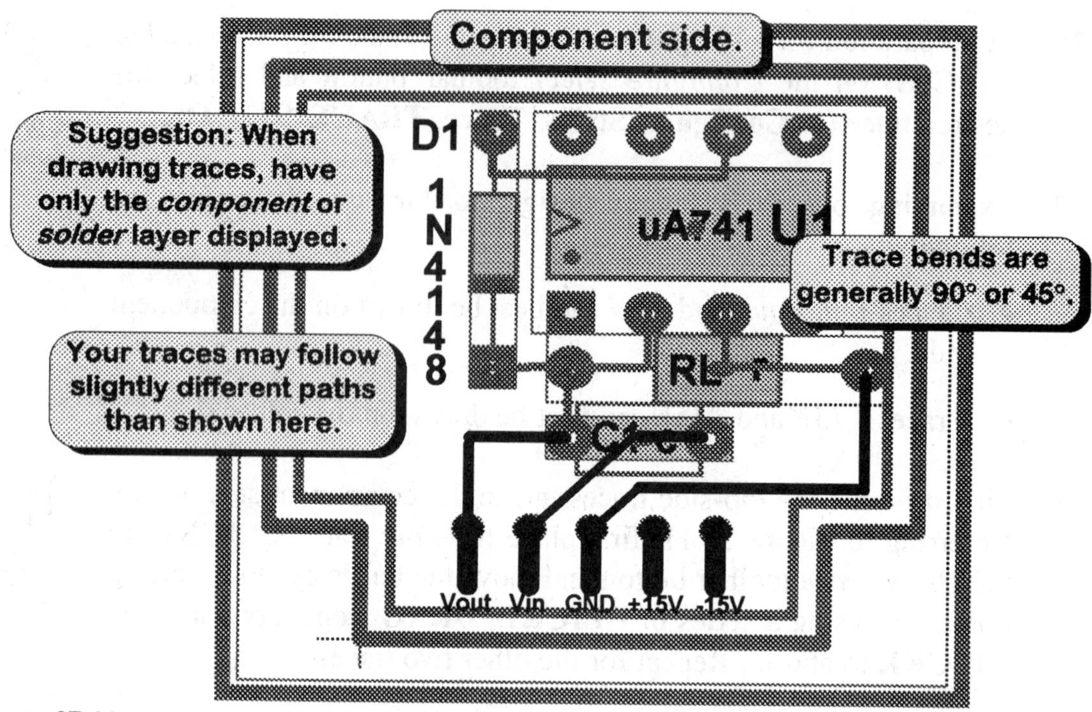

FIGURE 27.11
Adding the top-layer traces

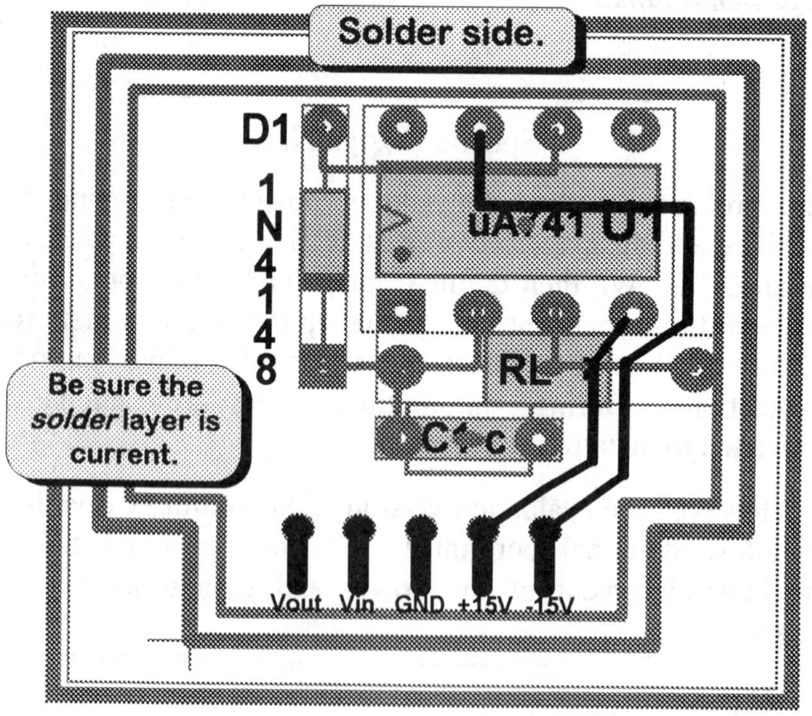

FIGURE 27.12
Adding the bottom-layer traces

30. Before we celebrate, let's check the electrical integrity of our design by using the *design rule check* (DRC) feature of PCBoards.

> The DRC system scans the design for *short circuits, clearance violations, signal keepout and keepin violations, trace width violations*, and *manufacturing requirement violations.*

If not already done, select the DRC layer. Did any DRC violations appear? (Are there any large yellow X's on the layout?) If yes, DCLICKL on each yellow X to determine the cause of each and correct the problem.

Advanced activities

31. Generate a bill of materials (list of components) for the clamper circuit of this chapter (from Schematics, **File**, **Reports**, **Display**).

    (a) Were all four components (R, C, 741, and D1) listed?

    Yes **No**

    (b) Do the attributes refer to each component's package type?

    Yes **No**

32. To help the selection process, MicroSim PCBoards offers *selection filters*. When one or more items are selected, they can be operated upon. To activate the selection filter, return to PCBoards and bring up the Selection Filter dialog box of Figure 27.13 by **Configure, Selection filter**. To select only pins: **Exclude All, Pins, OK**.

33. Select (**CLICKL**) various components. Is it true that only pins can be selected, and can they be selected at any layer?

    Yes **No**

34. Bring back the selection filter (**Configure, Selection filter**), enable *AutoExtend Selection*, **OK**. Return to the layout and select various pins. Is it true that when one pin is selected, all pins belonging to the same component are also selected?

    Yes **No**

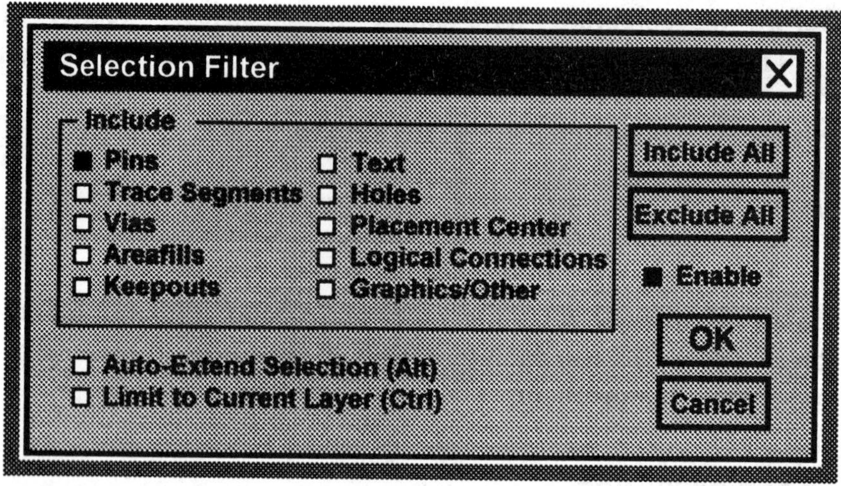

FIGURE 27.13
Selection filter dialog box

35. Try other selection filter combinations until you are familiar with this powerful feature.

36. Return to the schematic diagram, change the load resistor (RL) from 100k to 50k, and run MicroSim PCBoards again (**Tools**, **Run PCBoards**).

    (a) Did the *Compare to Netlist* dialog box come up, and did it indicate an incompatibility between the layout and the Schematic circuit?

    Yes **No**

    (b) **Preview** to bring up the *Forward ECO* dialog box. Did the ECO (engineering change order) refer to a change in RL?

    Yes **No**

    (c) **Apply** to make change. Was an ECO report issued?

    Yes **No**

EXERCISE

- Repeat the activities of this chapter on a circuit design from any previous chapter.

QUESTIONS AND PROBLEMS

1. What does it mean when a part (such as VSIN) is listed as *simulation only*?

2. What is the difference between the *component* and *solder* layers?

3. Name two ways of making the component layer current.

4. One inch is equivalent to how many *mils*?

5. Why are the connections of the *ratsnest* called *logical*?

6. One basic design rule is: Try to make all the traces on the component layer run vertically and those on the solder layer run horizontally. Why is this a good rule?

7. Why is the border definition usually one of the first design steps taken?

8. How does the *signal keepin* differ from the *border*?

9. What is the purpose of a *via*?

10. What is a *padstack*?

11. Which layer of the 19 choices do you believe most closely shows the footprint of a part? (Hint: See *MicroSim PCBoards Note 27.1.*)

12. Will an ECO be issued if a part is added or deleted?

APPENDIX A

Notes

Schematics

5.1 How do I simplify power connections?
12.1 How do I add a system bus?
12.2 How do I use the digital stimulus editor?

Probe

7.1 How do I create custom waveforms out of voltage segments?
11.1 How do I expand (zoom in on) digital waveforms?
11.2 How do I change the size of the digital display?

Filter Synthesis

20.1 What is the meaning of the various preferences options?

APPENDIX B

Probe's Mathematical Operators

Probe Function	Description
( )	Grouping
-	Logical complement
*/	Multiply/divide
+–	Add/subtract
&^|	AND, Exclusive OR, OR
AVGX(x,d)	Average (x to d)
RMS(x)	RMS average
DB(x)	x in dB
MIN(x)	Minimum real part of x
MAX(x)	Maximum real part of x
ABS(x)	Absolute value of x
SGN(x)	+1 if X>0, 0 if x = 0, –1 if x<0
SQRT(x)	$X^{1/2}$
EXP(x)	e^{X}
LOG(x)	Ln(x)
LOG10(x)	log(x)
M(x)	Magnitude of x
P(x)	Phase of x (degrees)
R(x)	Real part of x
IMG(x)	Imaginary part of x
G(x)	Group delay of x (sec)
PWR(x,y)	$(x)^{y}$
SIN(x)	sin(x)
COS(x)	cos(x)
TAN(x)	tan(x)
ATAN(x)	$\tan^{-1}(x)$
d(x)	Derivative of x with X-axis
s(x)	Integral of x with X-axis
AVG(x)	Average of x

APPENDIX C

Scale Suffixes

Symbol	Scale	Name
F	10E−15	*femto-*
P	10E−12	*pico-*
N	10E−9	*nano-*
U	10E−6	*micro-*
M	10E−3	*milli-*
K	10E+3	*kilo-*
MEG	10E+6	*mega-*
G	10E+9	*giga-*
T	10E+12	*tera-*

Appendix D

Spec Sheets

LM741

OPERATIONAL AMPLIFIER

Parameter	Conditions	Value			Unit
		Min	Typ	Max	
Input Offset Voltage	TA = 25°C		1.0	5.0	mV
Input Offset Current	TA = 25°C		20	200	nA
Input Bias Current	TA = 25°C		80	500	nA
Input Resistance	TA = 15°C, VS = ± 20V	.3	2.0		Mohms
Large Signal Voltage Gain	TA = 25°C, VS = ± 15V	50	200		V/mV
Output Short Circuit Current	TA = 25°C		25		mA
Common-Mode Rejection Ratio		70	90		DB
Bandwidth	TA = 25°C	.437	1.5		MHz
Slew Rate	TA = 25°C, Unity Gain		.5		V/us
Supply Current	TA = 25°C		1.7	2.8	mA
Power Consumption	TA = 25°C, VS = ± 15V		60	100	mW

TTL FAMILY CHARACTERISTICS

Standard 54/74

Parameter	Test Conditions	Min	Typ	Max	Unit
VIH Input HIGH voltage	**Guaranteed input HIGH voltage for all inputs**	**2.0**			**V**
VIL Input LOW voltage	**Guaranteed input LOW voltage for all inputs**			**.8**	**V**
VCD Input Clamp Diode Voltage	**VCC = Min lin = –12mA**		**–0.8**	**–1.5**	**V**
VOL Output LOW voltage	**VCC = Min IOL = 16mA**			**0.4**	**V**
VOH Output HIGH voltage	**VCC = Min IOH = –800uA**	**2.4**	**3.5**		**V**
IOH Output HIGH current (open collector)	**VCC = Max Vout = 5.5V**			**250**	**uA**
IOZH Output "off" current HIGH (3 state)	**VCC = Max Vout = 2.4V VOE = 2.0V**			**40**	**uA**
IOZL Output "off" current LOW	**VCC = Max Vout = .5V VOE = 2.0V**			**–40**	**uA**
IIH Input HIGH current	**VCC = Max Vin = 2.4V**			**40**	**uA**
IIH Input HIGH current at max input voltage	**VCC = Max Vin = 5.5V**			**1.0**	**mA**
IIL Input LOW current	**VCC = Max Vin = 0.4V**			**–1.6**	**mA**
IOS Output short circuit current	**VCC = Max Vout = .0V**	**–18**		**–55**	**mA**

Low Power Schottky 54LS/74LS

Parameter	Test Conditions	Min	Typ	Max	Unit
VIH Input HIGH voltage	**Guaranteed input HIGH voltage for all inputs**	**2.0**			**V**
VIL Input LOW voltage	**Guaranteed input LOW voltage for all inputs**			**.8**	**V**
VCD Input Clamp Diode Voltage	**VCC = Min lin = –12mA**		**–0.65**	**–1.5**	**V**
VOL Output LOW voltage	**VCC = Min IOL = 16mA**			**0.4**	**V**
VOH Output HIGH voltage	**VCC = Min IOH = –800uA**	**2.7**	**3.4**		**V**
IOH Output HIGH current (open collector)	**VCC = Max Vout = 5.5V**			**100**	**uA**
IOZH Output "off" current HIGH (3 state)	**VCC = Max Vout = 2.4V VOE = 2.0V**			**20**	**uA**
IOZL Output "off" current LOW	**VCC = Max Vout = .5V VOE = 2.0V**			**–20**	**uA**
IH Input HIGH current	**VCC = Max Vin = 2.4V**			**20**	**uA**
II Input HIGH current at max input voltage	**VCC = Max Vin = 5.5V**			**0.1**	**mA**
IIL Input LOW current	**VCC = Max Vin = 0.4V**			**–0.46**	**mA**
IOS Output short circuit current	**VCC = Max Vout = .0V**	**–18**		**–100**	**mA**

7442A BCD-to-Decimal Decoder

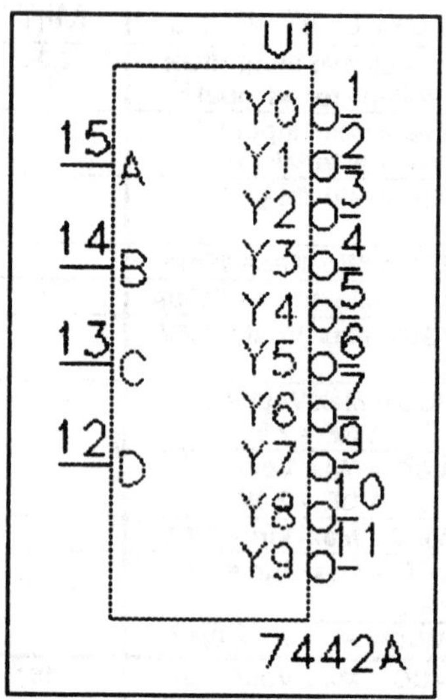

A3	A2	A1	A0	0	1	2	3	4	5	6	7	8	9
L	L	L	L	L	H	H	H	H	H	H	H	H	H
L	L	L	H	H	L	H	H	H	H	H	H	H	H
L	L	H	L	H	H	L	H	H	H	H	H	H	H
L	L	H	H	H	H	H	L	H	H	H	H	H	H
L	H	L	L	H	H	H	H	L	H	H	H	H	H
L	H	L	H	H	H	H	H	H	L	H	H	H	H
L	H	H	L	H	H	H	H	H	H	L	H	H	H
L	H	H	H	H	H	H	H	H	H	H	L	H	H
H	L	L	L	H	H	H	H	H	H	H	H	L	H
H	L	L	H	H	H	H	H	H	H	H	H	H	L
All others				All high									

7474 Dual D-Type FLIP-FLOP

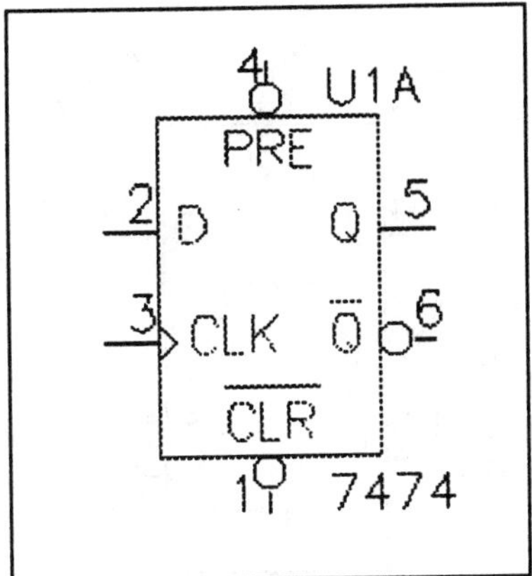

OPERATING MODE	INPUTS				OUTPUTS	
	SD	RD	CP	D	Q	NQ
Asynchronous Set	L	H	X	X	H	L
Asynchronous Reset	H	L	X	X	L	H
Undetermined	L	L	X	X	H	H
Load "1" (Set)	H	H	Rise	H	H	L
Load "0" (Reset)	H	H	Rise	L	L	H

7476 Dual JK FLIP-FLOP

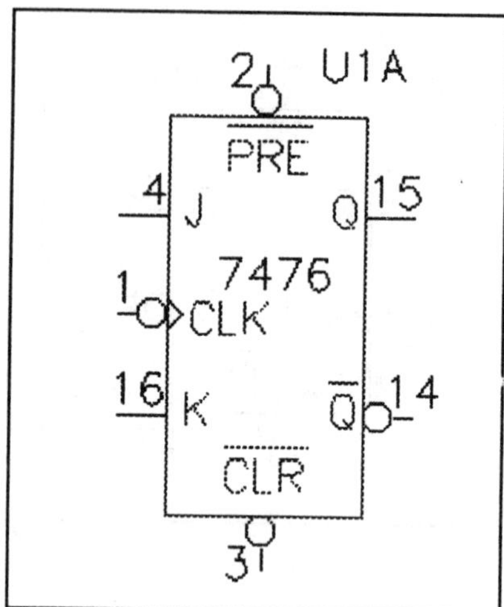

OPERATING MODE	INPUTS					OUTPUTS	
	SD	RD	CP	J	K	Q	NQ
Asynchronous Set	L	H	X	X	X	H	L
Asynchronous Reset	H	L	X	X	X	L	H
Undetermined	L	L	X	X	X	H	H
Toggle	H	H	Pulse	H	H	NQ	Q
Load "1" (Set)	H	H	Pulse	H	L	H	L
Load "0" (Reset)	H	H	Pulse	L	H	L	H
Hold (No Change)	H	H	Pulse	L	L	Q	NQ

7485 4-Bit Magnitude Comparator

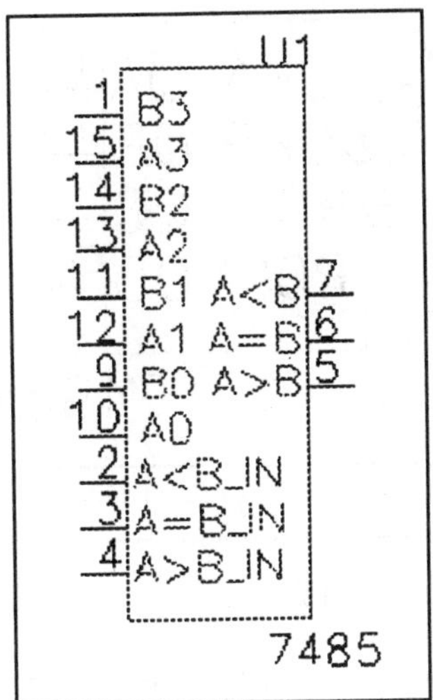

Input signals	A>B	A=B	A<B
A>B	H	L	L
A=B	L	H	L
A<B	L	L	H

7493A 4-Bit Binary Ripple Counter

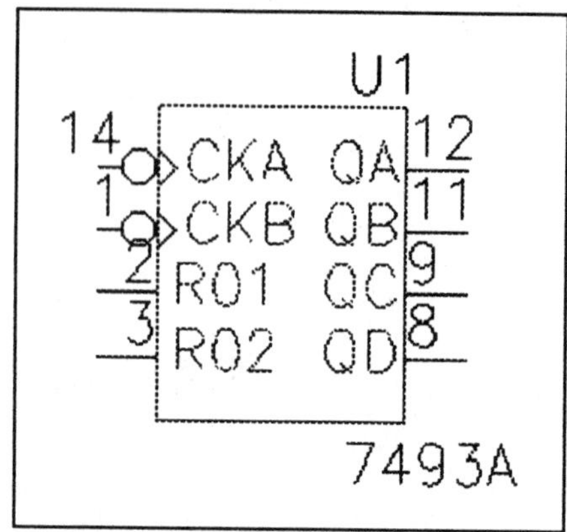

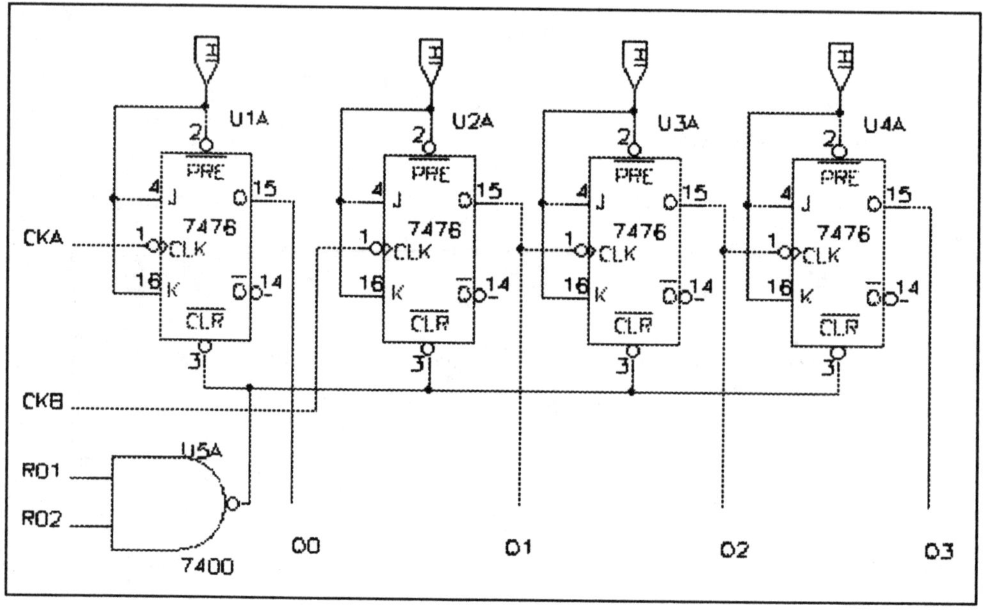

RESET INPUTS		OUTPUTS			
R01	**R02**	**Q0**	**Q1**	**Q2**	**Q3**
H	H	L	L	L	L
L	H	Count			
H	L	Count			
L	L	Count			

74147 10-Line-to-4-Line Priority Encoder

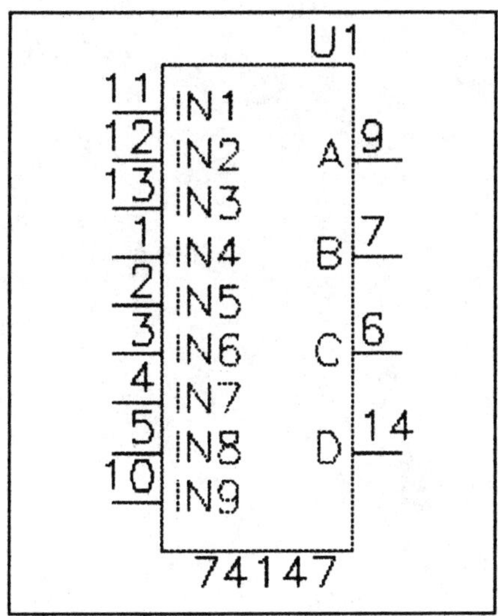

i1	I2	I3	I4	I5	I6	I7	I8	I9	A3	A2	A1	A0
H	H	H	H	H	H	H	H	H	H	H	H	H
X	X	X	X	X	X	X	X	L	L	H	H	L
X	X	X	X	X	X	X	L	H	L	H	H	H
X	X	X	X	X	X	L	H	H	H	L	L	L
X	X	X	X	X	L	H	H	H	H	L	L	H
X	X	X	X									
X	X	X										
X	X											
X												
L	H	H	H	H	H	H	H	H				

Index

T

U

V

W

Z